HANDBOOK OF MATHEMATICAL FORMULAS AND INTEGRALS

HANDBOOK OF MATHEMATICAL FORMULAS AND INTEGRALS

ALAN JEFFREY

Department of Engineering Mathematics
University of Newcastle upon Tyne
Newcastle upon Tyne
United Kingdom

Academic Press

San Diego   New York   Boston   London   Sydney   Tokyo   Toronto

This book is printed on acid-free paper. ∞

Copyright © 1995 by ACADEMIC PRESS, INC.

Academic Press, Inc.
A Division of Harcourt Brace & Company
525 B Street, Suite 1900, San Diego, California 92101-4495

United Kingdom Edition published by
Academic Press Limited
24-28 Oval Road, London NW1 7DX

Library of Congress Cataloging-in-Publication Data

Jeffrey, Alan.
    Handbook of mathematical formulas and integrals / by Alan Jeffrey.
        p.   cm.
    Includes index.
    ISBN 0-12-382580-6 (pbk.)
    1. Mathematics—Tables. 2. Mathematics—Formulas. I. Title.
QA47.J38   1995
510'.212—dc20   95-2344
                                                                 CIP

PRINTED IN THE UNITED STATES OF AMERICA
    96   97   98   99   00   MM   9   8   7   6   5   4   3   2

Contents

2 Functions and Identities

3 Derivatives of Elementary Functions

4 Indefinite Integrals of Algebraic Functions

5 Indefinite Integrals of Exponential Functions

6 Indefinite Integrals of Logarithmic Functions

7 Indefinite Integrals of Hyperbolic Functions

8 Indefinite Integrals Involving Inverse Hyperbolic Functions

9 Indefinite Integrals of Trigonometric Functions

10 Indefinite Integrals of Inverse Trigonometric Functions

11 The Gamma, Beta, Pi, and Psi Functions

12 Elliptic Integrals and Functions

13 Probability Integrals and the Error Function

14 Fresnel Integrals

15 Definite Integrals

16 Different Forms of Fourier Series

17 Bessel Functions

20 Fourier Transforms

21 Numerical Integration

22 Solutions of Standard Ordinary Differential Equations

23 Vector Analysis

24 Systems of Orthogonal Coordinates

25 Partial Differential Equations and Special Functions

Preface

This book contains a collection of general mathematical results, formulas, and integrals that occur throughout applications of mathematics. Many of the entries are based on the updated fifth edition of Gradshteyn and Ryzhik's "Tables of Integrals, Series, and Products," though during the preparation of the book, results were also taken from various other reference works. The material has been arranged in a straightforward manner, and for the convenience of the user a quick reference list of the simplest and most frequently used results is to be found in Chapter 0 at the front of the book. Tab marks have been added to pages to identify the twelve main subject areas into which the entries have been divided and also to indicate the main interconnections that exist between them. Keys to the tab marks are to be found inside the front and back covers.

The Table of Contents at the front of the book is sufficiently detailed to enable rapid location of the section in which a specific entry is to be found, and this information is supplemented by a detailed index at the end of the book. In the chapters listing integrals, instead of displaying them in their canonical form, as is customary in reference works, in order to make the tables more convenient to use, the integrands are presented in the more general form in which they are likely to arise. It is hoped that this will save the user the necessity of reducing a result to a canonical form before consulting the tables. Wherever it might be helpful, material has been added explaining the idea underlying a section or describing simple techniques that are often useful in the application of its results.

Standard notations have been used for functions, and a list of these together with their names and a reference to the section in which they occur or are defined is to be found at the front of the book. As is customary with tables of indefinite integrals, the additive arbitrary constant of integration has always been omitted. The result of an integration may take more than one form, often depending on the method used for its evaluation, so only the most common forms are listed.

A user requiring more extensive tables, or results involving the less familiar special functions, is referred to the short classified reference list at the end of the book. The list contains works the author found to be most useful and which a user is likely to find readily accessible in a library, but it is in no sense a comprehensive bibliography. Further specialist references are to be found in the bibliographies contained in these reference works.

Every effort has been made to ensure the accuracy of these tables and, whenever possible, results have been checked by means of computer symbolic algebra and integration programs, but the final responsibility for errors must rest with the author.

Alan Jeffrey

Index of Special Functions and Notations

Notation	Name	*Section or formula containing its definition*
$\lvert a \rvert$	Absolute value of the real number	1.1.2.1
am u	Amplitude of an elliptic function	12.2.1.1.2
$\sim$	Asymptotic relationship	1.14.2.1
α	Modular angle of an elliptic integral	12.1.2
arg z	Argument of complex number z	2.1.1.1
$A(x)$	$A(x) = 2P(x) - 1$; probability function	13.1.1.1.7
$\mathbf{A}$	Matrix	
$\mathbf{A}^{-1}$	Multiplicative inverse of a square matrix $\mathbf{A}$	1.5.1.1.9
$\mathbf{A}^{\mathrm{T}}$	Transpose of matrix $\mathbf{A}$	1.5.1.1.7
$\lvert \mathbf{A} \rvert$	Determinant associated with a square matrix $\mathbf{A}$	1.4.1.1
B_n	Bernoulli number	1.3.1.1
B_n^*	Alternative Bernoulli number	1.3.1.1.6
$B_n(x)$	Bernoulli polynomial	1.3.2.1.1
$B(x, y)$	Beta function	11.1.7.1
$\dbinom{n}{k}$	Binomial coefficient $$\binom{n}{k} = \frac{n!}{k!(n-k)!}, \qquad \binom{n}{0} = 1$$	1.2.1.1
$C(x)$	Fresnel cosine integral	14.1.1.1.1

Notation	Name	Section or formula containing its definition
C_{ij}	Cofactor of element a_{ij} in a square matrix $\mathbf{A}$	1.4.2
^nC_m or $_nC_m$	Combination symbol $^nC_m = \begin{pmatrix} n \\ m \end{pmatrix}$	1.6.2.1
cn u	Jacobian elliptic function	12.2.1.1.4
cn^{-1} u	Inverse Jacobian elliptic function	12.4.1.1.4
curl $\mathbf{F} = \nabla \times \mathbf{F}$	Curl of vector $\mathbf{F}$	23.8.1.1.6
$\delta(x)$	Dirac delta function	19.1.3
$D_n(x)$	Dirichlet kernel	1.13.1.10.3
dn u	Jacobian elliptic function	12.2.1.1.5
dn^{-1} u	Inverse Jacobian elliptic function	12.4.1.1.5
div $\mathbf{F} = \nabla \cdot \mathbf{F}$	Divergence of vector $\mathbf{F}$	23.8.1.1.4
$e^{i\theta}$	Euler formula; $e^{i\theta} = \cos\theta + i\sin\theta$	2.1.1.2.1
e	Euler's constant	0.3
$Ei(x)$	Exponential integral	5.1.2.2
$E(\varphi, k)$	Incomplete elliptic integral of the second kind	12.1.1.1.5
$E(k), E'(k)$	Complete elliptic integrals of the second kind	13.1.1.1.8, 13.1.1.1.10
e^{Az}	Matrix exponential	1.5.4.1
erf x	Error function	13.2.1.1
erfc x	Complementary error function	13.2.1.1.4
E_n	Euler number	1.3.1.1
E_n^*	Alternative Euler number	1.3.1.1.6
$E_n(x)$	Euler polynomial	1.3.2.3.1
$f(x)$	A function of x	
$f'(x)$	First derivative df/dx	1.15.1.1.6
$f^{(n)}(x)$	nth derivative $d^n f/dx^n$	1.12.1.1
$f^{(n)}(x_0)$	nth derivative $d^n f/dx^n$ at x_0	1.12.1.1
$F(\varphi, k)$	Incomplete elliptic integral of the first kind	12.1.1.1.4
$\|\Phi_n\|$	Norm of $\Phi_n(x)$	18.1.1.1
grad $\phi = \nabla\phi$	Gradient of the scalar function ϕ	23.8.1.6
$\Gamma(x)$	Gamma function	11.1.1.1
γ	Euler–Mascheroni constant	1.11.1.1.7
$H(x)$	Heaviside step function	19.1.2.5
$H_n(x)$	Hermite polynomial	18.5.3
i	Imaginary unit	1.1.1.1
Im$\{z\}$	Imaginary part of $z = x + iy$; Im$\{z\} = y$	1.1.1.2
$\mathbf{I}$	Unit (identity) matrix	1.5.1.1.3
i^n erfc x	nth repeated integral of erfc x	13.2.7.1.1
$I_{\pm v}(x)$	Modified Bessel function of the first kind of order v	17.6.1.1
$\int f(x)\,dx$	Indefinite integral (antiderivative) of $f(x)$	1.15.2
$\int_a^b f(x)\,dx$	Definite integral of $f(x)$ from $x = a$ to $x = b$	1.15.2.5

Notation	Name	Section or formula containing its definition
$J_{\pm \nu}(x)$	Bessel function of the first kind of order ν	17.1.1.1
k	Modulus of an elliptic integral	12.1.1.1
k'	Complementary modulus of an elliptic integral; $k' = \sqrt{1 - k^2}$	12.1.1.1
$\mathbf{K}(k), \mathbf{K}'(k)$	Complete elliptic integrals of the first kind	12.1.1.1.7, 12.1.1.1.9
$K_\nu(x)$	Modified Bessel function of the second kind of order ν	17.6.1.1
$\mathcal{L}[f(x); s]$	Laplace transform of $f(x)$	19.1.1
$L_n(x)$	Laguerre polynomial	18.4.1
$\log_a x$	Logarithm of x to the base a	2.2.1.1
$\ln x$	Natural logarithm of x (to the base e)	2.2.1.1
M_{ij}	Minor of element a_{ij} in a square matrix $\mathbf{A}$	1.4.2
$n!$	Factorial n; $n! = 1 \cdot 2 \cdot 3 \cdots n$; $\quad 0! = 1$	1.2.1.1
$(2n)!!$	Double factorial; $(2n)!! = 2 \cdot 4 \cdot 6 \cdots (2n)$	15.2.1
$(2n - 1)!!$	Double factorial; $(2n - 1)!! = 1 \cdot 3 \cdot 5 \cdots (2n - 1)$	15.2.1
$\left[\dfrac{n}{2}\right]$	Integral part of $n/2$	18.2.4.1.1
$^n P_m$ or $_n P_m$	Permutation symbol; $^n P_m = \dfrac{n!}{(n - m)!}$	1.6.1.1.3
$P_n(x)$	Legendre polynomial	18.2.1
$P(x)$	Normal probability distribution	13.1.1.1.5
$\displaystyle\prod_{k=1}^{n} u_k$	Product symbol; $\displaystyle\prod_{k=1}^{n} u_k = u_1 u_2 \cdots u_n$	1.9.1.1.1
P.V. $\int_{-\infty}^{\infty} f(x)dx$	Cauchy principal value of the integral	1.15.4.IV
π	Ratio of the circumference of a circle to its diameter	0.3
$\Pi(x)$	pi function	11.1.1.1
$\Pi(\varphi, n, k)$	Incomplete elliptic integral of the third kind	12.1.1.1.6
$\psi(z)$	psi (digamma) function	11.1.6.1
$Q(x)$	Probability function; $Q(x) = 1 - P(x)$	13.1.1.1.6
$Q(x)$	Quadratic form	1.5.2.1
$Q_n(x)$	Legendre function of the second kind	18.2.7
r	Modulus of $z = x + iy$; $r = (x^2 + y^2)^{1/2}$	2.1.1
$\text{Re}\{z\}$	Real part of $z = x + iy$; $\text{Re}\{z\} = x$	1.1.1.2
$\text{sgn}(x)$	sign of x defined as $x/|x|$	1.1.21
$\text{sn } u$	Jacobian elliptic function	12.2.1.1.3
$\text{sn}^{-1} u$	Inverse Jacobian elliptic function	12.4.1.1.3
$S(x)$	Fresnel sine integral	14.1.1.1.2
$\displaystyle\sum_{k=m}^{n} a_k$	Summation symbol; $\displaystyle\sum_{k=m}^{n} a_k = a_m + a_{m+1} + \cdots + a_n$	1.2.3
	and if $n < m$ we define $\displaystyle\sum_{k=m}^{n} a_k = 0$.	

Notation	Name	*Section or formula containing its definition*
$\displaystyle\sum_{k=m}^{\infty} a_k(x - x_0)^k$	Power series expanded about x_0	1.11.1.1.1
$T_n(x)$	Chebyshev polynomial	18.3.1.1
tr $\mathbf{A}$	Trace of a square matrix $\mathbf{A}$	15.1.1.10
$U_n(x)$	Chebyshev polynomial	18.3.11
$x = f^{-1}(y)$	Function inverse to $y = f(x)$	1.11.1.8
$Y_\nu(x)$	Bessel function of the second kind of order ν	17.1.1.1
z	Complex number $z = x + iy$	1.1.1.1
$|z|$	Modulus of $z = x + iy$; $r = |z| = (x^2 + y^2)^{1/2}$	1.1.1.1
$\bar{z}$	Complex conjugate of $z = x + iy$; $\bar{z} = x - iy$	1.1.1.1

0

Quick Reference List of Frequently Used Data

0.1 Useful Identities

0.1.1 Trigonometric identities

$$\sin^2 x + \cos^2 x = 1$$

$$\sec^2 x = 1 + \tan^2 x$$

$$\csc^2 x = 1 + \cot^2 x$$

$$\sin 2x = 2 \sin x \cos x$$

$$\cos 2x = \cos^2 x - \sin^2 x$$

$$= 1 - 2 \sin^2 x$$

$$= 2 \cos^2 x - 1$$

$$\sin^2 x = \frac{1}{2}(1 - \cos 2x)$$

$$\cos^2 x = \frac{1}{2}(1 + \cos 2x)$$

$$\sin(x + y) = \sin x \cos y + \cos x \sin y$$

$$\sin(x - y) = \sin x \cos y - \cos x \sin y$$

$$\cos(x + y) = \cos x \cos y - \sin x \sin y$$

$$\cos(x - y) = \cos x \cos y + \sin x \sin y$$

$$\tan(x + y) = \frac{\tan x + \tan y}{1 - \tan x \tan y}$$

$$\tan(x - y) = \frac{\tan x - \tan y}{1 + \tan x \tan y}$$

$2\sin^2 x = 1 - \cos 2x$

$\cos 2x + 1 = 2\cos^2 x$

0.1.2 Hyperbolic identities

$$\cosh^2 x - \sinh^2 x = 1$$

$$\text{sech}^2 x = 1 - \tanh^2 x$$

$$\text{csch}^2 x = \coth^2 x - 1$$

$$\sinh 2x = 2 \sinh x \cosh x$$

$$\cosh 2x = \cosh^2 x + \sinh^2 x$$
$$= 1 + 2 \sinh^2 x$$
$$= 2 \cosh^2 x - 1$$

$$\sinh^2 x = \frac{1}{2}(\cosh 2x - 1)$$

$$\cosh^2 x = \frac{1}{2}(\cosh 2x + 1)$$

$$\sinh(x + y) = \sinh x \cosh y + \cosh x \sinh y$$

$$\sinh(x - y) = \sinh x \cosh y - \cosh x \sinh y$$

$$\cosh(x + y) = \cosh x \cosh y + \sinh x \sinh y$$

$$\cosh(x - y) = \cosh x \cosh y - \sinh x \sinh y$$

$$\tanh(x + y) = \frac{\tanh x + \tanh y}{1 + \tanh x \tanh y}$$

$$\tanh(x - y) = \frac{\tanh x - \tanh y}{1 - \tanh x \tanh y}$$

$$\text{arcsinh} \frac{x}{a} = \ln\left[\frac{x + (a^2 + x^2)^{1/2}}{a}\right] \qquad [-\infty < x/a < \infty]$$

$$\text{arccosh} \frac{x}{a} = \ln\left[\frac{x + (x^2 - a^2)^{1/2}}{a}\right] \qquad [x/a > 1]$$

$$\text{arctanh} \frac{x}{a} = \frac{1}{2}\ln\left[\frac{a + x}{a - x}\right] \qquad [x^2 < a^2]$$

0.2 Complex Relationships

$$e^{ix} = \cos x + i \sin x$$

$$\sin x = \frac{e^{ix} - e^{-ix}}{2i}$$

$$\cos x = \frac{e^{ix} + e^{-ix}}{2}$$

$$(\cos x + i \sin x)^n = \cos nx + i \sin nx$$

$$\sin nx = \text{Im}\{(\cos x + i \sin x)^n\}$$

$$\cos nx = \text{Re}\{(\cos x + i \sin x)^n\}$$

$$\sinh x = \frac{e^x - e^{-x}}{2}$$

$$\cosh x = \frac{e^x + e^{-x}}{2}$$

$$\sin ix = i \sinh x$$

$$\cos ix = \cosh x$$

$$\sinh ix = i \sin x$$

$$\cosh ix = \cos x$$

0.3 Constants

$$e = 2.7182\ 81828\ 45904$$

$$\pi = 3.1415\ 92653\ 58979$$

$$\log_{10} e = 0.4342\ 94481\ 90325$$

$$\ln 10 = 2.3025\ 85092\ 99404$$
$$\gamma = 0.5772\ 15664\ 90153$$
$$(2\pi)^{-1/2} = 0.3989\ 42280\ 40143$$
$$\Gamma\left(\frac{1}{2}\right) = \pi^{1/2} = 1.7724\ 53850\ 90551$$

0.4 Derivatives of Elementary Functions

$f(x)$	$f'(x)$	$f(x)$	$f'(x)$
x^n	nx^{n-1}	$\sinh ax$	$a \cosh ax$
e^{ax}	ae^{ax}	$\cosh ax$	$a \sinh ax$
$\ln x$	$1/x$	$\tanh ax$	$a \operatorname{sech}^2 ax$
$\sin ax$	$a \cos ax$	$\operatorname{csch} ax$	$-a \operatorname{csch} ax \coth ax$
$\cos ax$	$-a \sin ax$	$\operatorname{sech} ax$	$-a \operatorname{sech} ax \tanh ax$
$\tan ax$	$a \sec^2 ax$	$\coth ax$	$-a \operatorname{csch}^2 ax$
$\csc ax$	$-a \csc ax \cot ax$	$\operatorname{arcsinh}\frac{x}{a}$	$1/\sqrt{x^2 + a^2}$
$\sec ax$	$a \sec ax \tan ax$		
$\cot ax$	$-a \csc^2 ax$	$\operatorname{arccosh}\frac{x}{a}$	$\begin{cases} 1/\sqrt{x^2 - a^2} \text{ for arccosh } \frac{x}{a} > 0, \ \frac{x}{a} > 1, \\ -1/\sqrt{x^2 - a^2} \text{ for arccosh } \frac{x}{a} < 0, \ \frac{x}{a} > 1. \end{cases}$
$\arcsin\frac{x}{a}$	$1/\sqrt{a^2 - x^2}$		
$\arccos\frac{x}{a}$	$-1/\sqrt{a^2 - x^2}$		
$\arctan\frac{x}{a}$	$a/(a^2 + x^2)$	$\operatorname{arctanh}\frac{x}{a}$	$a/(a^2 - x^2) \ [x^2 < a^2]$

0.5 Rules of Differentiation and Integration

1. $\dfrac{d}{dx}(u + v) = \dfrac{du}{dx} + \dfrac{dv}{dx}$ (sum)

2. $\dfrac{d}{dx}(uv) = u\dfrac{dv}{dx} + v\dfrac{du}{dx}$ (product)

3. $\dfrac{d}{dx}\left(\dfrac{u}{v}\right) = \left(v\dfrac{du}{dx} - u\dfrac{dv}{dx}\right)\bigg/ v^2$ for $v \neq 0$ (quotient)

4. $\dfrac{d}{dx}[f\{g(x)\}] = f'\{g(x)\}\dfrac{dg}{dx}$ (function of a function)

5. $\displaystyle\int (u + v)dx = \int u\,dx + \int v\,dx$ (sum)

6. $\displaystyle\int u\,dv = uv - \int v\,du$ (integration by parts)

7. $\dfrac{d}{d\alpha}\displaystyle\int_{\phi(\alpha)}^{\psi(\alpha)} f(x,\alpha)dx = \left(\dfrac{d\psi}{d\alpha}\right)f(\psi,\alpha) - \left(\dfrac{d\phi}{d\alpha}\right)f(\phi,\alpha) + \int_{\phi(\alpha)}^{\psi(\alpha)} \dfrac{\partial f}{\partial \alpha}dx$

(differentiation of an integral containing a parameter)

0.6 Standard Integrals

Common standard forms

1. $\displaystyle\int x^n dx = \frac{1}{n+1}x^{n+1}$ $[n \neq -1]$

2. $\displaystyle\int \frac{1}{x}dx = \ln|x| = \begin{cases} \ln x, & x > 0 \\ \ln(-x), & x < 0 \end{cases}$

3. $\displaystyle\int e^{ax} dx = \frac{1}{a}e^{ax}$

4. $\displaystyle\int a^x dx = \frac{a^x}{\ln a}$ $[a \neq 1, a > 0]$

5. $\displaystyle\int \ln x \, dx = x \ln x - x$

6. $\displaystyle\int \sin ax \, dx = -\frac{1}{a}\cos ax$

7. $\displaystyle\int \cos ax \, dx = \frac{1}{a}\sin ax$

8. $\displaystyle\int \tan ax \, dx = -\frac{1}{a}\ln|\cos ax|$

9. $\displaystyle\int \sinh ax \, dx = \frac{1}{a}\cosh ax$

10. $\displaystyle\int \cosh ax \, dx = \frac{1}{a}\sinh ax$

11. $\displaystyle\int \tanh ax \, dx = \frac{1}{a}\ln|\cosh ax|$

12. $\displaystyle\int \frac{1}{\sqrt{a^2 - x^2}} dx = \arcsin \frac{x}{a}$ $[x^2 \leq a^2]$

13. $\displaystyle\int \frac{1}{\sqrt{x^2 - a^2}} dx = \text{arccosh} \frac{x}{a}$
$= \ln|x + \sqrt{x^2 - a^2}|$ $[a^2 \leq x^2]$

14. $\displaystyle\int \frac{1}{\sqrt{a^2 + x^2}} dx = \text{arcsinh} \frac{x}{a}$
$= \ln|x + \sqrt{a^2 + x^2}|$

15. $\displaystyle\int \frac{1}{x^2 + a^2}\, dx = \frac{1}{a} \arctan \frac{x}{a}$

16. $\displaystyle\int \frac{dx}{a^2 - b^2 x^2} = \frac{1}{2ab} \ln \left| \frac{a + bx}{a - bx} \right|$

$\displaystyle\qquad = \frac{1}{ab} \operatorname{arctanh} \frac{bx}{a} \qquad [a^2 > b^2 x^2]$

Integrands involving algebraic functions

17. $\displaystyle\int (a + bx)^n\, dx = \frac{(a + bx)^{n+1}}{b(n + 1)} \qquad [n \neq -1]$

18. $\displaystyle\int \frac{1}{a + bx}\, dx = \frac{1}{b} \ln |a + bx|$

19. $\displaystyle\int x(a + bx)^n\, dx = \frac{(a + bx)^{n+1}}{b^2} \left(\frac{a + bx}{n + 2} - \frac{a}{n + 1} \right) \qquad [n \neq -1, -2]$

20. $\displaystyle\int \frac{x}{a + bx}\, dx = \frac{x}{b} - \frac{a}{b^2} \ln|a + bx|$

21. $\displaystyle\int \frac{x^2}{a + bx}\, dx = \frac{1}{b^3} \left[\frac{1}{2}(a + bx)^2 - 2a(a + bx) + a^2 \ln |a + bx| \right]$

22. $\displaystyle\int \frac{x}{(a + bx)^2}\, dx = \frac{1}{b^2} \left(\frac{a}{a + bx} + \ln |a + bx| \right)$

23. $\displaystyle\int \frac{x^2}{(a + bx)^2}\, dx = \frac{1}{b^3} \left(a + bx - \frac{a^2}{a + bx} - 2a \ln |a + bx| \right)$

24. $\displaystyle\int \frac{1}{x(a + bx)}\, dx = \frac{1}{a} \ln \left| \frac{x}{a + bx} \right|$

25. $\displaystyle\int \frac{1}{x^2(a + bx)}\, dx = -\frac{1}{ax} + \frac{b}{a^2} \ln \left| \frac{a + bx}{x} \right|$

26. $\displaystyle\int \frac{1}{x(a + bx)^2}\, dx = \frac{1}{a(a + bx)} + \frac{1}{a^2} \ln \left| \frac{x}{a + bx} \right|$

27. $\displaystyle\int \frac{1}{x\sqrt{a + bx}}\, dx = \begin{cases} \dfrac{1}{\sqrt{a}} \ln \left| \dfrac{\sqrt{a + bx} - \sqrt{a}}{\sqrt{a + bx} + \sqrt{a}} \right| & \text{if } a > 0 \\[2ex] \dfrac{2}{\sqrt{-a}} \arctan \sqrt{\dfrac{a+bx}{-a}} & \text{if } a < 0 \end{cases}$

28. $\displaystyle\int \frac{1}{x^2\sqrt{a + bx}}\, dx = -\frac{\sqrt{a + bx}}{ax} - \frac{b}{2a} \int \frac{1}{x\sqrt{a + bx}}\, dx$

29. $\displaystyle\int \frac{x}{\sqrt{a+bx}}\,dx = \frac{2}{3b^2}(bx - 2a)\sqrt{a+bx}$

30. $\displaystyle\int \frac{x^2 dx}{\sqrt{a+bx}} = \frac{2}{15b^3}(8a^2 + 3b^2 x^2 - 4abx)\sqrt{a+bx}$

31. $\displaystyle\int (\sqrt{a+bx})^n\,dx = \frac{2}{b}\frac{(a+bx)^{1+n/2}}{n+2}$ $[n \neq -2]$

32. $\displaystyle\int \frac{\sqrt{a+bx}}{x}\,dx = 2\sqrt{a+bx} + a\int \frac{1}{x\sqrt{a+bx}}\,dx$

33. $\displaystyle\int x\sqrt{a+bx}\,dx = \frac{2}{15b^2}(3bx - 2a)(a+bx)^{3/2}$

34. $\displaystyle\int \sqrt{a^2+x^2}\,dx = \frac{x}{2}\sqrt{a^2+x^2} + \frac{a^2}{2}\operatorname{arcsinh}\frac{x}{a}$

35. $\displaystyle\int x^2\sqrt{a^2+x^2}\,dx = \frac{x}{8}(a^2 + 2x^2)\sqrt{a^2+x^2} - \frac{a^4}{8}\operatorname{arcsinh}\frac{x}{a}$

36. $\displaystyle\int \frac{\sqrt{a^2+x^2}}{x}\,dx = \sqrt{a^2+x^2} - a\ln\left[\frac{(a^2+x^2)^{1/2}+a}{x}\right]$

37. $\displaystyle\int \frac{\sqrt{a^2+x^2}}{x^2}\,dx = \ln[(a^2+x^2)^{1/2}+x] - \frac{\sqrt{a^2+x^2}}{x}$

38. $\displaystyle\int \frac{1}{x\sqrt{a^2+x^2}}\,dx = -\frac{1}{a}\ln\left[\frac{(a^2+x^2)^{1/2}+a}{x}\right]$

39. $\displaystyle\int \frac{1}{x^2\sqrt{a^2+x^2}}\,dx = -\frac{\sqrt{a^2+x^2}}{a^2 x}$

40. $\displaystyle\int \sqrt{a^2-x^2}\,dx = \frac{x}{2}\sqrt{a^2-x^2} + \frac{a^2}{2}\arcsin\frac{x}{|a|}$ $[x^2 < a^2]$

41. $\displaystyle\int \frac{1}{x\sqrt{a^2-x^2}}\,dx = -\frac{1}{a}\ln\left[\frac{(a^2-x^2)^{1/2}+a}{x}\right]$ $[x^2 < a^2]$

42. $\displaystyle\int \sqrt{x^2-a^2}\,dx\,\frac{x}{2} = \sqrt{x^2-a^2} - \frac{a^2}{2}\ln[(x^2-a^2)^{1/2}+x]$ $[a^2 < x^2]$

43. $\displaystyle\int \frac{\sqrt{x^2-a^2}}{x}\,dx = \sqrt{x^2-a^2} - a\operatorname{arcsec}\left|\frac{x}{a}\right|$ $[a^2 \le x^2]$

44. $\displaystyle\int \frac{1}{x^2\sqrt{x^2-a^2}}\,dx = \frac{\sqrt{x^2-a^2}}{a^2 x}$ $[a^2 < x^2]$

45. $\displaystyle\int \frac{1}{(a^2 + x^2)^2}\, dx = \frac{x}{2a^2(a^2 + x^2)} + \frac{1}{2a^3}\arctan\frac{x}{a}$

46. $\displaystyle\int \frac{1}{(a^2 - x^2)^2}\, dx = \frac{x}{2a^2(a^2 - x^2)} - \frac{1}{4a^3}\ln\left[\frac{x-a}{x+a}\right]$ $[x^2 < a^2]$

Integrands involving trigonometric functions, powers of x, and exponentials

47. $\displaystyle\int \sin ax\, dx = -\frac{1}{a}\cos ax$

48. $\displaystyle\int \sin^2 ax\, dx = \frac{x}{2} - \frac{\sin 2ax}{4a}$

49. $\displaystyle\int \cos ax\, dx = \frac{1}{a}\sin ax$

50. $\displaystyle\int \cos^2 ax\, dx = \frac{x}{2} + \frac{\sin 2ax}{4a}$

51. $\displaystyle\int \sin ax \sin bx\, dx = \frac{\sin(a-b)x}{2(a-b)} - \frac{\sin(a+b)x}{2(a+b)}$ $[a^2 \neq b^2]$

52. $\displaystyle\int \cos ax \cos bx\, dx = \frac{\sin(a-b)x}{2(a-b)} + \frac{\sin(a+b)x}{2(a+b)}$ $[a^2 \neq b^2]$

53. $\displaystyle\int \sin ax \cos bx\, dx = -\frac{\cos(a+b)x}{2(a+b)} - \frac{\cos(a-b)x}{2(a-b)}$ $[a^2 \neq b^2]$

54. $\displaystyle\int \sin ax \cos ax\, dx = \frac{\sin^2 ax}{2a}$

55. $\displaystyle\int x \sin ax\, dx = \frac{\sin ax}{a^2} - \frac{x\cos ax}{a}$

56. $\displaystyle\int x^2 \sin ax\, dx = \frac{2x \sin ax}{a^2} - \frac{(a^2x^2 - 2)}{a^3}\cos ax$

57. $\displaystyle\int x \cos ax\, dx = \frac{x \sin ax}{a} + \frac{1}{a^2}\cos ax$

58. $\displaystyle\int x^2 \cos ax\, dx = \left(\frac{x^2}{a} - \frac{2}{a^3}\right)\sin ax + \frac{2x\cos ax}{a^2}$

59. $\displaystyle\int e^{ax} \sin bx\, dx = \frac{e^{ax}}{a^2 + b^2}(a \sin bx - b \cos bx)$

60. $\displaystyle\int e^{ax} \cos bx\, dx = \frac{e^{ax}}{a^2 + b^2}(a \cos bx + b \sin bx)$

61. $\displaystyle\int \sec ax\, dx = \frac{1}{a}\ln|\sec ax + \tan ax|$

62. $\displaystyle\int \csc ax\, dx = \frac{1}{a}\ln|\csc ax - \cot ax|$

63. $\displaystyle\int \cot ax\, dx = \frac{1}{a}\ln|\sin ax|$

64. $\displaystyle\int \tan^2 ax\, dx = \frac{1}{a}\tan ax - x$

65. $\displaystyle\int \sec^2 ax\, dx = \frac{1}{a}\tan ax$

66. $\displaystyle\int \csc^2 ax\, dx = -\frac{1}{a}\cot ax$

67. $\displaystyle\int \cot^2 ax\, dx = -\frac{1}{a}\cot ax - x$

Integrands involving inverse trigonometric functions

68. $\displaystyle\int \arcsin ax\, dx = x\arcsin ax + \frac{1}{a}\sqrt{1 - a^2x^2}$ $[a^2x^2 \le 1]$

69. $\displaystyle\int \arccos ax\, dx = x\arccos ax - \frac{1}{a}\sqrt{1 - a^2x^2}$ $[a^2x^2 \le 1]$

70. $\displaystyle\int \arctan ax\, dx = x\arctan ax - \frac{1}{2a}\ln(1 + a^2x^2)$

Integrands involving exponential and logarithmic functions

71. $\displaystyle\int e^{ax}\, dx = \frac{1}{a}e^{ax}$

72. $\displaystyle\int b^{ax}\, dx = \frac{1}{a}\frac{b^{ax}}{\ln b}$ $[b > 0, b \ne 1]$

73. $\displaystyle\int xe^{ax}\, dx = \frac{e^{ax}}{a^2}(ax - 1)$

74. $\displaystyle\int \ln ax\, dx = x\ln ax - x$

75. $\displaystyle\int \frac{\ln ax}{x}\, dx = \frac{1}{2}(\ln ax)^2$

76. $\displaystyle\int \frac{1}{x\ln ax}\, dx = \ln|\ln ax|$

Integrands involving hyperbolic functions

77. $\displaystyle\int \sinh ax \, dx = \frac{1}{a}\cosh ax$

78. $\displaystyle\int \sinh^2 ax \, dx = \frac{\sinh 2ax}{4a} - \frac{x}{2}$

79. $\displaystyle\int x \sinh ax \, dx = \frac{x}{a}\cosh ax - \frac{1}{a^2}\sinh ax$

80. $\displaystyle\int \cosh ax \, dx = \frac{1}{a}\sinh ax$

81. $\displaystyle\int \cosh^2 ax \, dx = \frac{\sinh 2ax}{4a} + \frac{x}{2}$

82. $\displaystyle\int x \cosh ax \, dx = \frac{x}{a}\sinh ax - \frac{1}{a^2}\cosh ax$

83. $\displaystyle\int e^{ax} \sinh bx \, dx = \frac{e^{ax}}{2}\left(\frac{e^{bx}}{a+b} - \frac{e^{-bx}}{a-b}\right) \qquad [a^2 \neq b^2]$

84. $\displaystyle\int e^{ax} \cosh bx \, dx = \frac{e^{ax}}{2}\left(\frac{e^{bx}}{a+b} + \frac{e^{-bx}}{a-b}\right) \qquad [a^2 \neq b^2]$

85. $\displaystyle\int e^{ax} \sinh ax \, dx = \frac{1}{4a}e^{2ax} - \frac{1}{2}x$

86. $\displaystyle\int e^{ax} \cosh ax \, dx = \frac{1}{4a}e^{2ax} + \frac{1}{2}x$

87. $\displaystyle\int \tanh ax \, dx = \frac{1}{a}\ln(\cosh ax)$

88. $\displaystyle\int \tanh^2 ax \, dx = x - \frac{1}{a}\tanh ax$

89. $\displaystyle\int \coth ax \, dx = \frac{1}{a}\ln|\sinh ax|$

90. $\displaystyle\int \coth^2 ax \, dx = x - \frac{1}{a}\coth ax$

91. $\displaystyle\int \operatorname{sech} ax \, dx = \frac{2}{a}\arctan e^{ax}$

92. $\displaystyle\int \operatorname{sech}^2 ax \, dx = \frac{1}{a}\tanh ax$

93. $\displaystyle \int \operatorname{csch} ax\,dx = \frac{1}{a}\ln\left|\tanh\frac{ax}{2}\right|$

94. $\displaystyle \int \operatorname{csch}^2 ax\,dx = -\frac{1}{a}\coth ax$

Integrands involving inverse hyperbolic functions

95. $\displaystyle \int \operatorname{arcsinh}\frac{x}{a}\,dx = x\operatorname{arcsinh}\frac{x}{a} - (a^2 + x^2)^{1/2}$

$$= x\ln\left[\frac{x + (a^2 + x^2)^{1/2}}{a}\right] - (a^2 + x^2)^{1/2} \qquad [a > 0]$$

96. $\displaystyle \int \operatorname{arccosh}\frac{x}{a}\,dx = x\operatorname{arccosh}\frac{x}{a} - (x^2 - a^2)^{1/2}$

$$= x\ln\left[\frac{x + (x^2 - a^2)^{1/2}}{a}\right] - (x^2 - a^2)^{1/2} \qquad \left[\operatorname{arccosh}\frac{x}{a} > 0,\, x^2 > a^2\right]$$

$$= x\operatorname{arccosh}\frac{x}{a} + (x^2 - a^2)^{1/2} \qquad \left[\operatorname{arccosh}\frac{x}{a} < 0,\, x^2 > a^2\right]$$

$$= x\ln\left[\frac{x + (x^2 - a^2)^{1/2}}{a}\right] + (x^2 - a^2)^{1/2}$$

97. $\displaystyle \int \operatorname{arctanh}\frac{x}{a}\,dx = x\operatorname{arctanh}\frac{x}{a} + \frac{a}{2}\ln(a^2 - x^2)$

$$= \frac{x}{2}\ln\left(\frac{a + x}{a - x}\right) + \frac{a}{2}\ln(a^2 - x^2) \qquad [x^2 < a^2]$$

98. $\displaystyle \int x\operatorname{arcsinh}\frac{x}{a}\,dx = \left(\frac{x^2}{2} + \frac{a^2}{4}\right)\operatorname{arcsinh}\frac{x}{a} - \frac{x}{4}\sqrt{a^2 + x^2}$

$$= \left(\frac{x^2}{2} + \frac{a^2}{4}\right)\ln\left[\frac{x + (a^2 + x^2)^{1/2}}{a}\right] - \frac{x}{4}\sqrt{a^2 + x^2} \qquad [a > 0]$$

99. $\displaystyle \int x\operatorname{arccosh}\frac{x}{a}\,dx = \left(\frac{x^2}{2} - \frac{a^2}{4}\right)\operatorname{arccosh}\frac{x}{a} - \frac{x}{4}\sqrt{x^2 - a^2}$

$$= \left(\frac{x^2}{2} - \frac{a^2}{4}\right)\ln\left[\frac{x + (x^2 - a^2)^{1/2}}{a}\right] - \frac{x}{4}\sqrt{x^2 - a^2}$$

$$\left[\operatorname{arccosh}\frac{x}{a} > 0,\, x^2 > a^2\right]$$

$$= \left(\frac{x^2}{2} - \frac{a^2}{4}\right)\operatorname{arccosh}\frac{x}{a} + \frac{x}{4}\sqrt{x^2 - a^2} \qquad \left[\operatorname{arccosh}\frac{x}{a} < 0,\, x^2 > a^2\right]$$

$$= \left(\frac{x^2}{2} - \frac{a^2}{4}\right)\ln\left[\frac{x + (x^2 - a^2)^{1/2}}{a}\right] + \frac{x}{4}\sqrt{x^2 - a^2}$$

100. $\displaystyle\int x \, \text{arctanh} \, \frac{x}{a} \, dx = \left(\frac{x^2 - a^2}{2}\right) \text{arctanh} \, \frac{x}{a} + \frac{1}{2}ax$

$$= \left(\frac{x^2 - a^2}{4}\right) \ln\left[\frac{a + x}{a - x}\right] + \frac{1}{2}ax \qquad [x^2 < a^2]$$

0.7 Standard Series

Power series

1. $(1 \pm x)^{-1} = 1 \mp x + x^2 \mp x^3 + x^4 \mp \cdots \qquad [|x| < 1]$

2. $(1 \pm x)^{-2} = 1 \mp 2x + 3x^2 \mp 4x^3 + 5x^4 \mp \cdots \qquad [|x| < 1]$

3. $(1 \pm x^2)^{-1} = 1 \mp x^2 + x^4 \mp x^6 + x^8 \mp \cdots \qquad [|x| < 1]$

4. $(1 \pm x^2)^{-2} = 1 \mp 2x^2 + 3x^4 \mp 4x^6 + 5x^8 \mp \cdots \qquad [|x| < 1]$

5. $(1 + x)^\alpha = 1 + \alpha x + \dfrac{\alpha(\alpha - 1)}{2!} x^2 + \dfrac{\alpha(\alpha - 1)(\alpha - 2)}{3!} x^3 + \cdots$

$$= 1 + \sum_{n=1}^{\infty} \frac{\alpha(\alpha - 1)(\alpha - 2) \cdots (\alpha - n + 1)}{n!} x^n, \, \alpha \text{ real and } |x| < 1.$$

(the binomial series)

These results may be extended by replacing x with $\pm x^k$ and making the appropriate modification to the convergence condition $|x| < 1$. Thus, replacing x with $\pm x^2/4$ and setting $\alpha = -1/2$ in power series 5 gives

$$\left(1 \pm \frac{x^2}{4}\right)^{-1/2} = 1 \mp \frac{1}{8}x^2 + \frac{3}{128}x^4 \mp \frac{5}{1024}x^6 + \cdots,$$

for $|x^2/4| < 1$, which is equivalent to $|x| < 2$.

Trigonometric series

6. $\sin x = x - \dfrac{x^3}{3!} + \dfrac{x^5}{5!} - \dfrac{x^7}{7!} + \cdots \qquad [|x| < \infty]$

7. $\cos x = 1 - \dfrac{x^2}{2!} + \dfrac{x^4}{4!} - \dfrac{x^6}{6!} + \cdots \qquad [|x| < \infty]$

8. $\tan x = x + \dfrac{x^3}{3} + \dfrac{2}{15}x^5 - \dfrac{17}{315}x^7 + \dfrac{62}{2835}x^9 + \cdots \qquad [|x| < \pi/2]$

Inverse trigonometric series

9. $\arcsin x = x + \dfrac{x^3}{2.3} + \dfrac{1.3}{2.4.5}x^5 + \dfrac{1.3.5}{2.4.6.7}x^7 + \cdots$

$$[|x| < 1, -\pi/2 < \arcsin x < \pi/2]$$

10. $\arccos x = \dfrac{\pi}{2} - \arcsin x \qquad [|x| < 1, \, 0 < \arccos x < \pi]$

11. $\arctan x = x - \dfrac{x^3}{3} + \dfrac{x^5}{5} - \dfrac{x^7}{7} + \cdots$ $[|x| < 1]$

$\qquad = \dfrac{\pi}{2} - \dfrac{1}{x} + \dfrac{1}{3x^3} - \dfrac{1}{5x^5} + \dfrac{1}{7x^7} - \cdots$ $[x > 1]$

$\qquad = -\dfrac{\pi}{2} - \dfrac{1}{x} + \dfrac{1}{3x^3} - \dfrac{1}{5x^5} + \dfrac{1}{7x^7} - \cdots$ $[x < -1]$

Exponential series

12. $e^x = 1 + x + \dfrac{x^2}{2!} + \dfrac{x^3}{3!} + \dfrac{x^4}{4!} + \cdots$ $[|x| < \infty]$

13. $e^{-x} = 1 - x + \dfrac{x^2}{2!} - \dfrac{x^3}{3!} + \dfrac{x^4}{4!} - \cdots$ $[|x| < \infty]$

Logarithmic series

14. $\ln(1 + x) = x - \dfrac{x^2}{2} + \dfrac{x^3}{3} - \dfrac{x^4}{4} + \dfrac{x^5}{5} - \cdots$ $[-1 < x \le 1]$

15. $\ln(1 + x) = -\left(x + \dfrac{x^2}{2} + \dfrac{x^3}{3} + \dfrac{x^4}{4}\right) + \cdots$ $[-1 \le x < 1]$

16. $\ln\left(\dfrac{1 + x}{1 - x}\right) = 2\left(x + \dfrac{x^3}{3} + \dfrac{x^5}{5} + \dfrac{x^7}{7} + \cdots\right) = 2\operatorname{arctanh} x$ $[|x| < 1]$

17. $\ln\left(\dfrac{1 - x}{1 + x}\right) = -2\left(x + \dfrac{x^3}{3} + \dfrac{x^5}{5} + \dfrac{x^7}{7} + \cdots\right) = -2\operatorname{arctanh} x$ $[|x| < 1]$

Hyperbolic series

18. $\sinh x = x + \dfrac{x^3}{3!} + \dfrac{x^5}{5!} + \dfrac{x^7}{7!} + \cdots$ $[|x| < \infty]$

19. $\cosh x = 1 + \dfrac{x^2}{2!} + \dfrac{x^4}{4!} + \dfrac{x^6}{6!} + \cdots$ $[|x| < \infty]$

20. $\tanh x = x - \dfrac{x^3}{3} + \dfrac{2}{15}x^5 - \dfrac{17}{315}x^7 + \dfrac{62}{2835}x^9 - \cdots$ $[|x| < \pi/2]$

Inverse hyperbolic series

21. $\operatorname{arcsinh} x = x - \dfrac{1}{2.3}x^3 + \dfrac{1.3}{2.4.5}x^5 - \dfrac{1.3.5}{2.4.6.7}x^7 + \cdots$ $[|x| < 1]$

22. $\operatorname{arccosh} x = \pm\left(\ln(2x) - \dfrac{1}{2.2x^2} - \dfrac{1.3}{2.4.4x^4} - \dfrac{1.3.5}{2.4.6.6x^6} - \cdots\right)$ $[x > 1]$

23. $\operatorname{arctanh} x = x + \dfrac{x^3}{3} + \dfrac{x^5}{5} + \dfrac{x^7}{7} + \cdots$ $[|x| < 1]$

0.8 Geometry

Triangle

Area $A = \frac{1}{2}ah = \frac{1}{2}ac\sin\theta$.

For the equilateral triangle in which $a = b = c$,

$$A = \frac{a^2\sqrt{3}}{4}, \qquad h = \frac{a\sqrt{3}}{2}.$$

The centroid C is located on the median RM (the line drawn from R to the midpoint M of PQ) with $MC = \frac{1}{3}RM$.

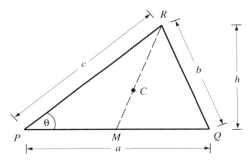

Parallelogram

Area $A = ah = ab\sin\alpha$.

The centroid C is located at the point of intersection of the diagonals.

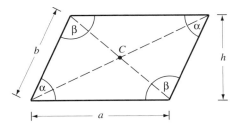

Trapezium A quadrilateral with two sides parallel, where h is the perpendicular distance between the parallel sides.

Area $A = \frac{1}{2}(a + b)h$.

The centroid C is located on PQ, the line joining the midpoints of AB and CD, with

$$QC = \frac{h}{3}\frac{(a + 2b)}{(a + b)}.$$

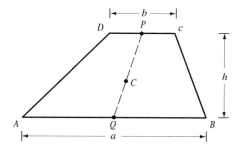

Rhombus   A parallelogram with all sides of equal length.

Area $A = a^2 \sin \alpha$.

The centroid C is located at the point of intersection of the diagonals.

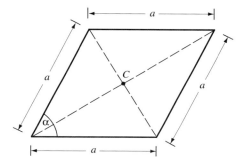

Cube

Area $A = 6a^2$.
Volume $V = a^3$.
Diagonal $d = a\sqrt{3}$.

The centroid C is located at the midpoint of a diagonal.

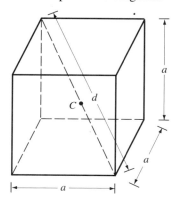

Rectangular parallelepiped

Area $A = 2(ab + ac + bc)$.
Volume $V = abc$.
Diagonal $d = \sqrt{a^2 + b^2 + c^2}$.

The centroid C is located at the midpoint of a diagonal.

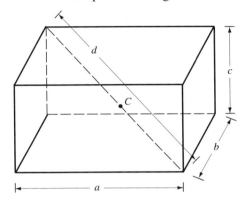

Pyramid   Rectangular base with sides of length a and b and four sides comprising pairs of identical isosceles triangles.

Area of sides $A_S = a\sqrt{h^2 + (b/2)^2} + b\sqrt{h^2 + (a/2)^2}$.
Area of base $A_B = ab$.
Total area $A = A_B + A_S$.
Volume $V = \dfrac{1}{3} abh$.

The centroid C is located on the axis of symmetry with $OC = h/4$.

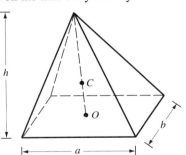

Rectangular (right) wedge   Base is rectangular, ends are isosceles triangles of equal size, and the remaining two sides are trapezia.

Area of sides $A_S = \frac{1}{2}(a + c)\sqrt{4h^2 + b^2} + \frac{1}{2}b\sqrt{4h^2 + 4(a - c)^2}$.

Area of base $A_B = ab$.

Total area $A = A_B + A_S$.

Volume $V = \dfrac{bh}{6}(2a + c)$.

The centroid C is located on the axis of symmetry with

$$OC = \frac{h}{2}\frac{(a + c)}{(2a + c)}.$$

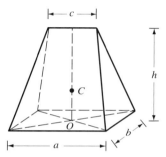

Tetrahedron   Formed by four equilateral triangles.

Surface area $A = a^2\sqrt{3}$.

Volume $V = \dfrac{a^3\sqrt{2}}{12}$.

The centroid C is located on the line from the centroid O of the base triangle to the vertex, with $OC = h/4$.

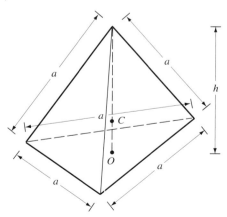

Oblique prism with plane end faces   If A_B is the area of a plane end face and h is the perpendicular distance between the parallel end faces, then

    Total area $=$ Area of plane sides $+\, 2A_B$.
    Volume $V = A_B h$.

The centroid C is located at the midpoint of the line $C_1 C_2$ joining the centroids of the parallel end faces.

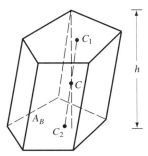

Circle

    Area $A = \pi r^2$,   Circumference $L = 2\pi r$,

where r is the radius of the circle. The centroid is located at the center of the circle.

Arc of circle

    Length of arc $AB : s = r\alpha\,(\alpha\,\text{radians})$.

The centroid C is located on the axis of symmetry with $OC = ra/s$.

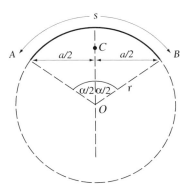

Sector of circle

$$\text{Area } A = \frac{sr}{2} = \frac{r^2\alpha}{2}\,(\alpha\text{ radians}).$$

The centroid C is located on the axis of symmetry with

$$OC = \frac{2}{3}\frac{ra}{s}.$$

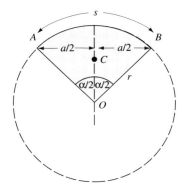

Segment of circle

$$a = 2\sqrt{2hr - h^2}.$$

$$h = r - \frac{1}{2}\sqrt{4r^2 - a^2} \qquad [h < r].$$

$$\text{Area } A = \frac{1}{2}[sr - a(r - h)].$$

The centroid C is located on the axis of symmetry with

$$OC = \frac{a^3}{12A}.$$

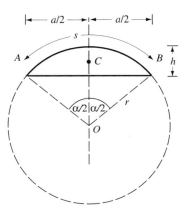

Annulus

$$\text{Area } A = \pi(R^2 - r^2) \qquad [r < R].$$

The centroid C is located at the center.

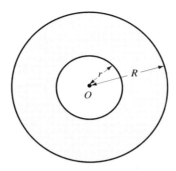

Right circular cylinder

Area of curved surface $A_S = 2\pi rh$.
Area of plane ends $A_B = 2\pi r^2$.
Total Area $A = A_B + A_S$.
Volume $V = \pi r^2 h$.

The centroid C is located on the axis of symmetry with $OC = h/2$.

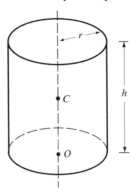

Right circular cylinder with an oblique plane face   Here, h_1 is the greatest height of a side of the cylinder, h_2 is the shortest height of a side of the cylinder, and r is the radius of the cylinder.

Area of curved surface $A_S = \pi r(h_1 + h_2)$.

Area of plane end faces $A_B = \pi r^2 + \pi r \sqrt{r^2 + \left(\dfrac{h_1 - h_2}{2}\right)^2}$.

Total area $A = \pi r \left[h_1 + h_2 + r + \sqrt{r^2 + \dfrac{(h_1 - h_2)^2}{2}} \right]$.

Volume $V = \dfrac{\pi r^2}{2}(h_1 + h_2)$.

The centroid C is located on the axis of symmetry with

$$OC = \frac{(h_1 + h_2)}{4} + \frac{1}{16}\frac{(h_1 - h_2)^2}{h_1 + h_2}.$$

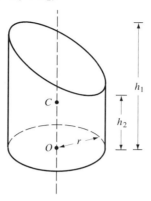

Cylindrical wedge   Here, r is radius of cylinder, h is height of wedge, $2a$ is base chord of wedge, b is the greatest perpendicular distance from the base chord to the wall of the cylinder measured perpendicular to the axis of the cylinder, and α is the angle subtended at the center O of the normal cross-section by the base chord.

Area of curved surface $A_S = \dfrac{2rh}{b}\left[(b - r)\dfrac{\alpha}{2} + a\right].$

Volume $V = \dfrac{h}{3b}\left[a(3r^2 - a^2) + 3r^2(b - r)\dfrac{\alpha}{2}\right].$

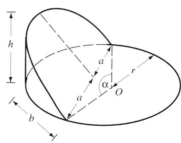

Right circular cone

Area of curved surface $A_S = \pi r s.$
Area of plane end $A_B = \pi r^2.$
Total area $A = A_B + A_S.$
Volume $V = \dfrac{1}{3}\pi r^2 h.$

The centroid C is located on the axis of symmetry with $OC = h/4$.

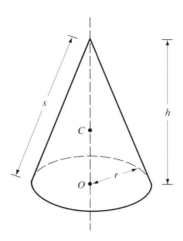

Frustrum of a right circular cone

$s = \sqrt{h^2 + (r_1 - r_2)^2}$.
Area of curved surface $A_S = \pi s(r_1 + r_2)$.
Area of plane ends $A_B = \pi\left(r_1^2 + r_2^2\right)$.
Total area $A = A_B + A_S$.

Volume $V = \dfrac{1}{3}\pi h\left(r_1^2 + r_1 r_2 + r_2^2\right)$.

The centroid C is located on the axis of symmetry with

$$OC = \frac{h}{4}\frac{\left(r_1^2 + 2r_1 r_2 + 3r_2^2\right)}{\left(r_1^2 + r_1 r_2 + r_2^2\right)}.$$

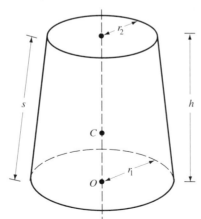

General cone If A is the area of the base and h is the perpendicular height, then

Volume $V = \dfrac{1}{3}Ah$.

The centroid C is located on the line joining the centroid O of the base to the vertex P with

$$OC = \frac{1}{4}OP.$$

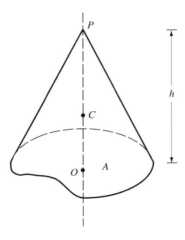

Sphere

Area $A = 4\pi r^2$ (r is radius of sphere).

Volume $V = \frac{4}{3}\pi r^3$.

The centroid is located at the center.

Spherical sector   Here, h is height of spherical segment cap, a is radius of plane face of spherical segment cap, and r is radius of sphere. For the area of the spherical cap and conical sides,

$$A = \pi r(2h + a).$$

Volume $V = \dfrac{2\pi r^2 h}{3}.$

The centroid C is located on the axis of symmetry with $OC = \frac{3}{8}(2r - h)$.

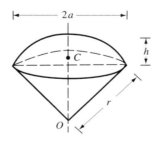

Spherical segment   Here, h is height of spherical segment, a is radius of plane face of spherical segment, r is radius of sphere, and $a = \sqrt{h(2r - h)}$.

Area of spherical surface $A_S = 2\pi rh$.
Area of plane face $A_B = \pi a^2$.
Total area $A = A_B + A_S$.

Volume $V = \dfrac{1}{3}\pi h^2(3r - h) = \dfrac{1}{6}\pi h(3a^2 + h^2)$.

The centroid C is located on the axis of symmetry with

$$OC = \frac{3}{4}\frac{(2r - h)^2}{(3r - h)}.$$

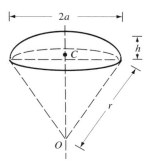

Spherical segment with two parallel plane faces   Here, a_1 and a_2 are the radii of the plane faces and h is height of segment.

Area of spherical surface $A_S = 2\pi rh$.
Area of plane end faces $A_B = \pi\left(a_1^2 + a_2^2\right)$.
Total area $A = A_B + A_S$.

Volume $V = \dfrac{1}{6}\pi h\left(3a_1^2 + 3a_2^2 + h^2\right)$.

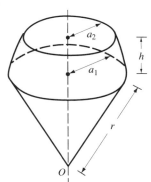

Ellipsoids of revolution   Let the ellipse have the equation

$$\frac{x^2}{a^2} + \frac{y^2}{b^2} = 1.$$

When rotated about the x-axis the volume is

$$V_x = \frac{4}{3}\pi a b^2.$$

When rotated about the y-axis the volume is

$$V_y = \frac{4}{3}\pi a^2 b.$$

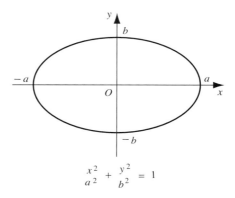

$$\frac{x^2}{a^2} + \frac{y^2}{b^2} = 1$$

Torus   Ring with circular cross-section:

Area $A = 4\pi^2 r R.$
Volume $V = 2\pi^2 r^2 R.$

(See Pappus's theorem 1.15.6.1.5.)

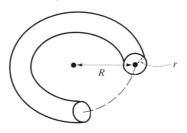

CHAPTER 1

Numerical, Algebraic, and Analytical Results for Series and Calculus

1.1 Algebraic Results Involving Real and Complex Numbers

1.1.1 Complex numbers

1.1.1.1 Basic definitions. If $a, b, c, d, \ldots$ are real numbers, the set $\mathbb{C}$ of complex numbers, in which individual complex numbers are denoted by $z, \zeta, \omega, \ldots$, is the set of all ordered number pairs $(a, b), (c, d), \ldots$ that obey the following rules defining the equality, sum, and product of complex numbers.

If $z_1 = (a, b)$ and $z_2 = (c, d)$, then:

(*Equality*) $z_1 = z_2$ implies, $a = c$ and $b = d$,

(*Sum*) $z_1 + z_2$ implies, $(a + c, b + d)$,

(*Product*) $z_1 z_2$ or $z_1 \cdot z_2$ implies, $(ac - bd, ad + bc)$.

Because $(a, 0) + (b, 0) = (a + b, 0)$ and $(a, 0) \cdot (b, 0) = (ab, 0)$ it is usual to set the complex number $(x, 0) = x$ and to call it a **purely real** number, since it has the properties of a real number. The complex number $i = (0, 1)$ is called the **imaginary unit,** and from the definition of multiplication $(0, 1) \cdot (0, 1) = (-1, 0)$, so it follows that

$$i^2 = -1 \qquad \text{or} \qquad i = \sqrt{-1}.$$

If $z = (x, y)$, the **real part** of z, denoted by $\text{Re}\{z\}$, is defined as $\text{Re}\{z\} = x$, while the **imaginary part** of z, denoted by $\text{Im}\{z\}$, is defined as $\text{Im}\{z\} = y$. A number of the form $(0, y) = (y, 0) \cdot (0, 1) = yi$ is called a **purely imaginary** number. The **zero** complex number $z = (0, 0)$ is also written $z = 0$. The **complex conjugate** $\bar{z}$ of a complex number $z = (x, y)$ is defined as

$$\bar{z} = (x, -y),$$

while its **modulus** (also called its **absolute value**) is the real number $|z|$ defined as

$$|z| = (x^2 + y^2)^{1/2},$$

so that

$$|z|^2 = z\bar{z}.$$

The **quotient** of the two complex numbers z_1 and z_2 is given by

$$\frac{z_1}{z_2} = \frac{z_1\bar{z}_2}{z_2\bar{z}_2} = \frac{z_1\bar{z}_2}{|z_2|^2} \qquad [z_2 \neq 0].$$

When working with complex numbers it is often more convenient to replace the ordered pair notation $z = (x, y)$ by the equivalent notation $z = x + iy$.

1.1.1.2 Properties of the modulus and complex conjugate.

1. If $z = (x, y)$ then
$$z + \bar{z} = 2\,\text{Re}\{z\} = 2x$$
$$z - \bar{z} = 2i\,\text{Im}\{z\} = 2iy$$

2. $|z| = |\bar{z}|$

3. $z = \overline{(\bar{z})}$

4. $\dfrac{1}{\bar{z}} = \overline{\left(\dfrac{1}{z}\right)} \qquad [z \neq 0]$

5. $\overline{(z^n)} = (\bar{z})^n$

6. $\left|\dfrac{\bar{z}_1}{\bar{z}_2}\right| = \dfrac{|\bar{z}_1|}{|\bar{z}_2|} \qquad [z_2 \neq 0]$

7. $\overline{(z_1 + z_2 + \cdots + z_n)} = \bar{z}_1 + \bar{z}_2 + \cdots + \bar{z}_n$

8. $\overline{z_1 z_2 \cdots z_n} = \bar{z}_1 \bar{z}_2 \cdots \bar{z}_n$

1.1.2 Algebraic inequalities involving real and complex numbers

1.1.2.1 The triangle and a related inequality.
If a, b are any two real numbers, then

$$|a + b| \leq |a| + |b| \qquad\qquad \text{(triangle inequality)}$$
$$|a - b| \geq ||a| - |b||,$$

where $|a|$, the absolute value of a is defined as

$$|a| = \begin{cases} a, & a \geq 0 \\ -a, & a < 0. \end{cases}$$

Analogously, if a, b are any two complex numbers, then

$$|a + b| \leq |a| + |b|$$
$$|a - b| \geq ||a| - |b||.$$

(triangle inequality)

1.1.2.2 Lagrange's identity. Let $a_1, a_2, \ldots, a_n$ and $b_1, b_2, \ldots, b_n$ be any two sets of real numbers; then

$$\left(\sum_{k=1}^{n} a_k b_k \right)^2 = \left(\sum_{k=1}^{n} a_k^2 \right) \left(\sum_{k=1}^{n} b_k^2 \right) - \sum_{1 \leq k < j \leq n} (a_k b_j - a_j b_k)^2.$$

1.1.2.3 Cauchy–Schwarz–Buniakowsky inequality. Let $a_1, a_2, \ldots, a_n$ and $b_1, b_2, \ldots, b_n$ be any two arbitrary sets of real numbers; then

$$\left(\sum_{k=1}^{n} a_k b_k \right)^2 \leq \left(\sum_{k=1}^{n} a_k^2 \right) \left(\sum_{k=1}^{n} b_k^2 \right).$$

The equality holds if, and only if, the sequences $a_1, a_2, \ldots, a_n$ and $b_1, b_2, \ldots, b_n$ are proportional. Analogously, let $a_1, a_2, \ldots, a_n$ and $b_1, b_2, \ldots, b_n$ be any two arbitrary sets of complex numbers; then

$$\left| \sum_{k=1}^{n} a_k b_k \right|^2 \leq \left(\sum_{k=1}^{n} |a_k|^2 \right) \left(\sum_{k=1}^{n} |b_k|^2 \right).$$

The equality holds if, and only if, the sequences $\overline{a}_1, \overline{a}_2, \ldots, \overline{a}_n$ and $b_1, b_2, \ldots, b_n$ are proportional.

1.1.2.4 Minkowski's inequality. Let $a_1, a_2, \ldots, a_n$ and $b_1, b_2, \ldots, b_n$ be any two sets of nonnegative real numbers and let $p > 1$; then

$$\left(\sum_{k=1}^{n} (a_k + b_k)^p \right)^{1/p} \leq \left(\sum_{k=1}^{n} a_k^p \right)^{1/p} + \left(\sum_{k=1}^{n} b_k^p \right)^{1/p}.$$

The equality holds if, and only if, the sequences $a_1, a_2, \ldots, a_n$ and $b_1, b_2, \ldots, b_n$ are proportional. Analogously, let $a_1, a_2, \ldots, a_n$ and $b_1, b_2, \ldots, b_n$ be any two arbitrary sets of complex numbers, and let the real number p be such that $p > 1$; then

$$\left(\sum_{k=1}^{n} |a_k + b_k|^p \right)^{1/p} \leq \left(\sum_{k=1}^{n} |a_k|^p \right)^{1/p} + \left(\sum_{k=1}^{n} |b_k|^p \right)^{1/p}.$$

1.1.2.5 Hölder's inequality. Let $a_1, a_2, \ldots, a_n$ and $b_1, b_2, \ldots, b_n$ be any two sets of nonnegative real numbers, and let $1/p + 1/q = 1$, with $p > 1$; then

$$\left(\sum_{k=1}^{n} a_k^p \right)^{1/p} \left(\sum_{k=1}^{n} b_k^q \right)^{1/q} \geq \sum_{k=1}^{n} a_k b_k.$$

The equality holds if, and only if, the sequences $a_1^p, a_2^p, \ldots, a_n^p$ and $b_1^q, b_2^q, \ldots, b_n^q$ are proportional. Analogously, let $a_1, a_2, \ldots, a_n$ and $b_1, b_2, \ldots, b_n$ be any two arbitrary sets of complex numbers, and let the real numbers p, q be such that $p > 1$ and $1/p + 1/q = 1$; then

$$\left(\sum_{k=1}^n |a_k|^p \right)^{1/p} \left(\sum_{k=1}^n |b_k|^p \right)^{1/q} \geq \left| \sum_{k=1}^n a_k b_k \right|.$$

1.1.2.6 Chebyshev's inequality.   Let $a_1, a_2, \ldots, a_n$ and $b_1, b_2, \ldots, b_n$ be two arbitrary sets of real numbers such that either $a_1 \geq a_2 \geq \cdots \geq a_n$ and $b_1 \geq b_2 \geq \cdots \geq b_n$, or $a_1 \leq a_2 \leq \cdots \leq a_n$ and $b_1 \leq b_2 \leq \cdots \leq b_n$; then

$$\left(\frac{a_1 + a_2 + \cdots + a_n}{n} \right) \left(\frac{b_1 + b_2 + \cdots + b_n}{n} \right) \leq \frac{1}{n} \sum_{k=1}^n a_k b_k.$$

The equality holds if, and only if, either $a_1 = a_2 = \cdots = a_n$ or $b_1 = b_2 = \cdots = b_n$.

1.1.2.7 Arithmetic–geometric inequality.   Let $a_1, a_2, \ldots, a_n$ be any set of positive numbers with arithmetic mean

$$A_n = \left(\frac{a_1 + a_2 + \cdots + a_n}{n} \right)$$

and geometric mean

$$G_n = (a_1 a_2 \cdots a_n)^{1/n};$$

then $A_n \geq G_n$ or, equivalently,

$$\left(\frac{a_1 + a_2 + \cdots + a_n}{n} \right) \geq (a_1 a_2 \cdots a_n)^{1/n}.$$

The equality holds only when $a_1 = a_2 = \cdots = a_n$.

1.1.2.8 Carleman's inequality.   If $a_1, a_2, \ldots, a_n$ is any set of positive numbers, then the geometric and arithmetic means satisfy the inequality

$$\sum_{k=1}^n G_k \leq e A_n$$

or, equivalently,

$$\sum_{k=1}^n (a_1 a_2 \cdots a_k)^{1/k} \leq e \left(\frac{a_1 + a_2 + \cdots + a_n}{n} \right),$$

where e is the best possible constant in this inequality.

The next inequality to be listed is of a somewhat different nature than that of the previous ones in that it involves a function of the type known as *convex*. When interpreted geometrically, a function $f(x)$ that is convex on an interval $I = [a, b]$ is one for which all points on the graph of $y = f(x)$ for $a < x < b$ lie below the chord joining the points $(a, f(a))$ and $(b, f(b))$.

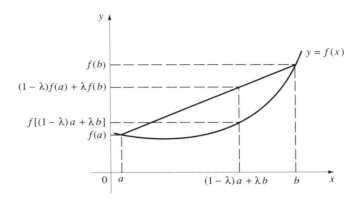

Definition of convexity. A function $f(x)$ defined on some interval $I = [a, b]$ is said to be **convex** on I if, and only if,

$$f[(1 - \lambda)a + \lambda f(b)] \leq (1 - \lambda)f(a) + \lambda f(b),$$

for $a \neq b$ and $0 < \lambda < 1$.

The function is said to be **strictly convex** on I if the preceding inequality is strict, so that

$$f[(1 - \lambda)a + \lambda f(b)] < (1 - \lambda)f(a) + \lambda f(b).$$

1.1.2.9 Jensen's inequality. Let $f(x)$ be convex on the interval $I = [a, b]$, let $x_1, x_2, \ldots, x_n$ be points in I, and take $\lambda_1, \lambda_2, \ldots, \lambda_n$ to be nonnegative numbers such that

$$\lambda_1 + \lambda_2 + \cdots + \lambda_n = 1.$$

Then the point $\lambda_1 x_1 + \lambda_2 x_2 + \cdots + \lambda_n x_n$ lies in I and

$$f(\lambda_1 x_1 + \lambda_2 x_2 + \cdots + \lambda_n x_n) \leq \lambda_1 f(x_1) + \lambda_2 f(x_2) + \cdots + \lambda_n f(x_n).$$

If all the λ_i's are strictly positive and $f(x)$ is strictly convex on I, then the equality holds if, and only if,

$$x_1 = x_2 = \cdots = x_n.$$

1.2 Finite Sums

1.2.1 The binomial theorem for positive integral exponents

1.2.1.1 Binomial theorem and binomial coefficients. If a, b are real or complex numbers, the **binomial theorem** for positive integral exponents n is

$$(a + b)^n = \sum_{k=0}^{n} \binom{n}{k} a^k b^{n-k} \qquad [n = 1, 2, 3, \ldots],$$

with

$$\binom{n}{0} = 1, \qquad \binom{n}{k} = \frac{n!}{k!(n-k)!} \qquad [k = 0, 1, 2, \ldots, \leq n],$$

and $k! = 1 \cdot 2 \cdot 3 \cdot \cdots \cdot k$ and, by definition $0! = 1$. The numbers $\binom{n}{k}$ are called **binomial coefficients.**

When expanded, the binomial theorem becomes

$$(a+b)^n = a^n + na^{n-1}b + \frac{n(n-1)}{2!}a^{n-2}b^2 + \frac{n(n-1)(n-2)}{3!}a^{n-3}b^3 + \cdots$$

$$+ \frac{n(n-1)}{2!}a^2b^{n-2} + nab^{n-1} + b^n.$$

An alternative form of the binomial theorem is

$$(a+b)^n = a^n\left(1 + \frac{b}{a}\right)^n = a^n \sum_{k=0}^{n} \binom{n}{k}\left(\frac{b}{a}\right)^{n-k} \qquad [n = 1, 2, 3, \ldots].$$

If n is a positive integer, the binomial expansion of $(a+b)^n$ contains $n+1$ terms, and so is a **finite sum**. However, if n is *not* a positive integer (it may be a positive or negative real number) the binomial expansion becomes an infinite series (see 0.7.5). (For a connection with probability see 1.6.1.1.4.)

1.2.1.2 Short table of binomial coefficients $\binom{n}{k}$.

						k							
n	0	1	2	3	4	5	6	7	8	9	10	11	12
1	1	1											
2	1	2	1										
3	1	3	3	1									
4	1	4	6	4	1								
5	1	5	10	10	5	1							
6	1	6	15	20	15	6	1						
7	1	7	21	35	35	21	7	1					
8	1	8	28	56	70	56	28	8	1				
9	1	9	36	84	126	126	84	36	9	1			
10	1	10	45	120	210	252	210	120	45	10	1		
11	1	11	55	165	330	462	462	330	165	55	11	1	
12	1	12	66	220	495	792	924	792	495	220	66	12	1

Reference to the first four rows of the table shows that

$$(a+b)^1 = a + b$$
$$(a+b)^2 = a^2 + 2ab + b^2$$
$$(a+b)^3 = a^3 + 3a^2b + 3ab^2 + b^3$$
$$(a+b)^4 = a^4 + 4a^3b + 6a^2b^2 + 4ab^3 + b^4.$$

The binomial coefficients $\binom{n}{k}$ can be generated in a simple manner by means of the following triangular array called **Pascal's triangle.**

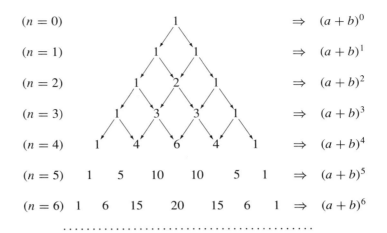

$(n = 0)$					1						$\Rightarrow$	$(a + b)^0$
$(n = 1)$				1		1					$\Rightarrow$	$(a + b)^1$
$(n = 2)$			1		2		1				$\Rightarrow$	$(a + b)^2$
$(n = 3)$		1		3		3		1			$\Rightarrow$	$(a + b)^3$
$(n = 4)$	1		4		6		4		1		$\Rightarrow$	$(a + b)^4$
$(n = 5)$	1	5		10		10		5	1		$\Rightarrow$	$(a + b)^5$
$(n = 6)$	1	6	15		20		15	6	1		$\Rightarrow$	$(a + b)^6$

The entries in the nth row are the binomial coefficients $\binom{n}{k}$ $(k = 0, 1, 2, \ldots, n)$. Each entry inside the triangle is obtained by summing the entries to its immediate left and right in the row above, as indicated by the arrows.

1.2.1.3 Relationships between binomial coefficients.

1. $\displaystyle \binom{n}{k} = \frac{n!}{k!(n-k)!} = \frac{n(n-1)(n-2)\cdots(n-k+1)}{k!} = \frac{n+1-k}{k}\binom{n}{k-1}$

2. $\displaystyle \binom{n}{0} = 1, \qquad \binom{n}{1} = n, \qquad \binom{n}{k} = \binom{n}{n-k}, \qquad \binom{n}{n} = 1$

3. $\displaystyle \binom{n}{k+1} = \frac{n-k}{k+1}\binom{n}{k}, \qquad \binom{n+1}{k} = \binom{n}{k} + \binom{n}{k-1},$

   $\displaystyle \binom{n+1}{k+1} = \binom{n}{k} + \binom{n}{k+1}$

4. $\displaystyle \binom{2n}{n} = \frac{(2n)!}{(n!)^2}, \qquad \binom{2n-1}{n} = \frac{n(2n-1)!}{(n!)^2}$

5. $\displaystyle \binom{-n}{k} = \frac{(-n)(-n-1)(-n-2)\cdots(-n-k+1)}{k!} = (-1)^k\binom{n+k-1}{k}$

1.2.1.4 Sums of binomial coefficients.   (n is an integer)

1. $\displaystyle \binom{n}{0} - \binom{n}{1} + \binom{n}{2} - \cdots + (-1)^m\binom{n}{m} = (-1)^m\binom{n-1}{m}$

$$[n \geq 1, \ m = 0, 1, 2, \ldots, n-1]$$

2. $\dbinom{n}{0} - \dbinom{n}{1} + \dbinom{n}{2} - \cdots + (-1)^n \dbinom{n}{n} = 0$

$$\left[\text{from 1.2.1.4.1 with } m = n - 1, \text{ because } \dbinom{n-1}{n-1} = 1 \right]$$

3. $\dbinom{n}{k} + \dbinom{n-1}{k} + \dbinom{n-2}{k} + \cdots + \dbinom{k}{k} = \dbinom{n+1}{k+1}$

4. $\dbinom{n}{0} + \dbinom{n}{1} + \dbinom{n}{2} + \cdots + \dbinom{n}{n} = 2^n$

5. $\dbinom{n}{0} + \dbinom{n}{2} + \dbinom{n}{4} + \cdots + \dbinom{n}{m} = 2^{n-1}$

$$[m = n \text{ (even } n\text{)}, m = n - 1 \text{ (odd } n\text{)}]$$

6. $\dbinom{n}{1} + \dbinom{n}{3} + \dbinom{n}{5} + \cdots + \dbinom{n}{m} = 2^{n-1}$

$$[m = n \text{ (even } n\text{)}, m = n - 1 \text{ (odd } n\text{)}]$$

7. $\dbinom{n}{0} + \dbinom{n}{3} + \dbinom{n}{6} + \cdots + \dbinom{n}{m} = \dfrac{1}{3}\left(2^n + 2\cos\dfrac{n\pi}{3} \right)$

$$[m = n \text{ (even } n\text{)}, m = n - 1 \text{ (odd } n\text{)}]$$

8. $\dbinom{n}{1} + \dbinom{n}{4} + \dbinom{n}{7} + \cdots + \dbinom{n}{m} = \dfrac{1}{3}\left(2^n + 2\cos\dfrac{(n-2)\pi}{3} \right)$

$$[m = n \text{ (even } n\text{)}, m = n - 1 \text{ (odd } n\text{)}]$$

9. $\dbinom{n}{2} + \dbinom{n}{5} + \dbinom{n}{8} + \cdots + \dbinom{n}{m} = \dfrac{1}{3}\left(2^n + 2\cos\dfrac{(n-4)\pi}{3} \right)$

$$[m = n \text{ (even } n\text{)}, m = n - 1 \text{ (odd } n\text{)}]$$

10. $\dbinom{n}{0} + \dbinom{n}{4} + \dbinom{n}{8} + \cdots + \dbinom{n}{m} = \dfrac{1}{2}\left(2^{n-1} + 2^{n/2}\cos\dfrac{n\pi}{4} \right)$

$$[m = n \text{ (even } n\text{)}, m = n - 1 \text{ (odd } n\text{)}]$$

11. $\dbinom{n}{1} + \dbinom{n}{5} + \dbinom{n}{9} + \cdots + \dbinom{n}{m} = \dfrac{1}{2}\left(2^{n-1} + 2^{n/2}\sin\dfrac{n\pi}{4} \right)$

$$[m = n \text{ (even } n\text{)}, m = n - 1 \text{ (odd } n\text{)}]$$

12. $\dbinom{n}{2} + \dbinom{n}{6} + \dbinom{n}{10} + \cdots + \dbinom{n}{m} = \dfrac{1}{2}\left(2^{n-1} - 2^{n/2}\cos\dfrac{n\pi}{4} \right)$

$$[m = n \text{ (even } n\text{)}, m = n - 1 \text{ (odd } n\text{)}]$$

13. $\dbinom{n}{3} + \dbinom{n}{7} + \dbinom{n}{11} + \cdots + \dbinom{n}{n} = \dfrac{1}{2}\left(2^{n-1} - 2^{n/2}\sin\dfrac{n\pi}{4} \right)$

14. $\dbinom{n}{0} + 2\dbinom{n}{1} + 3\dbinom{n}{2} + \cdots + (n+1)\dbinom{n}{n} = 2^{n-1}(n+2)$ $[n \geq 0]$

15. $\dbinom{n}{1} - 2\dbinom{n}{2} + 3\dbinom{n}{3} - \cdots + (-1)^{n+1}n\dbinom{n}{n} = 0$ $[n \geq 2]$

16. $\dbinom{N}{1} - 2^{n-1}\dbinom{N}{2} + 3^{n-1}\dbinom{N}{3} - \cdots + (-1)^N N^{n-1}\dbinom{N}{N} = 0$

$$[N \geq n; \, 0^0 \equiv 1]$$

17. $\dbinom{n}{1} - 2^n\dbinom{n}{2} + 3^n\dbinom{n}{3} - \cdots + (-1)^{n+1}n^n\dbinom{n}{n} = (-1)^{n+1}n!$

18. $\dbinom{n}{0} + \dfrac{1}{2}\dbinom{n}{1} + \dfrac{1}{3}\dbinom{n}{2} + \cdots + \dfrac{1}{n+1}\dbinom{n}{n} = \dfrac{2^{n+1}-1}{n+1}$

19. $\dbinom{n}{0}^2 + \dbinom{n}{1}^2 + \dbinom{n}{2}^2 + \cdots + \dbinom{n}{n}^2 = \dbinom{2n}{n}$

20. $\dbinom{n}{1} - \dfrac{1}{2}\dbinom{n}{2} + \dfrac{1}{3}\dbinom{n}{3} - \cdots + \dfrac{(-1)^{n+1}}{n}\dbinom{n}{n} = 1 + \dfrac{1}{2} + \dfrac{1}{3} + \cdots + \dfrac{1}{n}$

1.2.2 Arithmetic, geometric, and arithmetic–geometric series

1.2.2.1 Arithmetic series.

$$\sum_{k=0}^{n-1}(a+kd) = a + (a+d) + (a+2d) + \cdots + [a + (n-1)d]$$

$$= \frac{n}{2}[2a + (n-1)d] = \frac{n}{2}(a+l) \qquad [l - \text{the last term}]$$

1.2.2.2 Geometric series.

$$\sum_{k=1}^{n} ar^{k-1} = a + ar + ar^2 + \cdots + ar^{n-1}$$

$$= \frac{a(1-r^n)}{1-r} \qquad [r \neq 1]$$

1.2.2.3 Arithmetic–geometric series.

$$\sum_{k=1}^{n-1}(a+kd)r^k = a + (a+d)r + (a+2d)r^2 + \cdots + [a + (n-1)d]r^{n-1}$$

$$= \frac{r^n\{a(r-1) + [n(r-1) - r]d\}}{(r-1)^2} + \frac{[(d-a)r + a]}{(r-1)^2}$$

$$[r \neq 1, n > 1]$$

1.2.3 Sums of powers of integers

1.2.3.1

1. $\displaystyle\sum_{k=1}^{n} k = 1 + 2 + 3 + \cdots + n = \frac{n}{2}(n+1)$

2. $\displaystyle\sum_{k=1}^{n} k^2 = 1^2 + 2^2 + 3^2 + \cdots + n^2 = \frac{n}{6}(n+1)(2n+1)$

3. $\displaystyle\sum_{k=1}^{n} k^3 = 1^3 + 2^3 + 3^3 + \cdots + n^3 = \frac{n^2}{4}(n+1)^2$

4. $\displaystyle\sum_{k=1}^{n} k^4 = 1^4 + 2^4 + 3^4 + \cdots + n^4 = \frac{n}{30}(n+1)(2n+1)(3n^2+3n-1)$

5. $\displaystyle\sum_{k=1}^{n} k^q = \frac{n^{q+1}}{q+1} + \frac{n^q}{2} + \frac{1}{2}\binom{q}{1}B_2 n^{q-1} + \frac{1}{4}\binom{q}{3}B_4 n^{q-3}$

$$+ \frac{1}{6}\binom{q}{5}B_6 n^{q-5} + \frac{1}{8}\binom{q}{7}B_8 n^{q-7} + \cdots,$$

where $B_2 = 1/6$, $B_4 = -1/30$, $B_6 = 1/42,\ldots$, are the Bernoulli numbers (see 1.3.1) and the expansion terminates with the last term containing either n or n^2.

These results are useful when summing finite series with terms involving linear combinations of powers of integers. For example, using 1.2.3.1.1 and 1.2.3.1.2 leads to the result

$$\sum_{k=1}^{n}(3k-1)(k+2) = \sum_{k=1}^{n}(3k^2+5k-2) = 3\sum_{k=1}^{n}k^2 + 5\sum_{k=1}^{n}k - 2\sum_{k=1}^{n}1$$

$$= 3 \cdot \frac{n}{6}(n+1)(2n+1) + 5 \cdot \frac{n}{2}(n+1) - 2n$$

$$= n(n^2 + 4n + 1).$$

1.2.3.2

1. $\displaystyle\sum_{k=1}^{n}(km-1) = (m-1) + (2m-1) + (3m-1) + \cdots + (nm-1)$

$$= \frac{1}{2}mn(n+1) - n$$

2. $\displaystyle\sum_{k=1}^{n}(km-1)^2 = (m-1)^2 + (2m-1)^2 + (3m-1)^2 + \cdots + (nm-1)^2$

$$= \frac{1}{6}n[m^2(n+1)(2n+1) - 6m(n+1) + 6]$$

3. $\displaystyle\sum_{k=1}^{n}(km-1)^3 = (m-1)^3 + (2m-1)^3 + (3m-1)^3 + \cdots + (nm-1)^3$

$$= \frac{1}{4}n[m^3 n(n+1)^2 - 2m^2(n+1)(2n+1) + 6m(n+1) - 4]$$

4. $\displaystyle\sum_{k=1}^{n}(-1)^{k+1}(km-1)$

$$= (m-1) - (2m-1) + (3m-1) - \cdots + (-1)^{n+1}(nm-1)$$

$$= \frac{(-1)^n}{4}[2 - (2n+1)m] + \frac{(m-2)}{4}$$

5. $\displaystyle\sum_{k=1}^{n}(-1)^{k+1}(km-1)^2$

$$= (m-1)^2 - (2m-1)^2 + (3m-1)^2 - \cdots + (-1)^{n+1}(nm-1)^2$$

$$= \frac{(-1)^{n+1}}{2}[n(n+1)m^2 - (2n+1)m + 1] + \frac{(1-m)}{2}$$

6. $\displaystyle\sum_{k=1}^{n}(-1)^{k+1}(km-1)^3$

$$= (m-1)^3 - (2m-1)^3 + (3m-1)^3 - \cdots + (-1)^{n+1}(nm-1)^3$$

$$= \frac{(-1)^{n+1}}{8}[(4n^3 + 6n^2 - 1)m^3 - 12n(n+1)m^2$$

$$+ 6(2n+1)m - 4] - \frac{1}{8}(m^3 - 6m + 4)$$

7. $\displaystyle\sum_{k=1}^{n}(-1)^{k+1}(2k-1)^2 = 1^2 - 3^2 + 5^2 - \cdots + (-1)^{n+1}(2n-1)^2$

$$= \frac{(-1)^{n+1}}{2}(4n^2 - 1) - \frac{1}{2}$$

8. $\displaystyle\sum_{k=1}^{n}(-1)^{k+1}(2k-1)^3 = 1^3 - 3^3 + 5^3 - \cdots + (-1)^{n+1}(2n-1)^3$

$$= (-1)^{n+1}n(4n^2 - 3)$$

9. $\displaystyle\sum_{k=1}^{n}(-1)^{k+1}(3k-1) = 2 - 5 + 8 - \cdots + (-1)^{n+1}(3n-1)$

$$= \frac{(-1)^{n+1}}{4}(6n+1) + \frac{1}{4}$$

10. $\displaystyle\sum_{k=1}^{n}(-1)^{k+1}(3k-1)^2 = 2^2 - 5^2 + 8^2 - \cdots + (-1)^{n+1}(3n-1)^2$

$$= \frac{(-1)^{n+1}}{2}(9n^2 + 3n - 2) - 1$$

11. $\displaystyle\sum_{k=1}^{n}(-1)^{k+1}(3k-1)^3 = 2^3 - 5^3 + 8^3 - \cdots + (-1)^{n+1}(3n-1)^3$

$$= \frac{(-1)^{n+1}}{8}(108n^3 + 54n^2 - 72n - 13) - \frac{13}{8}$$

1.3 Bernoulli and Euler Numbers and Polynomials

1.3.1 Bernoulli and Euler numbers

1.3.1.1 Definitions and tables of Bernoulli and Euler numbers. The **Bernoulli numbers,** usually denoted by B_n, are rational numbers defined by the requirement that B_n is the coefficient of the term $t^n/n!$ on the right-hand side of the generating function

1. $$\frac{t}{e^t - 1} = \sum_{n=0}^{\infty} \frac{t^n}{n!} B_n.$$

The B_n are determined by multiplying the above identity in t by the Maclaurin series representation of $e^t - 1$, expanding the right-hand side, and then equating corresponding coefficients of powers of t on either side of the identity. This yields the result

$$t = B_0 t + \left(B_1 + \frac{1}{2} B_0 \right) t^2 + \left(\frac{1}{2} B_2 + \frac{1}{2} B_1 + \frac{1}{6} B_0 \right) t^3$$

$$+ \left(\frac{1}{6} B_3 + \frac{1}{4} B_2 + \frac{1}{6} B_1 + \frac{1}{24} B_0 \right) t^4 + \cdots.$$

Equating corresponding powers of t leads to the system of equations

$$B_0 = 1 \qquad \qquad \text{(coefficients of } t\text{)}$$

$$B_1 + \frac{1}{2} B_0 = 0 \qquad \qquad \text{(coefficients of } t^2\text{)}$$

$$\frac{1}{2} B_2 + \frac{1}{2} B_1 + \frac{1}{6} B_0 = 0 \qquad \qquad \text{(coefficients of } t^3\text{)}$$

$$\frac{1}{6} B_3 + \frac{1}{4} B_2 + \frac{1}{6} B_1 + \frac{1}{24} B_0 = 0 \qquad \qquad \text{(coefficients of } t^4\text{)}$$

$$\vdots$$

Solving these equations recursively generates $B_0, B_1, B_2, B_3, \ldots$, in this order, where

$$B_0 = 1, \quad B_1 = -\frac{1}{2}, \quad B_2 = \frac{1}{6}, \quad B_3 = 0, \quad B_4 = -\frac{1}{30}, \ldots.$$

Apart from $B_1 = \frac{1}{2}$, all Bernoulli numbers with an odd index are zero, so $B_{2n-1} = 0, n = 2, 3, \ldots$. The Bernoulli numbers with an even index B_{2n} follow by solving recursively these equations:

2. $$B_0 = 1; \quad \sum_{k=0}^{n-1} \binom{n}{k} B_{2k} = 0, \qquad k = 1, 2, \ldots,$$

which are equivalent to the system given above.

The Bernoulli number B_{2n} is expressible in terms of Bernoulli numbers of lower index by

3. $\quad B_{2n} = -\dfrac{1}{2n+1} + \dfrac{1}{2} - \displaystyle\sum_{k=2}^{2n-2} \dfrac{2n(2n-1)\cdots(2n-k+2)}{k!} B_k \qquad [n \geq 1].$

Short list of Bernoulli numbers.

$$
\begin{array}{lll}
B_0 = 1 & B_{12} = \dfrac{-691}{2730} & B_{24} = \dfrac{-236\,364\,091}{2730} \\[2mm]
B_1 = -\dfrac{1}{2} & B_{14} = \dfrac{7}{6} & B_{26} = \dfrac{8\,553\,103}{6} \\[2mm]
B_2 = \dfrac{1}{6} & B_{16} = \dfrac{-3617}{510} & B_{28} = \dfrac{-23\,749\,461\,029}{870} \\[2mm]
B_4 = -\dfrac{1}{30} & B_{18} = \dfrac{43\,867}{798} & B_{30} = \dfrac{8\,615\,841\,276\,005}{14322} \\[2mm]
B_6 = \dfrac{1}{42} & B_{20} = \dfrac{-174\,611}{330} & B_{32} = \dfrac{-7\,709\,321\,041\,217}{510} \\[2mm]
B_8 = -\dfrac{1}{30} & B_{22} = \dfrac{854\,513}{138} & B_{34} = \dfrac{2\,577\,687\,858\,367}{6} \\[2mm]
B_{10} = \dfrac{5}{66} & &
\end{array}
$$

4. $\quad B_{2n-1} = 0 \qquad [n = 2, 3, \ldots].$

The **Euler numbers** E_n, all of which are integers, are defined as the coefficients of $t^n/n!$ on the right-hand side of the generating function

5. $\quad \dfrac{2e^t}{e^t + 1} = \displaystyle\sum_{n=0}^{\infty} E_n \dfrac{t^n}{n!}.$

As $2e^t/(e^t+1) = 1/\cosh t$, it follows that $1/\cosh t$ is the generating function for Euler numbers. A procedure similar to the one described above for the determination of the Bernoulli numbers leads to the Euler numbers. Each Euler number with an odd index is zero.

Short list of Euler numbers.

$$
\begin{array}{ll}
E_0 = 1 & E_{12} = 2\,702\,765 \\
E_2 = -1 & E_{14} = -199\,360\,981 \\
E_4 = 5 & E_{16} = 19\,391\,512\,145 \\
E_6 = -61 & E_{18} = -2\,404\,879\,675\,441 \\
E_8 = 1385 & E_{20} = 370\,371\,188\,237\,525 \\
E_{10} = -50\,521 & E_{22} = -69\,348\,874\,393\,137\,901
\end{array}
$$

6. $E_{2n-1} = 0$ $[n = 1, 2, \ldots]$.

Alternative definitions of Bernoulli and Euler numbers are in use, in which the indexing and choice of sign of the numbers differ from the conventions adopted here. Thus when reference is made to other sources using Bernoulli and Euler numbers it is essential to determine which definition is in use.

The most commonly used alternative definitions lead to the following sequences of Bernoulli numbers B_n^* and Euler numbers E_n^*, where an asterisk has been added to distinguish them from the numbers used throughout this book.

$$B_1^* = \frac{1}{6} \qquad E_1^* = 1$$

$$B_2^* = \frac{1}{30} \qquad E_2^* = 5$$

$$B_3^* = \frac{1}{42} \qquad E_3^* = 61$$

$$B_4^* = \frac{1}{30} \qquad E_4^* = 1385$$

$$B_5^* = \frac{5}{66} \qquad E_5^* = 50521$$

$$\vdots \qquad\qquad \vdots$$

The relationships between these corresponding systems of Bernoulli and Euler numbers are as follows:

7. $B_{2n} = (-1)^{n+1} B_n, \quad E_{2n} = (-1)^n E_n^*, \qquad [n = 1, 2, \ldots]$.

The generating function for the B_n^* is

8. $2 - \dfrac{t}{2} \cot \dfrac{t}{2} = \displaystyle\sum_{k=0}^{\infty} B_k^* \dfrac{t^{2k}}{(2k)!}$ with $B_0^* = 1$, while the generating function for the E_n^* is

9. $\sec t = \displaystyle\sum_{k=0}^{\infty} E_k^* \dfrac{t^{2k}}{(2k)!}$ with $E_0^* = 1$.

1.3.1.2 Series representations for B_n and E_n.

1. $B_{2n} = \dfrac{(-1)^{n+1}(2n)!}{\pi^{2n} 2^{2n-1}} \left[1 + \dfrac{1}{2^{2n}} + \dfrac{1}{3^{2n}} + \dfrac{1}{4^{2n}} + \cdots \right]$

2. $B_{2n} = \dfrac{(-1)^{n+1}(2n)!}{\pi^{2n}(2^{2n-1} - 1)} \left[1 - \dfrac{1}{2^{2n}} + \dfrac{1}{3^{2n}} - \dfrac{1}{4^{2n}} + \cdots \right]$

3. $E_{2n} = \dfrac{(-1)^n 2^{2n+2}(2n)!}{\pi^{2n+1}} \left[1 - \dfrac{1}{3^{2n+1}} + \dfrac{1}{5^{2n+1}} - \dfrac{1}{7^{2n+1}} + \cdots \right]$

1.3.1.3 Relationships between B_n and E_n.

1. $E_{2n} = -\left[\dfrac{(2n)!}{(2n-2)!2!} E_{2n-2} + \dfrac{(2n)!}{(2n-4)!4!} E_{2n-4} \right.$

   $\left. + \dfrac{(2n)!}{(2n-6)!6!} E_{2n-6} + \cdots + E_0 \right]$

   $E_0 = 1 \qquad [n = 1, 2, \ldots].$

2. $B_{2n} = \dfrac{2n}{2^{2n}(2^{2n}-1)} \left[\dfrac{(2n-1)!}{(2n-2)!1!} E_{2n-2} + \dfrac{(2n-1)!}{(2n-4)!3!} E_{2n-4} \right.$

   $\left. + \dfrac{(2n-1)!}{(2n-6)!5!} E_{2n-6} + \cdots + E_0 \right]$

   $B_0 = 1, \quad B_1 = \dfrac{1}{2}, \quad E_0 = 1 \qquad [n = 1, 2, \ldots].$

1.3.1.4 The occurrence of Bernoulli numbers in series.

Bernoulli numbers enter into many summations, and their use can often lead to the form of the general term in a series expansion of a function that may be unobtainable by other means. For example, in terms of the Bernoulli numbers B_n, the generating function in 1.3.1.1 becomes

$$\frac{t}{e^t - 1} = 1 - \frac{1}{2}t + B_2 \frac{t^2}{2!} + B_4 \frac{t^4}{4!} + B_6 \frac{t^6}{6!} + \cdots .$$

However,

$$\frac{t}{e^t - 1} + \frac{t}{2} = \frac{t}{2} \coth \frac{t}{2},$$

so setting $t = 2x$ in the expansion gives

$$x \coth x = 1 + 2^2 B_2 \frac{x^2}{2!} + 2^4 B_4 \frac{x^4}{4!} + 2^6 B_6 \frac{x^6}{6!} + \cdots$$

$$= 1 + \sum_{k=1}^{\infty} 2^{2k} B_{2k} \frac{x^{2k}}{(2k)!}$$

or

1. $\coth x = \dfrac{1}{x} + \displaystyle\sum_{k=1}^{\infty} 2^{2k} B_{2k} \dfrac{x^{2k-1}}{(2k)!} \qquad [|x| < \pi].$

   Replacing x by ix this becomes

2. $\cot x = \dfrac{1}{x} + \displaystyle\sum_{k=1}^{\infty} (-1)^k 2^{2k} B_{2k} \dfrac{x^{2k-1}}{(2k)!} \qquad [|x| < \pi].$

   The following series for $\tan x$ is obtained by combining the previous result with the identity $\tan x = \cot x - 2 \cot 2x$ (see 2.4.1.5.7):

3. $\tan x = \sum_{k=1}^{\infty} (-1)^{k+1} 2^{2k} (2^{2k} - 1) B_{2k} \dfrac{x^{2k-1}}{(2k)!}$ $[|x| < \pi/2]$.

1.3.1.5 Sums of powers of integers with even or odd negative exponents.

1. $\sum_{k=1}^{\infty} \dfrac{1}{k^2} = \dfrac{1}{1^2} + \dfrac{1}{2^2} + \dfrac{1}{3^2} + \cdots = \dfrac{\pi^2}{6}$

2. $\sum_{k=1}^{\infty} \dfrac{1}{k^4} = \dfrac{1}{1^4} + \dfrac{1}{2^4} + \dfrac{1}{3^4} + \cdots = \dfrac{\pi^4}{90}$

3. $\sum_{k=1}^{\infty} \dfrac{1}{k^6} = \dfrac{1}{1^6} + \dfrac{1}{2^6} + \dfrac{1}{3^6} + \cdots = \dfrac{\pi^6}{945}$

4. $\sum_{k=1}^{\infty} \dfrac{1}{k^8} = \dfrac{1}{1^8} + \dfrac{1}{2^8} + \dfrac{1}{3^8} + \cdots = \dfrac{\pi^8}{9450}$

5. $\sum_{k=1}^{\infty} \dfrac{1}{k^{2n}} = (-1)^{n+1} \dfrac{(2\pi)^{2n} B_{2n}}{2 \cdot (2n)!}$ $[n = 1, 2, \ldots]$

6. $\sum_{k=1}^{\infty} (-1)^{k+1} \dfrac{1}{k^2} = \dfrac{1}{1^2} - \dfrac{1}{2^2} + \dfrac{1}{3^2} - \cdots = \dfrac{\pi^2}{12}$

7. $\sum_{k=1}^{\infty} (-1)^{k+1} \dfrac{1}{k^4} = \dfrac{1}{1^4} - \dfrac{1}{2^4} + \dfrac{1}{3^4} - \cdots = \dfrac{7\pi^4}{720}$

8. $\sum_{k=1}^{\infty} (-1)^{k+1} \dfrac{1}{k^6} = \dfrac{1}{1^6} - \dfrac{1}{2^6} + \dfrac{1}{3^6} - \cdots = \dfrac{31\pi^6}{30\,240}$

9. $\sum_{k=1}^{\infty} (-1)^{k+1} \dfrac{1}{k^8} = \dfrac{1}{1^8} - \dfrac{1}{2^8} + \dfrac{1}{3^8} - \cdots = \dfrac{127\pi^8}{1\,209\,600}$

10. $\sum_{k=1}^{\infty} (-1)^{k+1} \dfrac{1}{k^{2n}} = (-1)^{n+1} \dfrac{\pi^{2n}(2^{2n-1} - 1)}{(2n)!} B_{2n}$ $[n = 1, 2, \ldots]$

11. $\sum_{k=1}^{\infty} \dfrac{1}{(2k-1)^2} = \dfrac{1}{1^2} + \dfrac{1}{3^2} + \dfrac{1}{5^2} + \cdots = \dfrac{\pi^2}{8}$

12. $\sum_{k=1}^{\infty} \dfrac{1}{(2k-1)^4} = \dfrac{1}{1^4} + \dfrac{1}{3^4} + \dfrac{1}{5^4} + \cdots = \dfrac{\pi^4}{96}$

13. $\sum_{k=1}^{\infty} \dfrac{1}{(2k-1)^6} = \dfrac{1}{1^6} + \dfrac{1}{3^6} + \dfrac{1}{5^6} + \cdots = \dfrac{\pi^6}{960}$

14. $\sum_{k=1}^{\infty} \dfrac{1}{(2k-1)^8} = \dfrac{1}{1^8} + \dfrac{1}{3^8} + \dfrac{1}{5^8} + \cdots = \dfrac{17\pi^8}{161\,280}$

15. $\displaystyle\sum_{k=1}^{\infty}\frac{1}{(2k-1)^{2n}}=(-1)^{n+1}\frac{\pi^{2n}(2^{2n}-1)}{2(2n)!}B_{2n}$ $[n=1,2,\ldots]$

16. $\displaystyle\sum_{k=1}^{\infty}(-1)^{k+1}\frac{1}{(2k-1)}=1-\frac{1}{3}+\frac{1}{5}-\cdots=\frac{\pi}{4}$

17. $\displaystyle\sum_{k=1}^{\infty}(-1)^{k+1}\frac{1}{(2k-1)^3}=1-\frac{1}{3^3}+\frac{1}{5^3}-\cdots=\frac{5\pi^5}{1536}$

18. $\displaystyle\sum_{k=1}^{\infty}(-1)^{k+1}\frac{1}{(2k-1)^5}=1-\frac{1}{3^5}+\frac{1}{5^5}-\cdots=\frac{61\pi^7}{184\,320}$

19. $\displaystyle\sum_{k=1}^{\infty}(-1)^{k+1}\frac{1}{(2k-1)^7}=1-\frac{1}{3^7}+\frac{1}{5^7}-\cdots=\frac{277\pi^9}{8\,257\,536}$

20. $\displaystyle\sum_{k=1}^{\infty}(-1)^{k+1}\frac{1}{(2k-1)^{2n+1}}=(-1)^n\frac{\pi^{2n+1}}{2^{2n+2}(2n)!}E_{2n}$ $[n=0,1,\ldots]$

1.3.1.6 Asymptotic representations for B_{2n}.

1. $B_{2n}\sim(-1)^{n+1}\dfrac{2(2n)!}{(2\pi)^{2n}}$

2. $B_{2n}\sim(-1)^{n+1}4(\pi n)^{1/2}\left(\dfrac{n}{\pi e}\right)^{2n}$

3. $\dfrac{B_{2n}}{B_{2n+2}}\sim\dfrac{-4\pi^2}{(2n+1)(2n+2)}$

4. $\dfrac{B_{2n}}{B_{2n+2}}\sim-\left(\dfrac{\pi e}{n+1}\right)^2\left(\dfrac{n}{n+1}\right)^{2n}$

1.3.2 Bernoulli and Euler polynomials

1.3.2.1 The Bernoulli polynomials.   The Bernoulli polynomials $B_n(x)$ are defined by

1. $B_n(x)=\displaystyle\sum_{k=0}^{n}\binom{n}{k}B_k x^{n-k}$,

and they have as their generating function

2. $\dfrac{e^{xt}}{e^t-1}=\displaystyle\sum_{n=0}^{\infty}B_n(x)\dfrac{t^{n-1}}{n!}.$

The first eight Bernoulli polynomials are

3. $B_0(x)=1$

4. $B_1(x) = x - \dfrac{1}{2}$

5. $B_2(x) = x^2 - x + \dfrac{1}{6}$

6. $B_3(x) = x^3 - \dfrac{3}{2}x^2 + \dfrac{1}{2}x$

7. $B_4(x) = x^4 - 2x^3 + x^2 - \dfrac{1}{30}$

8. $B_5(x) = x^5 - \dfrac{5}{2}x^4 + \dfrac{5}{3}x^3 - \dfrac{1}{6}x$

9. $B_6(x) = x^6 - 3x^5 + \dfrac{5}{2}x^4 - \dfrac{1}{2}x^2 + \dfrac{1}{42}$

10. $B_7(x) = x^7 - \dfrac{7}{2}x^6 + \dfrac{7}{2}x^5 - \dfrac{7}{6}x^3 + \dfrac{1}{6}x$

The Bernoulli numbers B_n are related to the Bernoulli polynomials $B_n(x)$ by

11. $B_n = B_n(0)$ $[n = 0, 1, \ldots]$.

1.3.2.2 Functional relations and properties of Bernoulli polynomials.

1. $B_{m+1}(n) = B_{m+1} + (m+1)\sum\limits_{k=1}^{n-1} k^m$ $[m, n \text{ natural numbers}]$

2. $B_n(x+1) - B_n(x) = nx^{n-1}$ $[n = 0, 1, \ldots]$

3. $B_n(1-x) = (-1)^n B_n(x)$ $[n = 0, 1, \ldots]$

4. $(-1)^n B_n(-x) = B_n(x) + nx^{n-1}$ $[n = 0, 1, \ldots]$

5. $B_n(mx) = m^{n-1}\sum\limits_{k=0}^{m-1} B_n\left(x + \dfrac{k}{m}\right)$ $[m = 1, 2, \ldots, n = 0, 1, \ldots]$

6. $B_n'(x) = nB_{n-1}(x)$ $[n = 1, 2, \ldots]$

7. $\sum\limits_{k=1}^{m} k^n = \dfrac{B_{n+1}(m+1) - B_{n+1}}{n+1}$ $[m, n = 1, 2, \ldots]$

8. $B_n(x+h) = \sum\limits_{k=0}^{n} \binom{n}{k} B_k(x) h^{n-k}$ $[n = 0, 1, \ldots]$

1.3.2.3 The Euler polynomials.   The Euler polynomials $E_n(x)$ are defined by

1. $E_n(x) = \sum\limits_{k=0}^{n} \binom{n}{k} \dfrac{E_k}{2^k} \left(x - \dfrac{1}{2}\right)^{n-k}$,

and they have as their generating function

2. $\dfrac{2e^{xt}}{e^t + 1} = \sum\limits_{n=0}^{\infty} E_n(x) \dfrac{t^n}{n!}.$

The first eight Euler polynomials are

3. $E_0(x) = 1$

4. $E_1(x) = x - \dfrac{1}{2}$

5. $E_2(x) = x^2 - x$

6. $E_3(x) = x^3 - \dfrac{3}{2}x^2 + \dfrac{1}{4}$

7. $E_4(x) = x^4 - 2x^3 + x$

8. $E_5(x) = x^5 - \dfrac{5}{2}x^4 + \dfrac{5}{2}x^2 - \dfrac{1}{2}$

9. $E_6(x) = x^6 - 3x^5 + 5x^3 - 3x$

10. $E_7(x) = x^7 - \dfrac{7}{2}x^6 + \dfrac{35}{4}x^4 - \dfrac{21}{2}x^2 + \dfrac{17}{8}$

The Euler numbers E_n are related to the Euler polynomials $E_n(x)$ by

11. $E_n = 2^n E_n\left(\dfrac{1}{2}\right)$ (an integer) $[n = 0, 1, \ldots]$

1.3.2.4 Functional relations and properties of Euler polynomials.

1. $E_m(n + 1) = 2 \sum\limits_{k=1}^{n} (-1)^{n-k} k^m + (-1)^{n+1} E_m(0)$ [m, n natural numbers]

2. $E_n(x + 1) + E_n(x) = 2x^n$ $[n = 0, 1, \ldots]$

3. $E_n(1 - x) = (-1)^n E_n(x)$ $[n = 0, 1, \ldots]$

4. $(-1)^{n+1} E_n(-x) = E_n(x) - 2x^n$ $[n = 0, 1, \ldots]$

5. $E_n(mx) = m^n \sum\limits_{k=0}^{m-1} (-1)^k E_n\left(x + \dfrac{k}{m}\right)$ $[n = 0, 1, \ldots, m = 1, 3, \ldots]$

6. $E'_n(x) = n E_{n-1}(x)$ $[n = 1, 2, \ldots]$

7. $\sum\limits_{k=1}^{m} (-1)^{m-k} k^n = \dfrac{E_n(m + 1) + (-1)^m E_n(0)}{2}$ $[m, n = 1, 2, \ldots]$

8. $E_n(x + h) = \sum\limits_{k=0}^{n} \binom{n}{k} E_k(x) h^{n-k}$ $[n = 0, 1, \ldots]$

1.3.3 The Euler–Maclaurin summation formula

Let $f(x)$ have continuous derivatives of all orders up to and including $2m + 2$ for $0 \leq x \leq n$. Then if $a_k = f(k)$, the sum $\sum_{k=0}^{n} f(k) = \sum_{k=0}^{n} a_k$ determined by the **Euler–Maclaurin summation formula** is given by

$$\sum_{k=0}^{n} a_k = \int_0^n f(t)dt + \frac{1}{2}[f(0) + f(n)]$$

$$+ \sum_{k=1}^{m} \frac{B_{2k}}{(2k)!}[f^{(2k-1)}(n) - f^{(2k-1)}(0)] + R_m,$$

where the remainder term is

$$R_m = \frac{n B_{2m+2}}{(2m+2)!} f^{(2m+2)}(\theta n) \qquad \text{with } 0 < \theta < 1 \qquad [m, n = 1, 2, \ldots].$$

In special cases this formula yields an exact closed-form solution for the required sum, whereas in others it provides an asymptotic result. For example, if $f(x) = x^2$, the summation formula yields an exact result for $\sum_{k=1}^{n} n^2$, because every term after the one in B_2 is identically zero, including the remainder term R_1. The details of the calculations are as follows:

$$\sum_{k=0}^{n} k^2 = \sum_{k=1}^{n} k^2 = \int_0^n t^2 \, dt + \frac{1}{2}n^2 + \frac{B_2}{2!} \cdot 2n$$

$$= \frac{1}{3}n^3 + \frac{1}{2}n^2 + \frac{1}{6}n = \frac{1}{6}(n+1)(2n+1). \qquad \text{(see 1.2.3.1.2)}$$

However, no closed-form expression exists for $\sum_{k=1}^{n} 1/k^2$, so when applied to this case the formula can only yield an approximate sum. Setting $f(x) = 1/(x+1)^2$ in the summation formula gives

$$\sum_{k=0}^{n} \frac{1}{(k+1)^2} = \frac{1}{1^2} + \frac{1}{2^2} + \cdots + \frac{1}{(n+1)^2} \approx \int_0^n \frac{dt}{(t+1)^2}$$

$$+ \frac{1}{2}\left(1 + \frac{1}{(n+1)^2}\right) + \sum_{k=1}^{m} B_{2k}\left[1 - \frac{1}{(n+1)^{2k+1}}\right] + R_m$$

$$= \left(1 - \frac{1}{n+1}\right) + \frac{1}{2}\left(1 + \frac{1}{(n+1)^2}\right)$$

$$+ \sum_{k=1}^{m} B_{2k}\left[1 - \frac{1}{(n+1)^{2k+1}}\right] + R_m.$$

To illustrate matters, setting $n = 149$, $m = 1$, and neglecting R_1 gives the rather poor approximation $1.660\,022\,172$ to the actual result

$$\sum_{k=0}^{149} 1/(k+1)^2 = \sum_{k=1}^{150} 1/k^2 = 1.638\,289\,573 \cdots$$

obtained by direct calculation. Increasing m only yields a temporary improvement, because for large m, the numbers $|B_{2m}|$ increase like $(2m)!$ while alternating in sign, which causes the sum to oscillate unboundedly.

A more accurate result is obtained by summing, say, the first 9 terms numerically to obtain $\sum_{k=1}^{9} 1/k^2 = 1.539\,767\,731$, and then using the summation formula to estimate $\sum_{k=10}^{150} 1/k^2$. This is accomplished by setting $f(x) = 1/(x + 10)^2$ in the summation formula, because

$$\sum_{k=0}^{140} f(k) = \sum_{k=10}^{150} 1/k^2 = \frac{1}{10^2} + \frac{1}{11^2} + \cdots + \frac{1}{150^2}.$$

This time, again setting $m = 1$ and neglecting R_1 gives

$$\sum_{k=10}^{150} 1/k^2 \approx \left(\frac{1}{10} - \frac{1}{150} \right) + \frac{1}{2}\left(\frac{1}{10^2} + \frac{1}{150^2} \right) + \frac{1}{6}\left(\frac{1}{10^3} - \frac{1}{150^3} \right)$$

$$= 0.098\,522\,173.$$

Thus $\sum_{k=1}^{150} 1/k^2 \approx 1.539\,767\,731 + 0.098\,522\,173 = 1.638\,289\,904$, which is now accurate to six decimal places.

1.4 Determinants

1.4.1 Expansion of second- and third-order determinants

1.4.1.1 The **determinant** associated with the 2×2 matrix

1. $\mathbf{A} = \begin{bmatrix} a_{11} & a_{12} \\ a_{21} & a_{22} \end{bmatrix},$ (see 1.5.1.1.1)

with elements a_{ij} comprising real or complex numbers, or functions, denoted either by $|\mathbf{A}|$ or by $\det \mathbf{A}$, is defined by

2. $|\mathbf{A}| = a_{11}a_{22} - a_{12}a_{21}.$

This is called a **second-order** determinant.

The determinant associated with the 3×3 matrix

$$\mathbf{A} = \begin{bmatrix} a_{11} & a_{12} & a_{13} \\ a_{21} & a_{22} & a_{23} \\ a_{31} & a_{32} & a_{33} \end{bmatrix},$$

with elements a_{ij} comprising real or complex numbers, or functions, is defined by

3. $|\mathbf{A}| = a_{11}a_{22}a_{33} - a_{11}a_{23}a_{32} + a_{12}a_{23}a_{31} - a_{12}a_{21}a_{33} + a_{13}a_{21}a_{32} - a_{13}a_{22}a_{31}.$

This is called a **third-order** determinant.

1.4.2 Minors, cofactors, and the Laplace expansion

1.4.2.1 The $n \times n$ matrix

1. $\mathbf{A} = \begin{bmatrix} a_{11} & a_{12} & \cdots & a_{1n} \\ a_{21} & a_{22} & \cdots & a_{2n} \\ \vdots & \vdots & \vdots & \vdots \\ a_{n1} & a_{n2} & \cdots & a_{nn} \end{bmatrix}$

has associated with it the determinant

2. $|\mathbf{A}| = \begin{vmatrix} a_{11} & a_{12} & \cdots & a_{1n} \\ a_{21} & a_{22} & \cdots & a_{2n} \\ \vdots & \vdots & \vdots & \vdots \\ a_{n1} & a_{n2} & \cdots & a_{nn} \end{vmatrix}$

The **order** of a determinant is the number of elements in its leading diagonal (the diagonal from top left to bottom right), so an **n'th-order** determinant is associated with an $n \times n$ matrix. The Laplace expansion of a determinant of arbitrary order is given later.

The **minor** M_{ij} associated with the element a_{ij} in 1.4.2.1.1 is the $(n-1)$th-order determinant derived from 1.4.2.1.1 by deletion of its i'th row and j'th column. The **cofactor** C_{ij} associated with the element a_{ij} is defined as

3. $C_{ij} = (-1)^{i+j} M_{ij}$.

The **Laplace expansion** of determinant 1.4.2.1.1 may be either by elements of a row or of a column of $|\mathbf{A}|$.

Laplace expansion of $|\mathbf{A}|$ by elements of the i'th row.

4. $|\mathbf{A}| = \sum_{j=1}^{n} a_{ij} C_{ij} \qquad [i = 1, 2, \ldots, n]$.

Laplace expansion of $|\mathbf{A}|$ by elements of the j'th column.

5. $|\mathbf{A}| = \sum_{i=1}^{n} a_{ij} C_{ij} \qquad [j = 1, 2, \ldots, n]$.

A related property of determinants is that the sum of the products of the elements in any row and the cofactors of the corresponding elements in any other row is zero. Similarly, the sum of the products of the elements in any column and the cofactors of the corresponding elements in any other column is zero. Thus for a determinant of any order

6. $\sum_{j=1}^{n} a_{ij} C_{kj} = 0 \qquad [j \neq k, i = 1, 2, \ldots, n]$

and

7. $\sum_{i=1}^{n} a_{ij} C_{ik} = 0 \qquad [j \neq k, j = 1, 2, \ldots, n]$.

If the **Kronecker delta** symbol δ_{ij} is introduced, where

8. $\delta_{ij} = \begin{cases} 1, & i = j \\ 0, & i \neq j, \end{cases}$

results 1.4.2.1.4–7 may be combined to give

9. $\displaystyle\sum_{j=1}^{n} a_{ij}C_{kj} = \delta_{ik}|\mathbf{A}| \quad \text{and} \quad \sum_{i=1}^{n} a_{ij}C_{ik} = \delta_{jk}|\mathbf{A}|.$

These results may be illustrated by considering the matrix

$$\mathbf{A} = \begin{bmatrix} 1 & 2 & 1 \\ -2 & 4 & -1 \\ 2 & 1 & 3 \end{bmatrix}$$

and its associated determinant $|\mathbf{A}|$. Expanding $|\mathbf{A}|$ by elements of its second row gives

$$|\mathbf{A}| = -2C_{21} + 4C_{22} - C_{23},$$

but

$$C_{21} = (-1)^{2+1}\begin{vmatrix} 2 & 1 \\ 1 & 3 \end{vmatrix} = -5, \quad C_{22} = (-1)^{2+2}\begin{vmatrix} 1 & 1 \\ 2 & 3 \end{vmatrix} = 1,$$

$$C_{23} = (-1)^{2+3}\begin{vmatrix} 1 & 2 \\ 2 & 1 \end{vmatrix} = 3,$$

so

$$|\mathbf{A}| = (-2) \cdot (-5) + 4 \cdot (1) - 3 = 11.$$

Alternatively, expanding $|\mathbf{A}|$ by elements of its first column gives

$$|\mathbf{A}| = C_{11} - 2C_{21} + 2C_{31},$$

but

$$C_{11} = (-1)^{1+1}\begin{vmatrix} 4 & -1 \\ 1 & 3 \end{vmatrix} = 13, \quad C_{21} = (-1)^{2+1}\begin{vmatrix} 2 & 1 \\ 1 & 3 \end{vmatrix} = -5,$$

$$C_{31} = (-1)^{3+1}\begin{vmatrix} 2 & 1 \\ 4 & -1 \end{vmatrix} = -6,$$

so

$$|\mathbf{A}| = 13 - 2 \cdot (-5) + 2 \cdot (-6) = 11.$$

To verify 1.4.2.1.6 we sum the products of elements in the first row of $|\mathbf{A}|$ ($i = 1$) and the cofactors of the corresponding elements in the second row of $|\mathbf{A}|$ ($k = 2$) to obtain

$$\sum_{j=1}^{3} a_{ij}C_{2j} = a_{11}C_{21} + a_{12}C_{22} + a_{13}C_{23}$$
$$= 1 \cdot (-5) + 2 \cdot 1 + 1 \cdot 3 = 0.$$

1.4.3 Basic properties of determinants

1.4.3.1   Let $\mathbf{A} = [a_{ij}]$, $\mathbf{B} = [b_{ij}]$ be $n \times n$ matrices, when the following results are true.

1.   If any two adjacent rows (or columns) of $|\mathbf{A}|$ are interchanged, the sign of the resulting determinant is changed.

2.   If any two rows (or columns) of $|\mathbf{A}|$ are identical, then $|\mathbf{A}| = 0$.

3.   The value of a determinant is not changed if any multiple of a row (or column) is added to any other row (or column).

4. $|k\mathbf{A}| = k^n |\mathbf{A}|$ for any scalar k.

5. $|\mathbf{A}^T| = |\mathbf{A}|$,   where $\mathbf{A}^T$ is the transpose of $\mathbf{A}$.   (see 1.5.1.1.7)

6. $|\mathbf{A}||\mathbf{B}| = |\mathbf{AB}|$.   (see 1.5.1.1.27)

7. $|\mathbf{A}^{-1}| = 1/|\mathbf{A}|$,   where $\mathbf{A}^{-1}$ is the matrix inverse to $\mathbf{A}$.   (see 1.5.1.1.9)

8.   If the elements a_{ij} of $\mathbf{A}$ are functions of x, then

$$\frac{d|\mathbf{A}|}{dx} = \sum_{i,j=1}^{n} \frac{da_{ij}}{dx} C_{ij}, \qquad \text{where } C_{ij} \text{ is the cofactor of the element } a_{ij}.$$

1.4.4 Jacobi's theorem

1.4.4.1   Let M_r be an r-rowed minor of the nth-order determinant $|\mathbf{A}|$, associated with the $n \times n$ matrix $\mathbf{A} = [a_{ij}]$, in which the rows $i_1, i_2, \ldots, i_r$ are represented together with the columns $k_1, k_2, \ldots, k_r$.

Define the **complementary minor** to M_r to be the $(n - k)$-rowed minor obtained from $|\mathbf{A}|$ by deleting all the rows and columns associated with M_r, and the **signed complementary minor** $M^{(r)}$ to M_r to be

1. $M^{(r)} = (-1)^{i_1 + i_2 + \cdots + i_r + k_1 + k_2 + \cdots + k_r} \times$ (complementary minor to M_r).

Then, if Δ is the matrix of cofactors given by

2. $\Delta = \begin{vmatrix} C_{11} & C_{12} & \cdots & C_{1n} \\ C_{21} & C_{22} & \cdots & C_{2n} \\ \vdots & \vdots & \vdots & \vdots \\ C_{n1} & C_{n2} & \cdots & C_{nn} \end{vmatrix},$

and M_r and M_r' are corresponding r-rowed minors of $|\mathbf{A}|$ and Δ, it follows that

3. $M_r' = |\mathbf{A}|^{r-1} M^{(r)}$.

It follows that if $|\mathbf{A}| = 0$, then

4. $C_{pk} C_{nq} = C_{nk} C_{pq}$.

1.4.5 Hadamard's theorem

1.4.5.1   If $|\mathbf{A}|$ is an $n \times n$ determinant with elements a_{ij} that may be complex, then $|\mathbf{A}| \neq 0$ if

1. $|a_{ij}| > \sum_{j=1, j \neq i}^{n} |a_{ij}|.$

1.4.6 Hadamard's inequality

1.4.6.1   Let $\mathbf{A} = [a_{ij}]$ be an arbitrary $n \times n$ nonsingular matrix with real elements and determinant $|\mathbf{A}|$. Then

1. $\displaystyle |\mathbf{A}|^2 \le \prod_{i=1}^{n} \left(\sum_{k=1}^{n} a_{ik}^2 \right).$

This result remains true if $\mathbf{A}$ has complex elements but is such that $\mathbf{A} = \overline{\mathbf{A}}^{\mathrm{T}}$ ($\mathbf{A}$ is **hermitian**), where $\overline{\mathbf{A}}$ denotes the matrix obtained from $\mathbf{A}$ by replacing each element by its complex conjugate and T denotes the transpose operation (see 1.5.1.1.7).

Deductions.

1. If $M = \max|a_{ij}|$, then

    $$|\mathbf{A}| \le M^n n^{n/2}.$$

2. If the $n \times n$ matrix $\mathbf{A} = [a_{ij}]$ is positive definite (see 1.5.1.1.21), then

    $$|\mathbf{A}| \le a_{11} a_{22} \cdots a_{nn}.$$

3. If the real $n \times n$ matrix $\mathbf{A}$ is **diagonally dominant,** that is if

    $$\sum_{j \ne 1}^{n} |a_{ij}| < |a_{ii}| \qquad \text{for } i = 1, 2, \ldots, n,$$

    then $|\mathbf{A}| \ne 0$.

1.4.7 Cramer's rule

1.4.7.1   If n linear equations

1. $a_{11}x_1 + a_{12}x_2 + \cdots + a_{1n}x_n = b_1,$
    $a_{21}x_1 + a_{22}x_2 + \cdots + a_{2n}x_n = b_2,$
    $$\vdots \qquad \vdots \qquad \vdots \qquad \vdots \qquad \vdots$$
    $a_{n1}x_1 + a_{n2}x_2 + \cdots + a_{nn}x_n = b_n,$

    have a **nonsingular** coefficient matrix $\mathbf{A} = [a_{ij}]$, so that $|\mathbf{A}| \ne 0$, then there is a unique solution

2. $\displaystyle x_j = \frac{C_{1j}b_1 + C_{2j}b_2 + \cdots + C_{nj}b_j}{|\mathbf{A}|}$

    for $j = 1, 2, \ldots, n$, where C_{ij} is the cofactor of element a_{ij} in the coefficient matrix $\mathbf{A}$ (**Cramer's rule**).

1.4.8 Some special determinants

1.4.8.1 Vandermonde's determinant (alternant).

1. *Third order*

$$\begin{vmatrix} 1 & 1 & 1 \\ x_1 & x_2 & x_3 \\ x_1^2 & x_2^2 & x_3^2 \end{vmatrix} = (x_3 - x_2)(x_3 - x_1)(x_2 - x_1).$$

2. *n'th order*

$$\begin{vmatrix} 1 & 1 & \cdots & 1 \\ x_1 & x_2 & \cdots & x_n \\ x_1^2 & x_2^2 & \cdots & x_n^2 \\ \vdots & \vdots & \vdots & \vdots \\ x_1^{n-1} & x_2^{n-1} & \cdots & x_n^{n-1} \end{vmatrix} = \prod_{1 \le i < j \le n} (x_j - x_i),$$

where the right-hand side is the continued product of all the differences that can be formed from the $\frac{1}{2}n(n-1)$ pairs of numbers taken from $x_1, x_2, \ldots, x_n$, with the order of the differences taken in the reverse order of the suffixes that are involved.

1.4.8.2 Circulants.

3. *Second order*

$$\begin{vmatrix} x_1 & x_2 \\ x_2 & x_1 \end{vmatrix} = (x_1 + x_2)(x_1 - x_2).$$

4. *Third order*

$$\begin{vmatrix} x_1 & x_2 & x_3 \\ x_3 & x_1 & x_2 \\ x_2 & x_3 & x_1 \end{vmatrix} = (x_1 + x_2 + x_3)\big(x_1 + \omega x_2 + \omega^2 x_3\big)\big(x_1 + \omega^2 x_2 + \omega x_3\big),$$

where ω and ω^2 are the two complex cube roots of 1.

5. *n'th order*

$$\begin{vmatrix} x_1 & x_2 & x_3 & \cdots & x_n \\ x_n & x_1 & x_2 & \cdots & x_{n-1} \\ x_{n-1} & x_n & x_1 & \cdots & x_{n-2} \\ \vdots & \vdots & \vdots & \vdots & \vdots \\ x_2 & x_3 & x_4 & \cdots & x_1 \end{vmatrix} = \prod_{j=1} \big(x_1 + x_2\omega_j + x_3\omega_j^2 + \cdots + x_n\omega_j^{n-1}\big),$$

where ω_j is an nth root of 1.

The eigenvalues λ (see 1.5.11.18) of an $n \times n$ circulant matrix are

$$\lambda_j = x_1 + x_2\omega_j + x_3\omega_j^2 + \cdots + x_n\omega_j^{n-1},$$

where ω_j is again an n'th root of 1.

1.4.8.3 Jacobian determinants. If $f_1, f_2, \ldots, f_n$ are n real-valued functions that are differentiable with respect to $x_1, x_2, \ldots, x_n$, then the Jacobian $J_f(x)$ of the f_i with respect to the x_j is the determinant

1. $$J_f(x) = \begin{vmatrix} \dfrac{\partial f_1}{\partial x_1} & \dfrac{\partial f_1}{\partial x_2} & \cdots & \dfrac{\partial f_1}{\partial x_n} \\[2ex] \dfrac{\partial f_2}{\partial x_1} & \dfrac{\partial f_2}{\partial x_2} & \cdots & \dfrac{\partial f_2}{\partial x_n} \\[2ex] \vdots & \vdots & \vdots & \vdots \\[2ex] \dfrac{\partial f_n}{\partial x_1} & \dfrac{\partial f_n}{\partial x_2} & \cdots & \dfrac{\partial f_n}{\partial x_n} \end{vmatrix}.$$

The notation

2. $$\frac{\partial(f_1, f_2, \ldots, f_n)}{\partial(x_1, x_2, \ldots, x_n)}$$

is also used to denote the Jacobian $J_f(x)$.

1.4.8.4 Hessian determinants. The Jacobian of the derivatives $\partial\phi/\partial x_1$, $\partial\phi/\partial x_2$, $\ldots, \partial\phi/\partial x_n$ of a function $\phi(x_1, x_2, \ldots, x_n)$ with respect to $x_1, x_2, \ldots, x_n$ is called the **Hessian** H of ϕ, so that

1. $$H = \begin{vmatrix} \dfrac{\partial^2\phi}{\partial x_1^2} & \dfrac{\partial^2\phi}{\partial x_1\partial x_2} & \dfrac{\partial^2\phi}{\partial x_1\partial x_3} & \cdots & \dfrac{\partial^2\phi}{\partial x_1\partial x_n} \\[2ex] \dfrac{\partial^2\phi}{\partial x_2\partial x_1} & \dfrac{\partial^2\phi}{\partial x_2^2} & \dfrac{\partial^2\phi}{\partial x_2\partial x_3} & \cdots & \dfrac{\partial^2\phi}{\partial x_2\partial x_n} \\[2ex] \vdots & \vdots & \vdots & \vdots & \vdots \\[2ex] \dfrac{\partial^2\phi}{\partial x_n\partial x_1} & \dfrac{\partial^2\phi}{\partial x_n\partial x_2} & \dfrac{\partial^2\phi}{\partial x_n\partial x_3} & \cdots & \dfrac{\partial^2\phi}{\partial x_n^2} \end{vmatrix}.$$

1.4.8.5 Wronskian determinants. Let $f_1, f_2, \ldots, f_n$ be n functions, each n times differentiable with respect to x in some open interval (a, b). Then the Wronskian $W(x)$ of $f_1, f_2, \ldots, f_n$ is defined by

1. $$W(x) = \begin{vmatrix} f_1 & f_2 & \cdots & f_n \\[1ex] f_1^{(1)} & f_2^{(1)} & \cdots & f_n^{(1)} \\[1ex] \vdots & \vdots & \vdots & \vdots \\[1ex] f_1^{(n-1)} & f_2^{(n-1)} & \cdots & f_n^{(n-1)} \end{vmatrix},$$

where $f_i^{(r)} = d^r f_i/dx^r$.

Properties of Wronskian determinants.

2. dW/dx follows from $W(x)$ by replacing the last row of the determinant defining $W(x)$ by the n'th derivatives $f_1^{(n)}, f_2^{(n)}, \ldots, f_n^{(n)}$.

3. If constants $k_1, k_2, \ldots, k_n$ exist, not all zero, such that

$$k_1 f_1 + k_2 f_2 + \cdots + k_n f_n = 0$$

for all x in (a, b), then $W(x) = 0$ for all x in (a, b).

4. The vanishing of the Wronskian throughout (a, b) is necessary, but not sufficient, for the linear dependence of $f_1, f_2, \ldots, f_n$.

1.4.9 Routh–Hurwitz theorem

1.4.9.1   Let $P(\lambda)$ be the nth degree polynomial

1. $P(\lambda) \equiv \lambda^n + a_1 \lambda^{n-1} + a_2 \lambda^{n-2} + \cdots + a_n.$
    Form the n numbers:

2. $\Delta_1 = a_1, \quad \Delta_2 = \begin{vmatrix} a_1 & 1 \\ a_3 & a_2 \end{vmatrix}, \quad \Delta_3 = \begin{vmatrix} a_1 & 1 & 0 \\ a_3 & a_2 & a_1 \\ a_5 & a_4 & a_3 \end{vmatrix}, \cdots,$

$$\Delta_n = \begin{vmatrix} a_1 & 1 & 0 & 0 & \cdots & 0 \\ a_3 & a_2 & a_1 & 1 & \cdots & 0 \\ a_5 & a_4 & a_3 & a_2 & \cdots & 0 \\ \vdots & \vdots & \vdots & \vdots & \vdots & \vdots \\ a_{2n-1} & a_{2n-2} & a_{2n-3} & a_{2n-4} & \cdots & a_n \end{vmatrix},$$

and set $a_r = 0$ for $r > n$.

3. Then the necessary and sufficient conditions for the zeros of $P(\lambda)$ all to have negative real parts (the **Routh–Hurwitz conditions**) are

$$\Delta_i > 0 \qquad [i = 1, 2, \ldots, n].$$

1.5 Matrices

1.5.1 Special matrices

1.5.1.1 Basic definitions.

1. An $m \times n$ **matrix** is a rectangular array of elements (numbers or functions) with m rows and n columns. If a matrix is denoted by $\mathbf{A}$, the element (entry) in its i'th row and j'th column is denoted by a_{ij}, and we write $\mathbf{A} = [a_{ij}]$. A matrix with as many rows as columns is called a **square matrix**.

2. A square matrix $\mathbf{A}$ of the form

$$\mathbf{A} = \begin{bmatrix} \lambda_1 & 0 & 0 & \cdots & 0 \\ 0 & \lambda_2 & 0 & \cdots & 0 \\ 0 & 0 & \lambda_3 & \cdots & 0 \\ \vdots & \vdots & \vdots & \vdots & \vdots \\ 0 & 0 & 0 & \cdots & \lambda_n \end{bmatrix}$$

in which all entries away from the **leading diagonal** (the diagonal from top left to bottom right) are zero is called a **diagonal matrix.**

3. The **identity matrix,** or **unit matrix,** is a diagonal matrix $\mathbf{I}$ in which all entries in the leading diagonal are unity.

4. A **null matrix** is a matrix of any shape in which every entry is zero.

5. The $n \times n$ matrix $\mathbf{A} = [a_{ij}]$ is said to be **reducible** if the indices $1, 2, \ldots, n$ can be divided into two disjoint nonempty sets $i_1, i_2, \ldots, i_\mu; \ j_1, j_2, \ldots, j_\nu \, (\mu + \nu = n)$, such that

$$a_{i_\alpha j_\beta} = 0 \qquad [\alpha = 1, 2, \ldots, \mu; \ \beta = 1, 2, \ldots, \nu].$$

Otherwise $\mathbf{A}$ will be said to be **irreducible.**

6. An $m \times n$ matrix $\mathbf{A}$ is **equivalent** to an $m \times n$ matrix $\mathbf{B}$ if, and only if, $\mathbf{B} = \mathbf{PAQ}$ for suitable **nonsingular** $m \times m$ and $n \times n$ matrices $\mathbf{P}$ and $\mathbf{Q}$, respectively. A matrix $\mathbf{D}$ is said to be **nonsingular** if $|\mathbf{D}| \neq 0$.

7. If $\mathbf{A} = [a_{ij}]$ is an $m \times n$ matrix with element a_{ij} in its i'th row and the j'th column, then the **transpose** $\mathbf{A}^\mathrm{T}$ of $\mathbf{A}$ is the $n \times m$ matrix

$$\mathbf{A}^\mathrm{T} = [b_{ij}] \qquad \text{with } b_{ij} = a_{ji};$$

that is, the transpose $\mathbf{A}^\mathrm{T}$ of $\mathbf{A}$ is the matrix derived from $\mathbf{A}$ by interchanging rows and columns, so the i'th row of $\mathbf{A}$ becomes the i'th column of $\mathbf{A}^\mathrm{T}$ for $i = 1, 2, \ldots, m$.

8. If $\mathbf{A}$ is an $n \times n$ matrix, its **adjoint,** denoted by adj $\mathbf{A}$, is the transpose of the matrix of cofactors C_{ij} of $\mathbf{A}$, so that

$$\text{adj } \mathbf{A} = [C_{ij}]^\mathrm{T}. \qquad\qquad \text{(see 1.4.2.1.3)}$$

9. If $\mathbf{A} = [a_{ij}]$ is an $n \times n$ matrix with a nonsingular determinant $|\mathbf{A}|$, then its **inverse** $\mathbf{A}^{-1}$, also called its **multiplicative inverse,** is given by

$$\mathbf{A}^{-1} = \frac{\text{adj } \mathbf{A}}{|\mathbf{A}|}, \qquad \mathbf{A}^{-1}\mathbf{A} = \mathbf{A}\mathbf{A}^{-1} = \mathbf{I}.$$

10. The **trace** of an $n \times n$ matrix $\mathbf{A} = [a_{ij}]$, written tr $\mathbf{A}$, is defined to be the sum of the terms on the leading diagonal, so that

$$\text{tr } \mathbf{A} = a_{11} + a_{22} + \cdots + a_{nn}.$$

11. The $n \times n$ matrix $\mathbf{A} = [a_{ij}]$ is **symmetric** if $a_{ij} = a_{ji}$ for $i, j = 1, 2, \ldots, n$.

12. The $n \times n$ matrix $\mathbf{A} = [a_{ij}]$ is **skew-symmetric** if $a_{ij} = -a_{ji}$ for $i, j = 1, 2, \ldots, n$; so in a skew-symmetric matrix each element on the leading diagonal is zero.

13. An $n \times n$ matrix $\mathbf{A} = [a_{ij}]$ is of **upper triangular type** if $a_{ij} = 0$ for $i > j$ and of **lower triangular type** if $a_{ij} = 0$ for $j > i$.

14. A real $n \times n$ matrix $\mathbf{A}$ is orthogonal if, and only if, $\mathbf{AA}^\mathrm{T} = \mathbf{I}$.

15. If $\mathbf{A} = [a_{ij}]$ is an $n \times n$ matrix with complex elements, then its **Hermitian transpose** $\mathbf{A}^{\dagger}$ is defined to be

$$\mathbf{A}^{\dagger} = [\overline{a}_{ji}],$$

with the bar denoting the complex conjugate operation. The Hermitian transpose operation is also denoted by a superscript H, so that $\mathbf{A}^{\dagger} = \mathbf{A}^{H} = (\overline{\mathbf{A}})^{T}$.

A Hermitian matrix $\mathbf{A}$ is said to be **normal** if $\mathbf{A}$ and $\mathbf{A}^{\dagger}$ commute, so that $\mathbf{A}\mathbf{A}^{\dagger} = \mathbf{A}^{\dagger}\mathbf{A}$ or, in the equivalent notation, $\mathbf{A}\mathbf{A}^{H} = \mathbf{A}^{H}\mathbf{A}$.

16. An $n \times n$ matrix $\mathbf{A}$ is Hermitian if $\mathbf{A} = \mathbf{A}^{\dagger}$, or equivalently, if $\mathbf{A} = \overline{\mathbf{A}}^{T}$, with the overbar denoting the complex conjugate operation.

17. An $n \times n$ matrix $\mathbf{A}$ is **unitary** if $\mathbf{A}\mathbf{A}^{\dagger} = \mathbf{A}^{\dagger}\mathbf{A} = \mathbf{I}$.

18. If $\mathbf{A}$ is an $n \times n$ matrix, the **eigenvectors** $\mathbf{X}$ satisfy the equation

$$\mathbf{A}\mathbf{X} = \lambda\mathbf{X},$$

while the **eigenvalues** λ satisfy the characteristic equation

$$|\mathbf{A} - \lambda\mathbf{I}| = 0.$$

19. An $n \times n$ matrix $\mathbf{A}$ is **nilpotent** if $\mathbf{A}^{k} = 0$ for some k.

20. An $n \times n$ matrix $\mathbf{A}$ is **idempotent** if $\mathbf{A}^{2} = \mathbf{A}$.

21. An $n \times n$ matrix $\mathbf{A}$ is **positive definite** if $\mathbf{x}^{T}\mathbf{A}\mathbf{x} > 0$, for $\mathbf{x} \neq 0$ an n element column vector.

22. An $n \times n$ matrix $\mathbf{A}$ is **nonnegative definite** if $\mathbf{x}^{T}\mathbf{A}\mathbf{x} \geq \mathbf{0}$, for $\mathbf{x} \neq 0$ an n element column vector.

23. An $n \times n$ matrix $\mathbf{A}$ is **diagonally dominant** if $|a_{ii}| > \sum_{j \neq i} |a_{ij}|$ for all i.

24. Two matrices $\mathbf{A}$ and $\mathbf{B}$ are **equal** if, and only if, they are both of the same shape and corresponding elements are equal.

25. Two matrices $\mathbf{A}$ and $\mathbf{B}$ can be added (or subtracted) if, and only if, they have the same shape. If $\mathbf{A} = [a_{ij}]$, $\mathbf{B} = [b_{ij}]$, and $\mathbf{C} = \mathbf{A} + \mathbf{B}$, with $\mathbf{C} = [c_{ij}]$, then

$$c_{ij} = a_{ij} + b_{ij}.$$

Similarly, if $\mathbf{D} = \mathbf{A} - \mathbf{B}$, with $\mathbf{D} = [d_{ij}]$, then

$$d_{ij} = a_{ij} - b_{ij}.$$

26. If k is a scalar and $\mathbf{A} = [a_{ij}]$ is a matrix, then

$$k\mathbf{A} = [ka_{ij}].$$

27. If $\mathbf{A}$ is an $m \times n$ matrix and $\mathbf{B}$ is a $p \times q$ matrix, the matrix product $\mathbf{C} = \mathbf{A}\mathbf{B}$, in this order, is only defined if $n = p$, and then $\mathbf{C}$ is an $m \times q$ matrix. When the matrix product $\mathbf{C} = \mathbf{A}\mathbf{B}$ is defined, the entry c_{ij} in the i'th row and j'th column of $\mathbf{C}$ is $\mathbf{a}_i\mathbf{b}_j$, where $\mathbf{a}_i$ is the i'th row of $\mathbf{A}$, $\mathbf{C}_j$ is the j'th column of $\mathbf{B}$,

and if

$$\mathbf{a}_i = [a_{i1}, a_{i2}, \ldots, a_{in}], \quad \mathbf{b}_j = \begin{bmatrix} b_{1j} \\ b_{2j} \\ \vdots \\ b_{nj} \end{bmatrix},$$

then

$$\mathbf{a}_i \mathbf{b}_j = a_{i1} b_{1j} + a_{i2} b_{2j} + \cdots + a_{in} b_{nj}.$$

Thus if

$$\mathbf{A} = \begin{bmatrix} 3 & -1 & 4 \\ 1 & 0 & 2 \end{bmatrix}, \quad \mathbf{B} = \begin{bmatrix} 1 & 2 \\ 0 & -3 \\ 2 & 1 \end{bmatrix}, \quad \text{and} \quad \mathbf{C} = \begin{bmatrix} 1 & 2 & 1 \\ 0 & 1 & 2 \\ 4 & -1 & 1 \end{bmatrix},$$

$$\mathbf{AB} = \begin{bmatrix} 11 & 13 \\ 5 & 4 \end{bmatrix}, \quad \mathbf{BA} = \begin{bmatrix} 5 & -1 & 8 \\ -3 & 0 & -6 \\ 2 & -2 & 10 \end{bmatrix}, \quad \mathbf{AC} = \begin{bmatrix} 19 & 1 & 5 \\ 9 & 0 & 3 \end{bmatrix}$$

but $\mathbf{BC}$ is *not* defined.

28. Let the characteristic equation of the $n \times n$ matrix $\mathbf{A}$ determined by $|\mathbf{A} - \lambda\mathbf{I}| = 0$ have the form

$$\lambda^n + c_1\lambda^{n-1} + c_2\lambda^{n-2} + \cdots + c_n = 0.$$

Then $\mathbf{A}$ satisfies its characteristic equation, so

$$\mathbf{A}^n + c_1\mathbf{A}^{n-1} + c_2\mathbf{A}^{n-2} + \cdots + c_n\mathbf{I} = 0.$$

(Cayley–Hamilton theorem.)

29. If the matrix product $\mathbf{AB}$ is defined, then

$$(\mathbf{AB})^{\mathrm{T}} = \mathbf{B}^{\mathrm{T}}\mathbf{A}^{\mathrm{T}}.$$

30. If the matrix product $\mathbf{AB}$ is defined, then

$$(\mathbf{AB})^{-1} = \mathbf{B}^{-1}\mathbf{A}^{-1}.$$

31. The matrix $\mathbf{A}$ is **orthogonal** if $\mathbf{AA}^{\mathrm{T}} = \mathbf{I}$.

1.5.2 Quadratic forms

1.5.2.1 Definitions. A **quadratic form** involving the n real variables $x_1, x_2, \ldots, x_n$ that are associated with the real $n \times n$ matrix $\mathbf{A} = [a_{ij}]$ is the scalar expression

1. $Q(x_1, x_2, \ldots, x_n) = \sum_{i=1}^{n} \sum_{j=1}^{n} a_{ij} x_i x_j.$

In matrix notation, if $\mathbf{x}$ is the $n \times 1$ column vector with real elements $x_1, x_2, \ldots, x_n$, and $\mathbf{x}^{\mathrm{T}}$ is the transpose of $\mathbf{x}$, then

2. $Q(\mathbf{x}) = \mathbf{x}^{\mathrm{T}}\mathbf{A}\mathbf{x}.$

Employing the **inner product** notation, this same quadratic form may also be written

3. $Q(\mathbf{x}) \equiv (\mathbf{x}, \mathbf{A}\mathbf{x})$.

    If the $n \times n$ matrix $\mathbf{A}$ is Hermitian, so that $\overline{\mathbf{A}}^{\mathrm{T}} = \mathbf{A}$, where the bar denotes the complex conjugate operation, the quadratic form associated with the Hermitian matrix $\mathbf{A}$ and the vector $\mathbf{x}$, which may have complex elements, is the real quadratic form

4. $Q(\mathbf{x}) = (\mathbf{x}, \mathbf{A}\mathbf{x})$.

    It is always possible to express an arbitrary quadratic form

5. $Q(x) = \displaystyle\sum_{i=1}^{n} \sum_{j=1}^{n} \alpha_{ij} x_i x_j$

    in the form

6. $Q(\mathbf{x}) = (\mathbf{x}, \mathbf{A}\mathbf{x})$,

    in which $\mathbf{A} = [a_{ij}]$ is a symmetric matrix, by defining

7. $a_{ii} = \alpha_{ii} \qquad$ for $i = 1, 2, \ldots, n$

    and

8. $a_{ij} = \dfrac{1}{2}(\alpha_{ij} + \alpha_{ji}) \qquad$ for $i, j = 1, 2, \ldots, n \quad$ and $\quad i \neq j$.

9. When a quadratic form Q in n variables is reduced by a nonsingular linear transformation to the form

    $$Q = y_1^2 + y_2^2 + \cdots + y_p^2 - y_{p+1}^2 - y_{p+2}^2 - \cdots - y_r^2,$$

    the number p of positive squares appearing in the reduction is an invariant of the quadratic form Q, and does not depend on the method of reduction itself (**Sylvester's law of inertia**).

10. The **rank** of the quadratic form Q in the above canonical form is the total number r of squared terms (both positive and negative) appearing in its reduced from ($r \leq n$).

11. The **signature** of the quadratic form Q above is the number s of positive squared terms appearing in its reduced form. It is sometimes also defined to be $2s - r$.

12. The quadratic form $Q(\mathbf{x}) = (\mathbf{x}, \mathbf{A}\mathbf{x})$ is said to be **positive definite** when $Q(\mathbf{x}) > 0$ for $\mathbf{x} \neq 0$. It is said to be **positive semidefinite** if $Q(\mathbf{x}) \geq 0$ for $\mathbf{x} \neq 0$.

1.5.2.2 Basic theorems on quadratic forms.

1. Two real quadratic forms are **equivalent** under the group of linear transformations if, and only if, they have the same rank and the same signature.

2. A real quadratic form in n variables is positive definite if, and only if, its canonical form is

    $$Q = z_1^2 + z_2^2 + \cdots + z_n^2.$$

3. A real symmetric matrix $\mathbf{A}$ is positive definite if, and only if, there exists a real nonsingular matrix $\mathbf{M}$ such that $\mathbf{A} = \mathbf{M}\mathbf{M}^{\mathrm{T}}$.

4. Any real quadratic form in n variables may be reduced to the diagonal form

$$Q = \lambda_1 z_1^2 + \lambda_2 z_2^2 + \cdots + \lambda_n z_n^2, \quad \lambda_1 \geq \lambda_2 \geq \cdots \geq \lambda_n$$

by a suitable orthogonal point-transformation.

5. The quadratic form $Q = (\mathbf{x}, \mathbf{Ax})$ is positive definite if, and only if, every eigenvalue of $\mathbf{A}$ is positive; it is positive semidefinite if, and only if, all the eigenvalues of $\mathbf{A}$ are nonnegative; and it is indefinite if the eigenvalues of $\mathbf{A}$ are of both signs.

6. The necessary conditions for a Hermitian matrix $\mathbf{A}$ to be positive definite are

    (i) $a_{ii} > 0$ for all i,

    (ii) $a_{ii}a_{ij} > |a_{ij}|^2$ for $i \neq j$,

    (iii) the element of largest modulus must lie on the leading diagonal,

    (iv) $|\mathbf{A}| > 0$.

7. The quadratic form $Q = (\mathbf{x}, \mathbf{Ax})$ with $\mathbf{A}$ Hermitian will be positive definite if all the principal minors in the top left-hand corner of $\mathbf{A}$ are positive, so that

$$a_{11} > 0, \quad \begin{vmatrix} a_{11} & a_{12} \\ a_{21} & a_{22} \end{vmatrix} > 0, \quad \begin{vmatrix} a_{11} & a_{12} & a_{13} \\ a_{21} & a_{22} & a_{23} \\ a_{31} & a_{32} & a_{33} \end{vmatrix} > 0, \quad \cdots.$$

1.5.3 Differentiation and integration of matrices

1.5.3.1 If the $n \times n$ matrices $\mathbf{A}(t)$ and $\mathbf{B}(t)$ have elements that are differentiable functions of t, so that

$$\mathbf{A}(t) = [a_{ij}(t)], \quad \mathbf{B}(t) = [b_{ij}(t)],$$

then

1. $\dfrac{d}{dt}\mathbf{A}(t) = \left[\dfrac{d}{dt}a_{ij}(t)\right]$

2. $\dfrac{d}{dt}[\mathbf{A}(t) \pm \mathbf{B}(t)] = \left[\dfrac{d}{dt}a_{ij}(t) \pm \dfrac{d}{dt}b_{ij}(t)\right] = \dfrac{d}{dt}\mathbf{A}(t) \pm \dfrac{d}{dt}\mathbf{B}(t).$

3. If the matrix product $\mathbf{A}(t)\mathbf{B}(t)$ is defined, then

$$\frac{d}{dt}[\mathbf{A}(t)\mathbf{B}(t)] = \left[\frac{d}{dt}\mathbf{A}(t)\right]\mathbf{B}(t) + \mathbf{A}(t)\frac{d}{dt}\mathbf{B}(t).$$

4. If the matrix product $\mathbf{A}(t)\mathbf{B}(t)$ is defined, then

$$\frac{d}{dt}[\mathbf{A}(t)\mathbf{B}(t)]^{\mathrm{T}} = \left[\frac{d}{dt}\mathbf{B}(t)\right]^{\mathrm{T}}\mathbf{A}^{\mathrm{T}}(t) + \mathbf{B}^{\mathrm{T}}(t)\left(\frac{d}{dt}\mathbf{A}(t)\right)^{\mathrm{T}}.$$

5. If the square matrix $\mathbf{A}$ is nonsingular, so that $|\mathbf{A}| \neq 0$, then

$$\frac{d}{dt}[\mathbf{A}]^{-1} = -\mathbf{A}^{-1}(t)\left[\frac{d}{dt}\mathbf{A}(t)\right]\mathbf{A}^{-1}(t).$$

6. $\displaystyle\int_{t_0}^{t} \mathbf{A}(\tau)\, d\tau = \left[\int_{t_0}^{t} a_{ij}(\tau)\, d\tau \right].$

1.5.4 The matrix exponential

1.5.4.1 Definition.   If $\mathbf{A}$ is a square matrix, and z is any complex number, then the matrix exponential $e^{\mathbf{A}z}$ is defined to be

$$e^{\mathbf{A}z} = \mathbf{I} + \mathbf{A}z + \cdots + \frac{\mathbf{A}^n z^n}{n!} + \cdots = \sum_{r=0}^{\infty} \frac{1}{r!} \mathbf{A}^r z^r.$$

1.5.4.2 Basic properties of the matrix exponential.

1. $e^0 = \mathbf{I}, \quad e^{\mathbf{I}z} = \mathbf{I}e^z, \quad e^{\mathbf{A}(z_1+z_2)} = e^{\mathbf{A}z_1} \cdot e^{\mathbf{A}z_2}, \quad e^{-\mathbf{A}z} = \left(e^{\mathbf{A}z} \right)^{-1}, \quad e^{\mathbf{A}z} \cdot e^{\mathbf{B}z} = e^{(\mathbf{A}+\mathbf{B})z}$

   when $\mathbf{A} + \mathbf{B}$ is defined and $\mathbf{A}\mathbf{B} = \mathbf{B}\mathbf{A}$.

2. $\dfrac{d^r}{dz^r}\left(e^{\mathbf{A}z} \right) = \mathbf{A}^r e^{\mathbf{A}z} = e^{\mathbf{A}z} \mathbf{A}^r.$

3. If the square matrix $\mathbf{A}$ can be expressed in the form

   $$\mathbf{A} = \begin{bmatrix} \mathbf{B} & 0 \\ 0 & \mathbf{C} \end{bmatrix},$$

   with $\mathbf{B}$ and $\mathbf{C}$ square matrices, then

   $$e^{\mathbf{A}z} = \begin{bmatrix} e^{\mathbf{B}z} & 0 \\ 0 & e^{\mathbf{C}z} \end{bmatrix}.$$

1.5.5 The Gerschgorin circle theorem

1.5.5.1 The Gerschgorin circle theorem.   Each eigenvalue of an arbitrary $n \times n$ matrix $\mathbf{A} = [a_{ij}]$ lies in at least one of the circles $C_1, C_2, \ldots, C_n$ in the complex plane, where the circle C_r with radius ρ_r has its center at a_{rr}, where a_{rr} is the r'th element of the leading diagonal of $\mathbf{A}$, and

$$\rho_r = \sum_{\substack{j=1 \\ j \neq r}}^{n} |a_{rj}| = |a_{r1}| + |a_{r2}| + \cdots + |a_{r,r-1}| + |a_{r,r+1}| + \cdots + |a_{rn}|.$$

1.6 Permutations and Combinations

1.6.1 Permutations

1.6.1.1

1. A **permutation** of n mutually distinguishable elements is an arrangement or sequence of occurrence of the elements in which their *order* of appearance counts.

2. The number of possible mutually distinguishable permutations of n distinct elements is denoted either by ^nP_n or by $_nP_n$, where

$$^nP_n = n(n-1)(n-2)\cdots 3.2.1 = n!$$

3. The number of possible mutually distinguishable permutations of n distinct elements m at a time is denoted either by ^nP_m or by $_nP_m$, where

$$^nP_m = \frac{n!}{(n-m)!} \qquad [0! \equiv 1].$$

4. The number of possible identifiably different permutations of n elements of two different types, of which m are of type 1 and $n-m$ are of type 2 is

$$\frac{n!}{m!(n-m)!} = \binom{n}{m}. \qquad \textbf{(binomial coefficient)}$$

This gives the relationship between binomial coefficients and the number of m-element subsets of an n set. Expressed differently, this says that the coefficient of x^m in $(1+x)^n$ is the number of ways x's can be selected from precisely m of the n factors of the n-fold product being expanded.

5. The number of possible identifiably different permutations of n elements of m different types, in which m_r are of type r, with $m_1 + m_2 + \cdots + m_r = n$, is

$$\frac{n!}{m_1!m_2!\cdots m_r!}. \qquad \textbf{(multinomial coefficient)}$$

1.6.2 Combinations

1.6.2.1

1. A **combination** of n mutually distinguishable elements m at a time is a selection of m elements from the n *without regard to their order* of arrangement. The number of such combinations is denoted either by ^nC_m or by $_nC_m$, where

$$^nC_m = \binom{n}{m}.$$

2. The number of combinations of n mutually distinguishable elements in which each element may occur $0, 1, 2, \ldots, m$ times in any combination is

$$\binom{n+m-1}{m} = \binom{n+n-1}{n-1}.$$

3. The number of combinations of n mutually distinguishable elements in which each element must occur at least once in each combination is

$$\binom{m-1}{n-1}.$$

4. The number of distinguishable samples of m elements taken from n different elements, when each element may occur at most once in a sample, is

$$n(n-1)(n-2)\cdots(n-m+1).$$

5. The number of distinguishable samples of m elements taken from n different elements, when each element may occur $0, 1, 2, \ldots, m$ times in a sample is n^m.

1.7 Partial Fraction Decomposition

1.7.1 Rational functions

1.7.1.1   A function $R(x)$ of the form

1. $R(x) = \dfrac{N(x)}{D(x)},$

where $N(x)$ and $D(x)$ are polynomials in x, is called a **rational function** of x. The replacement of $R(x)$ by an equivalent expression involving a sum of simpler rational functions is called a decomposition of $R(x)$ into **partial fractions**. This technique is of use in the integration of arbitrary rational functions. Thus in the identity

$$\frac{3x^2 + 2x + 1}{x^3 + x^2 + x} = \frac{1}{x} + \frac{2x + 1}{x^2 + x + 1},$$

the expression on the right-hand side is a partial fraction expansion of the rational function $(3x^2 + 2x + 1)/(x^3 + x^2 + x)$.

1.7.2 Method of undetermined coefficients

1.7.2.1

1. The general form of the simplest possible partial fraction expansion of $R(x)$ in 1.7.1.1 depends on the respective degrees of $N(x)$ and $D(x)$, and on the decomposition of $D(x)$ into *real* factors. The form of the partial fraction decomposition to be adopted is determined as follows.

2. *Case 1.*   (Degree of $N(x)$ less than degree of $D(x)$).

    (i)   Let the degree of $N(x)$ be less than the degree of $D(x)$, and factor $D(x)$ into the simplest possible set of *real* factors. There may be linear factors with multiplicity 1, such as $(ax+b)$; linear factors with multiplicity r, such as $(ax+b)^r$; quadratic factors with multiplicity 1, such as (ax^2+bx+c); or quadratic factors with multiplicity m such as $(ax^2 + bx + c)^m$, where $a, b, \ldots,$ are real numbers, and the quadratic factors cannot be expressed as the product of *real* linear factors.

    (ii) To each linear factor with multiplicity 1, such as $(ax + b)$, include in the partial fraction decomposition a term such as

    $$\frac{A}{ax + b},$$

where A is an undetermined constant.

(iii) To each linear factor with multiplicity r, such as $(ax + b)^r$, include in the partial fraction decomposition terms such as

$$\frac{B_1}{ax + b} + \frac{B_2}{(ax + b)^2} + \cdots + \frac{B_r}{(ax + b)^r},$$

where $B_1, B_2, \ldots, B_r$ are undetermined constants.

(iv) To each quadratic factor with multiplicity 1, such as $(ax^2 + bx + c)$, include in the partial fraction decomposition a term such as

$$\frac{C_1 x + D_1}{ax^2 + bx + c},$$

where C_1, D_1 are undetermined constants.

(v) To each quadratic factor with multiplicity m, such as $(ax^2 + bx + c)^m$, include in the partial fraction decomposition terms such as

$$\frac{E_1 x + F_1}{ax^2 + bx + c} + \frac{E_2 x + F_2}{(ax^2 + bx + c)^2} + \cdots + \frac{E_m x + F_m}{(ax^2 + bx + c)^m},$$

where $E_1, D_1, \ldots, E_m, F_m$ are undetermined constants.

(vi) The final general form of the partial fraction decomposition of $R(x)$ in 1.7.1.1.1 is then the sum of all the terms generated in (ii) to (vi) containing the **undetermined coefficients** $A, B_1, B_2, \ldots, E_m, F_m$.

(vii) The specific values of the undetermined coefficients $A_1, B_1, \ldots, E_m, F_m$ are determined by equating $N(x)/D(x)$ to the sum of terms obtained in (vi), multiplying this identity by the factored form of $D(x)$, and equating the coefficients of corresponding powers of x on each side of the identity. If, say, there are N undetermined coefficients, an alternative to deriving N equations satisfied by them by equating coefficients of corresponding powers of x is to obtain N equations by substituting N convenient different values for x.

3. *Case 2.* (Degree of $N(x)$ greater than or equal to degree of $D(x)$).

(i) If the degree m of $N(x)$ is greater than or equal to the degree n of $D(x)$, first use long division to divide $D(x)$ into $N(x)$ to obtain an expression of the form

$$\frac{N(x)}{D(x)} = P(x) + \frac{M(x)}{D(x)},$$

where $P(x)$ is a known polynomial of degree $m - n$, and the degree of $M(x)$ is less than the degree of $D(x)$.

(ii) Decompose $M(x)/D(x)$ into partial fractions as in Case 1.

(iii) The required partial fraction decomposition is then the sum of $P(x)$ and the terms obtained in (ii) above.

An example of Case 1 is provided by considering the following rational function and applying the above rules to the factors x and $(x^2 + x + 1)$ of the denominator to obtain

$$\frac{3x^2 + 2x + 1}{x^3 + x^2 + x} = \frac{3x^2 + 2x + 1}{x(x^2 + x + 1)} = \frac{A}{x} + \frac{Bx + C}{x^2 + x + 1}.$$

Multiplication by $x(x^2 + x + 1)$ yields

$$3x^2 + 2x + 1 = A(x^2 + x + 1) + x(Bx + C).$$

Equating coefficients of corresponding powers of x gives

$$1 = A \qquad\qquad\qquad\qquad\text{(coefficients of } x^0)$$
$$2 = A + C \qquad\qquad\qquad\text{(coefficients of } x)$$
$$3 = A + B \qquad\qquad\qquad\text{(coefficients of } x^2)$$

so $A = 1$, $B = 2$, $C = 1$, and the required partial fraction decomposition becomes

$$\frac{3x^2 + 2x + 1}{x^3 + x^2 + x} = \frac{1}{x} + \frac{2x + 1}{x^2 + x + 1}.$$

An example of Case 2 is provided by considering $(2x^3 + 5x^2 + 7x + 5)/(x + 1)^2$. The degree of the numerator exceeds that of the denominator, so division by $(x + 1)^2$ is necessary. We have

$$
\begin{array}{r}
2x + 1 \\
x^2 + 2x + 1 \enclose{longdiv}{2x^3 + 5x^2 + 7x + 5} \\
\underline{2x^3 + 4x^2 + 2x} \\
x^2 + 5x + 5 \\
\underline{x^2 + 2x + 1} \\
3x + 4
\end{array}
$$

and so

$$\frac{2x^3 + 5x^2 + 7x + 5}{(x + 1)^2} = 1 + 2x + \frac{3x + 4}{(x + 1)^2}.$$

An application of the rules of Case 1 to the rational function $(3x + 4)/(x + 1)^2$ gives

$$\frac{3x + 4}{(x + 1)^2} = \frac{A}{x + 1} + \frac{B}{(x + 1)^2},$$

or

$$3x + 4 = A(x + 1) + B.$$

Equating coefficients of corresponding powers of x gives

$$4 = A + B \qquad\qquad\qquad\text{(coefficients of } x^0)$$
$$3 = A \qquad\qquad\qquad\qquad\text{(coefficients of } x)$$

so $A = 3$, $B = 1$ and the required partial fraction decomposition becomes

$$\frac{2x^3 + 5x^2 + 7x + 5}{(x + 1)^2} = 1 + 2x + \frac{3}{x + 1} + \frac{1}{(x + 1)^2}.$$

1.8 Convergence of Series

1.8.1 Types of convergence of numerical series

1.8.1.1   Let $\{u_k\}$ with $k = 1, 2, \ldots$, be an infinite sequence of numbers, then the **infinite** numerical series

1. $\displaystyle\sum_{k=1}^{\infty} u_k = u_1 + u_2 + u_3 + \cdots$

    is said to **converge** to the sum S if the sequence $\{S_n\}$ of **partial** sums

2. $S_n = \displaystyle\sum_{k=1}^{n} u_k = u_1 + u_2 + \cdots + u_n$ has a finite limit S, so that

3. $S = \lim\limits_{n \to \infty} S_n$.

    If S is infinite, or the sequence $\{S_n\}$ has no limit, the series 1.8.1.1.1 is said to **diverge**.

4. The series 1.8.1.1.1 is convergent if for each $\varepsilon > 0$ there is a number $N(\varepsilon)$ such that

    $$|S_m - S_n| < \varepsilon \qquad \text{for all } m > n > N.$$

    (Cauchy criterion for convergence)

5. The series 1.8.1.1.1 is said to be **absolutely convergent** if the series of absolute values

    $$\sum_{k=1}^{n} |u_k| = |u_1| + |u_2| + |u_3| + \cdots$$

    converges. Every absolutely convergent series is convergent. If series 1.8.1.1.1 is such that it is convergent, but it is not absolutely convergent, it is said to be **conditionally convergent**.

1.8.2 Convergence tests

1.8.2.1   Let the series 1.8.1.1.1 be such that $u_k \neq 0$ for any k and

1. $\lim\limits_{k \to \infty} \left| \dfrac{u_{k+1}}{u_k} \right| = r$.

Then series 1.8.1.1.1 is absolutely convergent if $r < 1$ and it **diverges** if $r > 1$. The test fails to provide information about convergence or divergence if $r = 1$.

(d'Alembert's ratio test)

1.8.2.2   Let the series 1.8.1.1.1 be such that $u_k \neq 0$ for any k and

1. $\lim\limits_{k \to \infty} |u_k|^{1/k} = r$.

The series 1.8.1.1.1 is absolutely convergent if $r < 1$ and it diverges if $r > 1$. The test fails to provide information about convergence or divergence if $r = 1$.

(Cauchy's n'th root test)

1.8.2.3   Let the series $\sum_{k=1}^{\infty} u_k$ be such that

$$\lim_{k \to \infty} k \left\{ \left| \frac{u_k}{u_{k+1}} \right| - 1 \right\} = r.$$

Then the series is absolutely convergent if $r > 1$ and it is divergent if $r < 1$. The test fails to provide information about convergence or divergence if $r = 1$. (**Raabe's test:** This test is a more delicate form of ratio test and it is often useful when the ratio test fails because $r = 1$.)

1.8.2.4   Let the series in 1.8.1.1.1 be such that $u_k \geq 0$ for $k = 1, 2, \ldots$, and let $\sum_{k=1}^{\infty} a_k$ be a convergent series of positive terms such that $u_k \leq a_k$ for all k. Then

1. $\sum_{k=1}^{\infty} u_k$ is convergent and

$$\sum_{k=1}^{\infty} u_k \leq \sum_{k=1}^{\infty} a_k \qquad \textbf{(comparison test for convergence)}$$

2. If $\sum_{k=1}^{\infty} a_k$ is a divergent series of nonnegative terms and $u_k \geq a_k$ for all k, then $\sum_{k=1}^{\infty} u_k$ is divergent.   **(comparison test for divergence)**

1.8.2.5   Let $\sum_{k=1}^{\infty} u_k$ be a series of positive terms whose convergence is to be determined, and let $\sum_{k=1}^{\infty} a_k$ be a comparison series of positive terms known to be either convergent or divergent. Let

1. $\lim_{k \to \infty} \dfrac{u_k}{a_k} = L,$

    where L is either a nonnegative number or infinity.
2. If $\sum_{k=1}^{\infty} a_k$ converges and $0 \leq L < \infty$, $\sum_{k=1}^{\infty} u_k$ converges.
3. If $\sum_{k=1}^{\infty} a_k$ diverges and $0 < L \leq \infty$, $\sum_{k=1}^{\infty} u_k$ diverges.

    (limit comparison test)

1.8.2.6   Let $\sum_{k=1}^{\infty} u_k$ be a series of positive nonincreasing terms, and let $f(x)$ be a nonincreasing function defined for $k \geq N$ such that

1. $f(k) = u_k$.

    Then the series $\sum_{k=1}^{\infty} u_k$ converges or diverges according as the improper integral

2. $\int_N^{\infty} f(x)\, dx$

    converges or diverges.   **(Cauchy integral test)**

1.8.2.7   Let the sequence $\{a_k\}$ with $a_k > a_{k+1} > 0$, for $n = 1, 2, \ldots$, be such that

1. $\lim_{k \to \infty} a_k = 0$.

    Then the **alternating series** $\sum_{k=1}^{\infty} (-1)^{k+1} a_k$ converges, and

2. $|S - S_N| \leq a_{N+1}$,

    where

$$S_N = a_1 - a_2 + a_3 - \cdots + (-1)^{N+1} a_N.$$

1.8.2.8   If the series

1. $\displaystyle\sum_{k=1}^{\infty} v_k = v_1 + v_2 + \cdots$

converges and the sequence of numbers $\{u_k\}$ form a monotonic bounded sequence, that is, if $|u_k| < M$ for some number M and all k, the series

2. $\displaystyle\sum_{k=1}^{\infty} u_k v_k = u_1 v_1 + u_2 v_2 + \cdots$

converges.   **(Abel's test)**

1.8.2.9   If the partial sums of the series 1.8.2.8.1 are bounded and if the numbers u_k constitute a monotonic decreasing sequence with limit zero, that is if

$$\left|\sum_{k=1}^{n} u_k\right| < M \qquad [n = 1, 2, \ldots] \quad \text{and} \quad \lim_{k \to \infty} u_k = 0,$$

then the series 1.8.2.8.1 converges.   **(Dirichlet's test)**

1.8.3 Examples of infinite numerical series

1.8.3.1 Geometric and arithmetic–geometric series.

1. $\displaystyle\sum_{k=0}^{\infty} ar^{k-1} = \frac{a}{1-r} \qquad [|r| < 1].$ (see 1.2.2.2)

2. $\displaystyle\sum_{k=0}^{\infty} (a + kd)r^k = \frac{a}{1-r} + \frac{rd}{(1-r)^2} \qquad [|r| < 1].$ (see 1.2.2.3)

1.8.3.2 Binomial expansion.

1. $(1 + a)^q = 1 + qa + \dfrac{q(q-1)}{2!}a^2 + \dfrac{q(q-1)(q-2)}{3!}a^3 + \cdots$

$\qquad + \dfrac{q(q-1)(q-2)\cdots(q-r+1)}{r!}a^r + \cdots$

$\qquad\qquad\qquad\qquad\qquad\qquad\qquad$ [any real q, $|a| < 1$].

2. $(a + b)^q = a^q \left(1 + \dfrac{b}{a}\right)^q$

$\qquad = a^q\left[1 + q\left(\dfrac{b}{a}\right) + \dfrac{q(q-1)}{2!}\left(\dfrac{b}{a}\right)^2 + \dfrac{q(q-1)(q-2)}{3!}\left(\dfrac{b}{a}\right)^3 + \cdots\right.$

$\qquad\qquad \left. + \dfrac{q(q-1)(q-2)\cdots(q-r+1)}{r!}\left(\dfrac{b}{a}\right)^r + \cdots\right]$

$\qquad\qquad\qquad\qquad\qquad$ [any real q, $|b/a| < 1$].   (see 0.7.5)

1.8.3.3 Series with rational sums.

1. $\displaystyle\sum_{k=1}^{\infty} \frac{1}{k(k+1)} = 1$

2. $\displaystyle\sum_{k=1}^{\infty} \frac{1}{k(k+2)} = \frac{3}{4}$

3. $\displaystyle\sum_{k=1}^{\infty} \frac{1}{(2k-1)(2k+1)} = \frac{1}{2}$

4. $\displaystyle\sum_{k=1}^{\infty} \frac{k}{(4k^2-1)^2} = \frac{1}{8}$

5. $\displaystyle\sum_{k=1}^{\infty} \frac{k}{(k+1)!} = 1$

1.8.3.4 Series involving π.

1. $\displaystyle\sum_{k=1}^{\infty} (-1)^{k+1} \frac{1}{(2k-1)} = \frac{\pi}{4}$

2. $\displaystyle\sum_{k=1}^{\infty} \frac{1}{k^2} = \frac{\pi^2}{6}$

3. $\displaystyle\sum_{k=1}^{\infty} (-1)^{k+1} \frac{1}{k^2} = \frac{\pi^2}{12}$

4. $\displaystyle\sum_{k=1}^{\infty} \frac{(-1)^{k+1}}{(2k-1)^3} = \frac{\pi^3}{32}$

5. $\displaystyle\sum_{k=1}^{\infty} \frac{1}{(2k-1)^4} = \frac{\pi^4}{96}$

6. $\displaystyle\sum_{k=1}^{\infty} \frac{1}{(4k-1)(4k+1)} = \frac{1}{2} - \frac{\pi}{8}$

1.8.3.5 Series involving e.

1. $\displaystyle\sum_{k=0}^{\infty} \frac{1}{k!} = e = 2.71828\ldots$

2. $\displaystyle\sum_{k=0}^{\infty} \frac{(-1)^k}{k!} = \frac{1}{e} = 0.36787\ldots$

3. $\displaystyle\sum_{k=1}^{\infty} \frac{2k}{(2k+1)!} = \frac{1}{e} = 0.36787\ldots$

4. $\displaystyle\sum_{k=0}^{\infty} \frac{1}{(2k)!} = \frac{1}{2}\left(e + \frac{1}{e}\right) = 1.54308\ldots$

5. $\displaystyle\sum_{k=0}^{\infty} \frac{1}{(2k+1)!} = \frac{1}{2}\left(e - \frac{1}{e}\right) = 1.17520\ldots$

1.8.3.6 Series involving a logarithm.

1. $\displaystyle\sum_{k=1}^{\infty}(-1)^{k+1}\frac{1}{k} = \ln 2$

2. $\displaystyle\sum_{k=1}^{\infty}(-1)^{k+1}\frac{1}{k \cdot m^k} = \ln\left(\frac{1+m}{m}\right) \qquad [m = 1, 2, \ldots]$

3. $\displaystyle\sum_{k=1}^{\infty}\frac{1}{k(4k^2 - 1)} = 2\ln 2 - 1$

4. $\displaystyle\sum_{k=1}^{\infty}\frac{1}{k(9k^2 - 1)} = \frac{3}{2}(\ln 3 - 1)$

5. $\displaystyle\sum_{k=1}^{\infty}\frac{1}{k(4k^2 - 1)^2} = \frac{3}{2} - 2\ln 2$

6. $\displaystyle\sum_{k=1}^{\infty}\frac{12k^2 - 1}{k(4k^2 - 1)^2} = 2\ln 2$

7. $\displaystyle\sum_{k=1}^{\infty}\frac{1}{(2k-1)k(2k+1)} = 2\ln 2 - 1$

8. $\displaystyle\sum_{k=1}^{\infty}\frac{(-1)^{k+1}}{(2k-1)k(2k+1)} = (1 - \ln 2)$

9. $\displaystyle\sum_{k=1}^{\infty}\frac{1}{2^k k} = \ln 2$

1.9 Infinite Products

1.9.1 Convergence of infinite products

1.9.1.1   Let $\{u_k\}$ be an infinite sequence of numbers and denote the product of the first n elements of the sequence by $\prod_{k=1}^{n} u_k$, so that

1. $\displaystyle\prod_{k=1}^{n} u_k = u_1 u_2 \cdots u_n.$

Then if the limit $\lim_{n\to\infty}\prod_{k=1}^{n} u_k$ exists, whether finite or infinite, but of definite sign, this limit is called the value of the **infinite product** $\prod_{k=1}^{\infty} u_k$, and we write

2. $\lim\limits_{n\to\infty} \prod\limits_{k=1}^{n} u_k = \prod\limits_{k=1}^{\infty} u_k.$

If an infinite product has a finite *nonzero* value it is said to **converge.** An infinite product that does not converge is said to **diverge.**

1.9.1.2   If $\{a_k\}$ is an infinite sequence of numbers, in order that the infinite product

1. $\prod\limits_{k=1}^{\infty}(1 + a_k)$

should converge it is necessary that $\lim_{n\to\infty} a_k = 0$.

1.9.1.3   If $a_k > 0$ or $a_k < 0$ for all k starting with some particular value, then for the infinite product 1.9.1.2.1 to converge, it is necessary and sufficient for $\sum_{k=1}^{\infty} a_k$ to converge.

1.9.1.4   The infinite product $\prod_{k=1}^{\infty}(1 + a_k)$ is said to **converge absolutely** if the infinite product $\prod_{k=1}^{\infty}(1 + |a_k|)$ converges.

1.9.1.5   Absolute convergence of an infinite product implies its convergence.

1.9.1.6   The infinite product $\prod_{k=1}^{\infty}(1 + a_k)$ converges absolutely if, and only if, the series $\sum_{k=1}^{\infty} a_k$ converges absolutely.

1.9.1.7   The infinite product $\prod_{k=1}^{\infty}|a_k| < \infty$ if, and only if, $\sum_{k=1}^{\infty} \ln|a_k| < \infty$.

1.9.2 Examples of infinite products

1. $\prod\limits_{k=1}^{\infty}\left(1 + \dfrac{(-1)^{k+1}}{2k - 1}\right) = \sqrt{2}$

2. $\prod\limits_{k=2}^{\infty}\left(1 - \dfrac{1}{k^2}\right) = \dfrac{1}{2}$

3. $\prod\limits_{k=1}^{\infty}\left(1 - \dfrac{1}{(2k + 1)^2}\right) = \dfrac{\pi}{4}$

4. $\prod\limits_{k=2}^{\infty}\left(1 - \dfrac{2}{k(k + 1)}\right) = \dfrac{1}{3}$

5. $\prod\limits_{k=2}^{\infty}\left(1 - \dfrac{2}{k^3 + 1}\right) = \dfrac{2}{3}$

6. $\prod\limits_{k=2}^{\infty}\left(1 + \dfrac{1}{2^k - 2}\right) = 2$

7. $\prod\limits_{k=2}^{\infty}\left(1 + \dfrac{1}{k^2 - 1}\right) = 2$

8. $\displaystyle\prod_{k=1}^{\infty}\left(1-\frac{1}{4k^2}\right)=\frac{2}{\pi}$

9. $\displaystyle\prod_{k=1}^{\infty}\left(1-\frac{1}{9k^2}\right)=\frac{3^{3/2}}{2\pi}$

10. $\displaystyle\prod_{k=1}^{\infty}\left(1-\frac{1}{16k^2}\right)=\frac{2^{3/2}}{\pi}$

11. $\displaystyle\prod_{k=1}^{\infty}\left(1-\frac{1}{36k^2}\right)=\frac{3}{\pi}$

12. $\displaystyle\prod_{k=0}^{\infty}\left(1+\left(\frac{1}{2}\right)^{2^k}\right)=2$

13. $\displaystyle\frac{2}{1}\cdot\left(\frac{4}{3}\right)^{1/2}\left(\frac{6\cdot 8}{5\cdot 7}\right)^{1/4}\left(\frac{10\cdot 12\cdot 14\cdot 16}{9\cdot 11\cdot 13\cdot 15}\right)^{1/8}\cdots=e$

14. $\displaystyle\sqrt{\frac{1}{2}}\cdot\sqrt{\frac{1}{2}+\frac{1}{2}\sqrt{\frac{1}{2}}}\cdot\sqrt{\frac{1}{2}+\frac{1}{2}\sqrt{\frac{1}{2}+\frac{1}{2}\sqrt{\frac{1}{2}}}}\cdots=\frac{2}{\pi}$ **(Vieta's formula)**

15. $\displaystyle\prod_{k=1}^{\infty}\left(\frac{2k}{2k-1}\right)\left(\frac{2k}{2k+1}\right)=\frac{2}{1}\cdot\frac{2}{3}\cdot\frac{4}{3}\cdot\frac{4}{5}\cdot\frac{6}{5}\cdot\frac{6}{7}\cdots=\frac{\pi}{2}$ **(Wallis's formula)**

16. $\displaystyle\prod_{k=1}^{\infty}\left(1-\frac{x^2}{k^2\pi^2}\right)=\frac{\sin x}{x}$

17. $\displaystyle\prod_{k=0}^{\infty}\left(1-\frac{4x^2}{(2k+1)^2\pi^2}\right)=\cos x$

18. $\displaystyle\prod_{k=1}^{\infty}\left(1+\frac{x^2}{k^2\pi^2}\right)=\frac{\sinh x}{x}$

19. $\displaystyle\prod_{k=0}^{\infty}\left(1+\frac{4x^2}{(2k+1)^2\pi^2}\right)=\cosh x$

1.10 Functional Series

1.10.1 Uniform convergence

1.10.1.1   Let $\{f_k(x)\}$, $k=1,2,\ldots$, be an infinite sequence of functions. Then a series of the form

1. $\displaystyle\sum_{k=1}^{\infty} f_k(x),$

is called a **functional series**. The set of values of the independent variable x for which the series converges is called the **region of convergence** of the series.

1.10.1.2   Let D be a region in which the functional series 1.10.1.1.1 converges for each value of x. Then the series is said to **converge uniformly** in D if, for every $\varepsilon > 0$, there exists a number $N(\varepsilon)$ such that, for $n > N$, it follows that

1. $$\left| \sum_{k=n+1}^{\infty} f_k(x) \right| < \varepsilon,$$

for all x in D.

The **Cauchy criterion** for the uniform convergence of series 1.10.1.1.1 requires that

2. $|f_m(x) + f_{m+1}(x) + \cdots + f_n(x)| < \varepsilon$,

for every $\varepsilon > 0$, all x in D and all $n > m > N$.

1.10.1.3   Let $\{f_k(x)\}$, $k = 1, 2, \ldots$, be an infinite sequence of functions, and let $\{M_k\}$, $k = 1, 2, \ldots$, be a sequence of positive numbers such that $\sum_{k=1}^{\infty} M_k$ is convergent. Then, if

1. $|f_k(x)| \leq M_k$,

for all x in a region D and all $k = 1, 2, \ldots$, the functional series in 1.10.1.1.1 converges uniformly for all x in D.   **(Weierstrass's M test)**

1.10.1.4   Let the series 1.10.1.1.1 converge for all x in some region D, in which it defines a function

1. $$f(x) = \sum_{k=1}^{\infty} f_k(x).$$

Then the series is said to **converge uniformly** to $f(x)$ in D if, for every $\varepsilon > 0$, there exists a number $N(\varepsilon)$ such that, for $n > N$, it follows that

2. $$\left| f(x) - \sum_{k=0}^{n} f_k(x) \right| < \varepsilon$$

for all x in D.

1.10.1.5   Let the infinite sequence of functions $\{f_k(x)\}$, $k = 1, 2, \ldots$, be continuous for all x in some region D. Then if the functional series $\sum_{k=1}^{\infty} f_k(x)$ is uniformly convergent to the function $f(x)$ for all x in D, the function $f(x)$ is continuous in D.

1.10.1.6   Suppose the series 1.10.1.1.1 converges uniformly in a region D, and that for each x in D the sequence of functions $\{g_k(x)\}$, $k = 1, 2, \ldots$ is monotonic and uniformly bounded, so that for some number $L > 0$

1. $|g_k(x)| \leq L$

for each $k = 1, 2, \ldots$, and all x in D. Then the series

2. $$\sum_{k=1}^{\infty} f_k(x) g_k(x)$$

converges uniformly in D.   **(Abel's theorem)**

1.10.1.7   Suppose the partial sums $S_n(x) = \sum_{k=1}^{n} f_k(x)$ of 1.10.1.1.1 are uniformly bounded, so that

1. $\left| \sum_{k=1}^{n} f_k(x) \right| < L$

for some L, all $n = 1, 2, \ldots$, and all x in the region of convergence D. Then, if $\{g_k(x)\}, k = 1, 2, \ldots$, is a monotonic decreasing sequence of functions that approaches zero uniformly for all x in D, the series

2. $\sum_{k=1}^{\infty} f_k(x) g_k(x)$

converges uniformly in D.   **(Dirichlet's theorem)**

1.10.1.8   If each function in the infinite sequence of functions $\{f_k(x)\}, k = 1, 2, \ldots$, is integrable on the interval $[a, b]$, and if the series 1.10.1.1.1 converges uniformly on this interval, the series may be **integrated term by term (termwise),** so that

1. $\int_a^b \left[\sum_{k=1}^{\infty} f_k(x) \right] dx = \sum_{k=1}^{\infty} \int_a^b f_k(x)\, dx \qquad [a \leq x \leq b].$

1.10.1.9   Let each function in the infinite sequence of functions $\{f_k(x)\}, k = 1, 2, \ldots$, have a continuous derivative $f_k'(x)$ on the interval $[a, b]$. Then if series 1.10.1.1.1 converges on this interval, and if the series $\sum_{k=1}^{\infty} f_k'(x)$ converges uniformly on the same interval, the series 1.10.1.1.1 may be **differentiated term by term (termwise),** so that

$$\frac{d}{dx} \left[\sum_{k=1}^{\infty} f_k(x) \right] dx = \sum_{k=1}^{\infty} f_k'(x).$$

1.11 Power Series

1.11.1 Definition

1.11.1.1   A functional series of the form

1. $\sum_{k=0}^{\infty} a_k (x - x_0)^k = a_0 + a_1(x - x_0) + a_2(x - x_0)^2 + \cdots$

is called a **power series** in x expanded about the point x_0 with **coefficients** a_k. The following is true of any power series: If it is not everywhere convergent, the region of convergence (in the complex plane) is a circle of radius R with its center at the point x_0; at every interior point of this circle the power series 1.11.1.1.1 converges absolutely, and outside this circle it diverges. The circle is called the **circle of convergence** and its radius R is called the **radius of convergence**. A series that converges at all points of the complex plane is said to have an infinite radius of convergence ($R = +\infty$).

1.11.1.2   The radius of convergence R of the power series in 1.11.1.1.1 may be determined by

1. $R = \lim\limits_{k \to \infty} \left| \dfrac{a_k}{a_{k+1}} \right|$,

    when the limit exists; by

2. $R = \dfrac{1}{\lim_{k \to \infty} |a_k|^{1/k}}$,

    when the limit exists; or by the **Cauchy–Hadamard formula**

3. $R = \dfrac{1}{\lim \sup |a_k|^{1/k}}$,

    which is always defined (though the result is difficult to apply).

The **circle of convergence** of the power series in 1.11.1.1.1 is $|x - x_0| = R$, so the series is absolutely convergent in the open disk

$$|x - x_0| < R$$

and divergent outside it where x and x_0 are points in the complex plane.

1.11.1.3   The power series 1.11.1.1.1 may be integrated and differentiated term by term inside its circle of convergence; thus

1. $\displaystyle \int_{x_0}^{x} \left[\sum_{k=0}^{\infty} a_k (x - x_0)^k \right] dx = \sum_{k=0}^{\infty} \frac{a_k}{k+1} (x - x_0)^{k+1}$,

2. $\displaystyle \frac{d}{dx} \left[\sum_{k=0}^{\infty} a_k (x - x_0)^k \right] = \sum_{k=1}^{\infty} k a_k (x - x_0)^{k-1}$.

The radii of convergence of the series 1.11.1.3.1 and 1.11.1.3.2 just given are both the same as that of the original series 1.11.1.1.1.

Operations on power series.

1.11.1.4   Let $\sum_{k=0}^{\infty} a_k (x - x_0)^k$ and $\sum_{k=0}^{\infty} b_k (x - x_0)^k$ be two power series expanded about x_0. Then the quotient of these series

1. $\displaystyle \frac{\sum_{k=0}^{\infty} b_k (x - x_0)^k}{\sum_{k=0}^{\infty} a_k (x - x_0)^k} = \frac{1}{a_0} \sum_{k=0}^{\infty} c_k (x - x_0)^k$,

    where the c_k follow from the equations

2. $b_0 = c_0$
    $a_0 b_1 = a_0 c_1 + a_1 c_0$
    $a_0 b_2 = a_0 c_2 + a_1 c_1 + a_2 c_0$
    $a_0 b_3 = a_0 c_3 + a_1 c_2 + a_2 c_1 + a_3 c_0$
    $$\vdots$$
    $$a_0 c_n + \sum_{k=1}^{n} a_k c_{n-k} - a_0 b_n = 0$$

or from

$$3. \quad c_n = \frac{(-1)^n}{a_0^n} \begin{vmatrix} a_1 b_0 - a_0 b_1 & a_0 & 0 & \cdots & 0 \\ a_2 b_0 - a_0 b_2 & a_1 & a_0 & \cdots & 0 \\ a_3 b_0 - a_0 b_3 & a_2 & a_1 & \cdots & 0 \\ \vdots & \vdots & \vdots & \vdots & \vdots \\ a_{n-1} b_0 - a_0 b_{n-1} & a_{n-2} & a_{n-3} & \cdots & a_0 \\ a_n b_0 - a_0 b_n & a_{n-1} & a_{n-2} & \cdots & a_1 \end{vmatrix}, \quad c_0 = b_0/a_0.$$

For example, if $a_k = 1/k!$, $b_k = 2^k/k!$, $x_0 = 0$, it follows that $\sum_{k=0}^{\infty} a_k(x-x_0)^k = e^x$ and $\sum_{k=0}^{\infty} b_k(x-x_0)^k = e^{2x}$, so in this case

$$\sum_{k=0}^{\infty} b_k(x-x_0)^k \Big/ \sum_{k=0}^{\infty} a_k(x-x_0)^k = e^{2x}/e^x = e^x.$$

This is confirmed by the above method, because from 1.11.1.4.2, $c_0 = 1$, $c_1 = 1$, $c_2 = \frac{1}{2}$, $c_3 = \frac{1}{6}$, ..., so, as expected,

$$\left(\sum_{k=0}^{\infty} x^k/k! \right) \Big/ \left[\sum_{k=0}^{\infty} (2x)^k/k! \right]$$

$$= 1 + x + \frac{x^2}{2} + \frac{x^3}{6} + \cdots = 1 + x + \frac{x^2}{2!} + \frac{x^3}{3!} + \cdots = e^x.$$

1.11.1.5 Let $\sum_{k=0}^{\infty} a_k(x-x_0)^k$ be a power series expanded about x_0, and let n be a natural number. Then, when this power series is **raised to the power n,** we have

$$1. \quad \left[\sum_{k=0}^{\infty} a_k(x-x_0)^k \right]^n = \sum_{k=0}^{\infty} c_k(x-x_0)^k,$$

where

$$c_0 = a_0^n, \quad c_m = \frac{1}{m a_0} \sum_{k=1}^{m} (kn - m + k) a_k c_{m-k} \qquad [m \geq 1].$$

For example, if $a_k = 1/k!$, $x_0 = 0$, $n = 3$, it follows that

$$\sum_{k=0}^{\infty} a_k(x-x_0)^k = e^x, \quad \text{so} \quad \left[\sum_{k=0}^{\infty} a_k(x-x_0)^k \right]^n = e^{3x}.$$

This is confirmed by the above method, because from 1.11.1.5.2, $c_0 = 1$, $c_1 = 3$, $c_2 = 9/2$, $c_3 = 9/2$, ..., so, as expected,

$$\left(\sum_{k=0}^{\infty} x^k/k! \right)^3 = 1 + 3x + \frac{9}{2}x^2 + \frac{9}{2}x^3 + \cdots$$

$$= 1 + 3x + \frac{(3x)^2}{2!} + \frac{(3x)^3}{3!} + \cdots = e^{3x}.$$

1.11.1.6 Let $y = \sum_{k=1}^{\infty} a_k(x-x_0)^k$ and $\sum_{k=1}^{\infty} b_k y^k$ be two power series. Then substituting for y in the second power series gives

1. $\sum_{k=1}^{\infty} b_k y^k = \sum_{k=1}^{\infty} c_k (x - x_0)^k$, where

2. $c_1 = a_1 b_1, \quad c_2 = a_2 b_1 + a_1^2 b_2, \quad c_3 = a_3 b_1 + 2a_1 a_2 b_2 + a_1^3 b_3,$
   $c_4 = a_4 b_1 + a_2^2 b_2 + 2a_1 a_3 b_2 + 3a_1^2 a_2 b_3 + a_1^4 b_4, \dots$

   For example, if $a_k = (-1)^k/(k+1)$, $b_k = 1/k!$, $x_0 = 0$, it follows that $y = \sum_{k=1}^{\infty} a_k (x - x_0)^k = \ln(1 + x)$, and $\sum_{k=1}^{\infty} b_k y^k = e^y - 1$, so the result of substituting for y is to give $\sum_{k=1}^{\infty} b_k y^k = \exp\{\ln(1 + x)\} - 1 = x$. This is confirmed by the above method, because from 1.11.1.6.2, $c_1 = 1$, $c_2 = c_3 = c_4 = \dots = 0$ so, as expected,

   $$\sum_{k=1}^{\infty} b_k y^k = \sum_{k=1}^{\infty} c_k (x - x_0)^k = x.$$

1.11.1.7   Let $\sum_{k=0}^{\infty} a_k (x - x_0)^k$ and $\sum_{k=0}^{\infty} b_k (x - x_0)^k$ be two power series expanded about x_0. Then the **product** of these series is given by

1. $$\left[\sum_{k=0}^{\infty} a_k (x - x_0)^k \right] \left[\sum_{k=0}^{\infty} b_k (x - x_0)^k \right] = \sum_{k=0}^{\infty} c_k (x - x_0)^k,$$
   where

2. $$c_n = \sum_{k=0}^{n} a_k b_{n-k}, \quad c_0 = a_0 b_0.$$

For example, if $a_k = 1/k!$, $b_k = 1/k!$, $x_0 = 0$, it follows that

$$\sum_{k=0}^{\infty} a_k (x - x_0)^k = \sum_{k=0}^{\infty} b_k (x - x_0)^k = e^x,$$

so in this case

$$\left[\sum_{k=0}^{\infty} a_k (x - x_0)^k \right] \left[\sum_{k=0}^{\infty} b_k (x - x_0)^k \right] = e^x \cdot e^x = e^{2x}.$$

This is confirmed by the above method, because from 1.11.1.7.2 $c_0 = 1$, $c_1 = 2$, $c_2 = 2$, $c_3 = 4/3$, $\dots$, so, as expected,

$$\left(\sum_{k=0}^{\infty} x^k/k! \right) \left(\sum_{k=0}^{\infty} x^k/k! \right) = 1 + 2x + 2x^2 + \frac{4}{3}x^3 + \cdots$$

$$= 1 + 2x + \frac{(2x)^2}{2!} + \frac{(2x)^3}{3!} + \cdots = e^{2x}.$$

1.11.1.8   Let

1. $$y - y_0 = \sum_{k=1}^{\infty} a_k (x - x_0)^k$$

   be a power series expanded about x_0. Then the **reversion** of this power series corresponds to finding a power series in $y - y_0$, expanded about y_0, that represents the function inverse to the one defined by the power series in 1.11.1.8.1. Thus, if the original series is written concisely as $y = f(x)$, reversion of the series corresponds to finding the power series for the inverse function $x = f^{-1}(y)$. The reversion of the series in 1.11.1.8.1 is given by

2. $\quad x - x_0 = \sum_{k=1}^{\infty} A_k (y - y_0)^k,$

where

3. $\quad A_1 = \dfrac{1}{a_1}, \quad A_2 = \dfrac{-a_2}{a_1^3}, \quad A_3 = \dfrac{2a_2^2 - a_1 a_3}{a_1^3}$

$A_4 = \dfrac{5a_1 a_2 a_3 - a_1^2 a_4 - 5a_2^3}{a_1^7},$

$A_5 = \dfrac{6a_1^2 a_2 a_4 + 3a_1^2 a_3^2 + 14a_2^4 - a_1^3 a_5 - 21a_1 a_2^2 a_3}{a_1^9}$

$A_6 = \dfrac{7a_1^3 a_2 a_5 + 7a_1^3 a_3 a_4 + 84a_1 a_2^3 a_3 - a_1^4 a_6 - 28a_1^2 a_2 a_3^2 - 42a_2^5 - 28a_1^2 a_2^2 a_4}{a_1^{11}}.$

$\vdots$

For example, given the power series

$$y = \text{arc sinh } x = x - \frac{1}{6}x^3 + \frac{3}{40}x^5 - \frac{15}{336}x^7 + \cdots,$$

setting $x_0 = 0$, $y_0 = 0$, $a_1 = 1$, $a_2 = 0$, $a_3 = -1/6$, $a_4 = 0$, $a_5 = 3/40, \ldots$, it follows from 1.11.1.8.3 that $A_1 = 1$, $A_2 = 0$, $A_3 = 1/6$, $A_4 = 0$, $A_5 = 1/120, \ldots$, so, as would be expected,

$$x = \sinh y = y + \frac{y^3}{6} + \frac{y^5}{120} + \cdots = y + \frac{y^3}{3!} + \frac{y^5}{5!} + \cdots.$$

1.12 Taylor Series

1.12.1 Definition and forms of remainder term

If a function $f(x)$ has derivatives of all orders in some interval containing the point x_0, the power series in $x - x_0$ of the form

1. $\quad f(x_0) + (x - x_0) f^{(1)}(x_0) + \dfrac{(x - x_0)^2}{2!} f^{(2)}(x_0) + \dfrac{(x - x_0)^3}{3!} f^{(3)}(x_0) + \cdots,$

where $f^{(n)}(x_0) = (d^n f / dx^n)_{x=x_0}$, is called the **Taylor series** expansion of $f(x)$ about x_0.

The Taylor series expansion converges to the function $f(x)$ if the **remainder**

2. $\quad R_n(x) = f(x) - f(x_0) - \sum_{k=1}^{n} \dfrac{(x - x_0)^k}{k!} f^{(k)}(x_0)$

approaches zero as $n \to \infty$.

The remainder term $R_n(x)$ can be expressed in a number of different forms, including the following:

3. $R_n(x) = \dfrac{(x - x_0)^{n+1}}{(n + 1)!} f^{(n+1)}[x_0 + \theta(x - x_0)]$ $[0 < \theta < 1]$.

(Lagrange form)

4. $R_n(x) = \dfrac{(x - x_0)^{n+1}}{n!} (1 - \theta)^n f^{(n+1)}[x_0 + \theta(x - x_0)]$ $[0 < \theta < 1]$.

(Cauchy form)

5. $R_n(x) = \dfrac{\psi(x - x_0) - \psi(0)}{\psi'[(x - x_0)(1 - \theta)]} \dfrac{(x - x_0)^n (1 - \theta)^n}{n!} f^{(n+1)}[\xi + \theta(x - x_0)]$

$[0 < \theta < 1]$,   **(Schlömilch form)**

where $\psi(x)$ is an arbitrary function with the properties that (i) it and its derivative $\psi'(x)$ are continuous in the interval $(0, x - x_0)$ and (ii) the derivative $\psi'(x)$ does not change sign in that interval.

6. $R_n(x) = \dfrac{(x - x_0)^{n-1}(1 - \theta)^{n-p-1}}{(p + 1)n!} f^{(n+1)}[x_0 + \theta(x - x_0)]$

$[0 < p \le n; 0 < \theta < 1]$.

[Rouché form obtained from 1.12.1.5 with $\psi(x) = x^{p+1}]$.

7. $R_n(x) = \dfrac{1}{n!} \displaystyle\int_{x_0}^{x} f^{(n+1)}(t)(x - t)^n dt.$ **(integral form)**

1.12.2

The Taylor series expansion of $f(x)$ is also written as:

1. $f(a + x) = \displaystyle\sum_{k=0}^{\infty} \dfrac{x^k}{k!} f^{(k)}(a) = f(a) + \dfrac{x}{1!} f^{(1)}(a) + \dfrac{x^2}{2!} f^{(2)}(a) + \cdots$

2. $f(x) = \displaystyle\sum_{k=0}^{\infty} \dfrac{x^k}{k!} f^{(k)}(0) = f(0) + \dfrac{x}{1!} f^{(1)}(0) + \dfrac{x^2}{2!} f^{(2)}(0) + \cdots$

(Maclaurin series: a Taylor series expansion about the origin).

1.12.3

The Taylor series expansion of a function $f(x, y)$ of the two variables that possesses partial derivatives of all orders in some region D containing the point (x_0, y_0) is:

1. $f(x, y) = f(x_0, y_0) + (x - x_0)\left(\dfrac{\partial f}{\partial x}\right)_{(x_0, y_0)} + (y - y_0)\left(\dfrac{\partial f}{\partial y}\right)_{(x_0, y_0)}$

$+ \dfrac{1}{2!}\left[(x - x_0)^2\left(\dfrac{\partial^2 f}{\partial x^2}\right)_{(x_0, y_0)} + 2(x - x_0)(y - y_0)\left(\dfrac{\partial^2 f}{\partial x \partial y}\right)_{(x_0, y_0)}\right.$

$\left. + (y - y_0)^2\left(\dfrac{\partial^2 f}{\partial y^2}\right)_{(x_0, y_0)}\right] + \cdots.$

In its simplest form the remainder term $R_n(x, y)$ satisfies a condition analogous to 1.12.1.3, so that

2. $R_n(x, y) = \dfrac{1}{(n+1)!}(D_{n+1}f)_{(x_0+\theta_1(x-x_0), y_0+\theta_2(y-y_0))}$

$$[0 < \theta_1 < 1, 0 < \theta_2 < 1]$$

where

3. $D_n \equiv \left((x-x_0)\dfrac{\partial}{\partial x} + (y-y_0)\dfrac{\partial}{\partial y}\right)^n$.

1.13 Fourier Series

1.13.1 Definitions

1.13.1.1 Let $f(x)$ be a function that is defined over the interval $[-l, l]$, and by periodic extension outside it; that is,

1. $f(x - 2l) = f(x)$, for all x.

Suppose also that $f(x)$ is absolutely integrable (possibly improperly) over the interval $(l, -l)$; that is, $\int_{-l}^{l} |f(x)|\, dx$ is finite. Then the **Fourier series** of $f(x)$ is the trigonometric series

2. $\dfrac{1}{2}a_0 + \sum\limits_{k=1}^{\infty}\left(a_k \cos\dfrac{k\pi x}{l} + b_k \sin\dfrac{k\pi x}{l}\right)$,

where the **Fourier coefficients** a_k, b_k are given by the formulas:

3. $a_k = \dfrac{1}{l}\int_{-l}^{l} f(t)\cos\dfrac{k\pi t}{l}\, dt = \dfrac{1}{l}\int_{\alpha}^{\alpha+2l} f(t)\cos\dfrac{k\pi t}{l}\, dt$

[any real α, $k = 0, 1, 2, \ldots$],

4. $b_k = \dfrac{1}{l}\int_{-l}^{l} f(t)\sin\dfrac{k\pi t}{l}\, dt = \dfrac{1}{l}\int_{\alpha}^{\alpha+2l} f(t)\sin\dfrac{k\pi t}{l}\, dt$

[any real α, $k = 1, 2, 3, \ldots$].

Convergence of Fourier series.

1.13.1.2 It is important to know in what sense the Fourier series of $f(x)$ represents the function $f(x)$ itself. This is the question of the **convergence** of Fourier series, which is discussed next. The Fourier series of a function $f(x)$ at a point x_0 converges to the number

1. $\dfrac{f(x_0 + 0) + f(x_0 - 0)}{2}$,

if, for some $h > 0$, the integral

2. $\displaystyle\int_0^h \dfrac{|f(x_0 + t) + f(x_0 - t) - f(x_0 + 0) - f(x_0 - 0)|}{t}\, dt$

exists, where it is assumed that $f(x)$ is either continuous at x_0 or it has a finite jump discontinuity at x_0 (a **saltus**) at which both the one-sided limits $f(x_0 - 0)$ and $f(x_0 + 0)$ exist. Thus, if $f(x)$ is continuous at x_0, the Fourier series of $f(x)$ converges to the value $f(x_0)$ at the point x_0, while if a finite jump discontinuity occurs at x_0 the Fourier series converges to the average of the values $f(x_0 + 0)$ and $f(x_0 - 0)$ of $f(x)$ to the immediate left and right of x_0.   **(Dini's condition)**

1.13.1.3   A function $f(x)$ is said to satisfy **Dirichlet conditions** on the interval $[a, b]$ if it is bounded on the interval, and the interval $[a, b]$ can be partitioned into a finite number of subintervals inside each of which the function $f(x)$ is continuous and monotonic (either increasing or decreasing).

The Fourier series of a periodic function $f(x)$ that satisfies Dirichlet conditions on the interval $[a, b]$ converges at every point x_0 of $[a, b]$ to the value $\frac{1}{2}\{f(x_0 + 0) + f(x_0 - 0)\}$.   **(Dirichlet's result)**

1.13.1.4   Let the function $f(x)$ be defined on the interval $[a, b]$, where $a < b$, and let the interval be partitioned into subintervals in an arbitrary manner with the ends of the intervals at

1. $a = x_0 < x_1 < x_2 < \cdots < x_{n-1} < x_n = b$.
     Form the sum

2. $\displaystyle\sum_{k=1}^{n} |f(x_k) - f(x_{k-1})|$.

Then different partitions of the interval $[a, b]$, that is, different choices of the points x_k, will give rise to different sums of the form in 1.13.1.4.2. If the set of these sums is bounded above, the function $f(x)$ is said to be of **bounded variation** on the interval $[a, b]$. The least upper bound of these sums is called the **total variation** of the function $f(x)$ on the interval $[a, b]$.

1.13.1.5   Let the function $f(x)$ be piecewise-continuous on the interval $[a, b]$, and let it have a piecewise-continuous derivative within each such interval in which it is continuous. Then, at every point x_0 of the interval $[a, b]$, the Fourier series of the function $f(x)$ converges to the value $\frac{1}{2}\{f(x_0 + 0) + f(x_0 - 0)\}$.

1.13.1.6   A function $f(x)$ defined in the interval $[0, l]$ can be expanded in a cosine series (**half-range Fourier cosine series**) of the form

1. $\displaystyle\frac{1}{2}a_0 + \sum_{k=1}^{\infty} a_k \cos \frac{k\pi x}{l}$,
     where

2. $a_k = \dfrac{2}{l} \displaystyle\int_0^l f(t) \cos \frac{k\pi t}{l}\, dt \qquad [k = 0, 1, 2, \ldots]$.

1.13.1.7   A function $f(x)$ defined in the interval $[0, l]$ can be expanded in a sine series (**half-range Fourier sine series**):

1. $\displaystyle\sum_{k=1}^{\infty} b_k \sin \frac{k\pi x}{l}$,

   where

2. $b_k = \dfrac{2}{l} \displaystyle\int_0^l f(t) \sin \frac{k\pi t}{l}\, dt \qquad [k = 1, 2, \ldots].$

The convergence tests for these half-range Fourier series are analogous to those given in 1.13.1.2 to 1.13.1.5.

1.13.1.8 The Fourier coefficients a_k, b_k defined in 1.13.1.1 for a function $f(x)$ that is absolutely integrable over $[-l, l]$ are such that

$$\lim_{k\to\infty} a_k = 0 \quad \text{and} \quad \lim_{k\to\infty} b_k = 0. \qquad \textbf{(Riemann–Lebesgue lemma)}$$

1.13.1.9 Let $f(x)$ be a piecewise-smooth or piecewise-continuous function defined on the interval $[-l, l]$, and by periodic extension outside it. Then for all real x the **complex Fourier series** for $f(x)$ is

1. $\displaystyle\lim_{m\to\infty} \sum_{k=-m}^{m} c_k \exp[ik\pi x/l]$,

   where

2. $c_k = \dfrac{1}{2l} \displaystyle\int_{-l}^{l} f(x) e^{-ik\pi x/l}\, dx \qquad [k = 0, \pm1, \pm2, \ldots].$

The convergence properties and conditions for convergence of complex Fourier series are analogous to those already given for Fourier series.

1.13.1.10 The **n'th partial sum** $s_n(x)$ of a Fourier series for $f(x)$ on the interval $(-\pi, \pi)$, defined by

1. $s_n(x) = \frac{1}{2}a_0 + \sum_{k=1}^{n}(a_k \cos kx + b_k \sin kx)$ is given by the **Dirichlet integral representation:**

2. $s_n(x) = \dfrac{1}{\pi} \displaystyle\int_{-\pi}^{\pi} f(\tau) \frac{\sin\left[\left(n + \frac{1}{2}\right)(x - \tau)\right]}{2 \sin \frac{1}{2}(x - \tau)}\, d\tau,$

   where

3. $D_n(x) = \dfrac{1}{2\pi} \dfrac{\sin\left[\left(n + \frac{1}{2}\right)(x - \tau)\right]}{\sin \frac{1}{2}(x - \tau)}$

   is called the **Dirichlet kernel**.

4. $\dfrac{1}{\pi} \displaystyle\int_{-\pi}^{\pi} D_n(x)\, dx = 1.$

1.13.1.11 Let $f(x)$ be continuous and piecewise-smooth on the interval $(-l, l)$, with $f(-l) = f(l)$. Then the Fourier series of $f(x)$ converges uniformly to $f(x)$ for x in the interval $(-l, l)$.

1.13.1.12 Let $f(x)$ be piecewise-continuous on $[-l, l]$ with the Fourier series given in 1.13.1.1.2. Then term-by-term integration of the Fourier series for $f(x)$

yields a series representation of the function

$$F(x) = \int_\alpha^x f(t)\,dt, \qquad [-l \le \alpha < x < l].$$

When expressed differently, this result is equivalent to

$$\int_{-l}^x f(t)\,dt = \int_{-l}^x \frac{a_0}{2}\,dt + \sum_{k=1}^\infty \int_{-l}^x \left(a_k \cos\frac{k\pi t}{l} + b_k \sin\frac{k\pi t}{l} \right) dt.$$

(integration of Fourier series)

1.13.1.13   Let $f(x)$ be a continuous function over the interval $[-l, l]$ and such that $f(-l) = f(l)$. Suppose also that the derivative $f'(x)$ is piecewise-continuous over this interval. Then at every point at which $f''(x)$ exists, the Fourier series for $f(x)$ may be differentiated term by term to yield a Fourier series that converges to $f'(x)$. Thus, if $f(x)$ has the Fourier series given in 1.13.1.1.2,

$$f'(x) = \sum_{k=1}^\infty \frac{d}{dx}\left(a_k \cos\frac{k\pi x}{l} + b_k \sin\frac{k\pi x}{l} \right) \qquad [-l \le x \le l].$$

(differentiation of Fourier series)

1.13.1.14   Let $f(x)$ be a piecewise-continuous over $[-l, l]$ with the Fourier series given in 1.13.1.1.2. Then

$$\frac{1}{2}a_0^2 + \sum_{k=1}^\infty \left(a_k^2 + b_k^2 \right) \le \frac{1}{l} \int_{-l}^l [f(x)]^2\,dx.$$

(Bessel's inequality)

1.13.1.15   Let $f(x)$ be continuous over $[-l, l]$, and periodic with period $2l$, with the Fourier series given in 1.13.1.2. Then

$$\frac{1}{2}a_0^2 + \sum_{k=1}^\infty \left(a_k^2 + b_k^2 \right) = \frac{1}{l} \int_{-l}^l [f(x)]^2\,dx.$$

(Parseval's identity)

1.13.1.16   Let $f(x)$ and $g(x)$ be two functions defined over the interval $[-l, l]$ with respective Fourier coefficients a_k, b_k and α_k, β_k, and such that the integrals $\int_{-l}^l [f(x)]^2\,dx$ and $\int_{-l}^l [g(x)]^2\,dx$ are both finite [$f(x)$ and $g(x)$ are **square integrable**]. Then the **Parseval identity** for the product function $f(x)g(x)$ becomes

$$\frac{a_0\alpha_0}{2} + \sum_{k=1}^\infty (a_k\alpha_k + b_k\beta_k) = \frac{1}{l} \int_{-l}^l f(x)g(x)\,dx.$$

(generalized Parseval identity)

1.14 Asymptotic Expansions

1.14.1 Introduction

1.14.1.1   Among the set of all divergent series is a special class known as the **asymptotic series**. These are series which, although divergent, have the property that the sum of a suitable number of terms provides a good approximation to the functions they represent. In the case of an alternating asymptotic series, the greatest accuracy is obtained by truncating the series at the term of smallest absolute value. When working with an alternating asymptotic series representing a function $f(x)$, the magnitude of the error involved when $f(x)$ is approximated by summing only the first n terms of the series does not exceed the magnitude of the $(n+1)$'th term (the first term to be discarded).

An example of this type is provided by the asymptotic series for the function

$$f(x) = \int_x^\infty \frac{e^{x-t}}{t} \, dt.$$

Integrating by parts n times gives

$$f(x) = \frac{1}{x} - \frac{1}{x^2} + \frac{2!}{x^3} - \cdots + (-1)^{n-1}\frac{(n-1)!}{x^n} + (-1)^n n! \int_x^\infty \frac{e^{x-t}}{t^{n+1}} \, dt.$$

It is easily established that the infinite series

$$\frac{1}{x} - \frac{1}{x^2} + \frac{2!}{x^3} - \cdots + (-1)^{n-1}\frac{(n-1)!}{x^n} + \cdots$$

is divergent for all x, so if $f(x)$ is expanded as an infinite series, the series diverges for all x.

The remainder after the n'th term can be estimated by using the fact that

$$n! \int_x^\infty \frac{e^{x-t}}{t^{n+1}} \, dt < \frac{n!}{x^{n+1}} \int_x^\infty e^{x-t} \, dt = \frac{n!}{x^{n+1}},$$

from which it can be seen that when $f(x)$ is approximated by the first n terms of this divergent series, the magnitude of the error involved is less than the magnitude of the $(n+1)$'th term (see 1.8.2.7.2).

For any fixed value of x, the terms in the divergent series decrease in magnitude until the N'th term, where N is the integral part of x, after which they increase again. Thus, for any fixed x, truncating the series after the N'th term will yield the best approximation to $f(x)$ for that value of x.

In general, even when x is only moderately large, truncating the series after only a few terms will provide an excellent approximation to $f(x)$. In the above case, if $x = 30$ and the series is truncated after only two terms, the magnitude of the error involved when evaluating $f(30)$ is less than $2!/30^3 = 7.4 \times 10^{-5}$.

1.14.2 Definition and properties of asymptotic series

1. Let $S_n(z)$ be the sum of the first $(n+1)$ terms of the series

$$S(z) = A_0 + \frac{A_1}{z} + \frac{A_2}{z^2} + \cdots + \frac{A_n}{z^n} + \cdots,$$

where, in general, z is a complex number. Let $R_n(z) = S(z) - S_n(z)$. Then the series $S(z)$ is said to be an **asymptotic expansion** of $f(z)$ if for arg z in some interval $\alpha \le \arg z \le \beta$, and for each fixed n,

$$\lim_{|z| \to \infty} z^n R_n(z) = 0.$$

The relationship between $f(z)$ and its asymptotic expansion $S(z)$ is indicated by writing $f(z) \sim S(z)$.

2. The operations of addition, subtraction, multiplication, and raising to a power can be performed on asymptotic series just as on absolutely convergent series. The series obtained as a result of these operations will also be an asymptotic series.

3. Division of asymptotic series is permissible and yields another asymptotic series provided the first term A_0 of the divisor is not equal to zero.

4. Term-by-term integration of an asymptotic series is permissible and yields another asymptotic series.

5. Term-by-term differentiation of an asymptotic series is *not*, in general, permissible.

6. For arg z in some interval $\alpha \leq \arg z \leq \beta$, the asymptotic expansion of a function $f(x)$ is unique.

7. A series can be the asymptotic expansion of more than one function.

1.15 Basic Results from the Calculus

1.15.1 Rules for differentiation

1.15.1.1   Let u, v be differentiable functions of their arguments and let k be a constant.

1. $\dfrac{dk}{dx} = 0$

2. $\dfrac{d(ku)}{dx} = k\dfrac{du}{dx}$

3. $\dfrac{d(u+v)}{dx} = \dfrac{du}{dx} + \dfrac{dv}{dx}$ **(differentiation of a sum)**

4. $\dfrac{d(uv)}{dx} = u\dfrac{dv}{dx} + v\dfrac{du}{dx}$ **(differentiation of a product)**

5. $\dfrac{d}{dx}(u/v) = \dfrac{v\dfrac{du}{dx} - u\dfrac{dv}{dx}}{v^2}$ $[v \neq 0]$ **(differentiation of a quotient)**

6. Let v be differentiable at some point x, and u be differentiable at the point $v(x)$; then the composite function $(u \circ v)(x)$ is differentiable at the point x. In particular, if $z = u(y)$ and $y = v(x)$, so that $z = u(v(x)) = (u \circ v)(x)$, then

$$\frac{dz}{dx} = \frac{dz}{dy} \cdot \frac{dy}{dx}$$

or, equivalently,

$$\frac{dz}{dx} = u'(y)v'(x) = u'(v(x))v'(x).$$ **(chain rule)**

7. If the n'th derivatives $u^{(n)}(x)$ and $v^{(n)}(x)$ exist, so also does $d^n(uv)/dx^n$ and

$$\frac{d^n(uv)}{dx^n} = \sum_{k=0}^{n} \binom{n}{k} u^{(n-k)}(x)v^{(k)}(x).$$ **(Leibnitz's formula)**

In particular (for $n = 1$; see 1.15.1.1.3),

$$\frac{d^2(uv)}{dx^2} = \frac{d^2u}{dx^2}v + 2\frac{du}{dx}\frac{dv}{dx} + u\frac{d^2v}{dx^2} \qquad [n = 2],$$

$$\frac{d^3(uv)}{dx^3} = \frac{d^3u}{dx^3}v + 3\frac{d^2u}{dx^2}\frac{dv}{dx} + 3\frac{du}{dx}\frac{d^2v}{dx^2} + u\frac{d^3v}{dx^3} \qquad [n = 3].$$

8. Let the function $f(x)$ be continuous at each point of the closed interval $[a, b]$ and differentiable at each point of the open interval $[a, b]$. Then there exists a number ξ, with $a < \xi < b$, such that

$$f(b) - f(a) = (b - a)f'(\xi).$$ **(mean-value theorem for derivatives)**

9. Let $f(x)$, $g(x)$ be functions such that $f(x_0) = g(x_0) = 0$, but $f'(x_0)$ and $g'(x_0)$ are not both zero. Then

$$\lim_{x \to x_0} \left[\frac{f(x)}{g(x)} \right] = \frac{f'(x_0)}{g'(x_0)}.$$ **(L'Hôpital's rule)**

Let $f(x)$, $g(x)$ be functions such that $f(x_0) = g(x_0) = 0$, and let their first n derivatives all vanish at x_0, so that $f^{(r)}(x_0) = g^{(r)}(x_0) = 0$, for $r = 1, 2, \ldots, n$. Then, provided that not both of the $(n + 1)$'th derivatives $f^{(n+1)}(x)$ and $g^{(n+1)}(x)$ vanish at x_0

$$\lim_{x \to x_0} \left[\frac{f(x)}{g(x)} \right] = \frac{f^{(n+1)}(x_0)}{g^{(n+1)}(x_0)}.$$ **(generalized L'Hôpital's rule)**

1.15.2 Integration

1.15.2.1 A function $F(x)$ is called an **antiderivative** of a function $f(x)$ if

1. $\frac{d}{dx}[F(x)] = f(x).$

   The operation of determining an antiderivative is called **integration**. Functions of the form

2. $F(x) + C$

   are antiderivatives of $f(x)$ in 1.15.2.1.1 and this is indicated by writing

3. $$\int f(x)\,dx = F(x) + C,$$

where C is an arbitrary constant.

The expansion on the right-hand side of 1.15.2.1.3 is called an **indefinite integral**. The term *indefinite* is used because of the arbitrariness introduced by the arbitrary additive **constant of integration** C.

4. $$\int u\left(\frac{dv}{dx}\right) dx = uv - \int v\left(\frac{du}{dx}\right) dx. \qquad \textbf{(formula for integration by parts)}$$

This is often abbreviated to

$$\int u\,dv = uv - \int v\,du.$$

5. A **definite integral** involves the integration of $f(x)$ from a **lower limit** $x = a$ to an upper limit $x = b$, and it is written

$$\int_a^b f(x)\,dx.$$

The **first fundamental theorem of calculus** asserts that if $F(x)$ is an antiderivative of $f(x)$, then

$$\int_a^b f(x)\,dx = F(b) - F(a).$$

Since the variable of integration in a definite integral does not enter into the final result it is called a **dummy variable,** and it may be replaced by any other symbol. Thus

$$\int_a^b f(x)\,dx = \int_a^b f(t)\,dt = \cdots = \int_a^b f(s)\,ds = F(b) - F(a).$$

6. If

$$F(x) = \int_a^x f(t)\,dt,$$

then $dF/dx = f(x)$ or, equivalently,

$$\frac{d}{dx}\int_a^x f(t)\,dt = f(x). \qquad \textbf{(second fundamental theorem of calculus)}$$

7. $$\int_a^b f(x)\,dx = -\int_b^a f(x)\,dx. \qquad \textbf{(reversal of limits changes the sign)}$$

8. Let $f(x)$ and $g(x)$ be continuous in the interval $[a, b]$, and let $g'(x)$ exist and be continuous in this same interval. Then, with the substitution $u = g(x)$,

(i) $$\int f[g(x)]g'(x)\,dx = \int f(u)\,du,$$

$$\textbf{(integration of indefinite integral by substitution)}$$

(ii) $\displaystyle\int_a^b f[g(x)]g'(x)\,dx = \int_{g(a)}^{g(b)} f(u)\,du.$

(integration of a definite integral by substitution)

9. Let $f(x)$ be finite and integrable over the interval $[a, b]$ and let ξ be such that $a < \xi < b$, then

$$\int_{\xi-0}^{\xi+0} f(x)\,dx = 0.$$ **(integral over an interval of zero length)**

10. If ξ is a point in the closed interval $[a, b]$, then

$$\int_a^b f(x)\,dx = \int_a^{\xi} f(x)\,dx \int_{\xi}^b f(x)\,dx.$$

(integration over contiguous intervals)

This result, in conjunction with 1.15.2.1.9, is necessary when integrating a piecewise-continuous function $f(x)$ over the interval $[a, b]$. Let

$$f(x) = \begin{cases} \varphi(x), & a \le x \le \xi \\ \psi(x), & \xi < x \le b \end{cases},$$

with $\varphi(\xi - 0) \ne \psi(\xi + 0)$, so that $f(x)$ has a finite discontinuity (**saltus**) at $x = \xi$. Then

$$\int_a^b f(x)\,dx = \int_a^{\xi-0} f(x)\,dx + \int_{\xi+0}^b f(x)\,dx.$$

(integration of a discontinuous function)

11. $\displaystyle\int_a^b u\left(\frac{dv}{dx}\right)dx = u(b)v(b) - u(a)v(a) - \int_a^b v\left(\frac{du}{dx}\right)dx.$

(formula for integration by parts of a definite integral)

Using the notation $(uv)|_a^b = u(b)v(b) - u(a)v(a)$ this last result is often contracted to

$$\int_a^b u\,dv = (uv)\Big|_a^b - \int_a^b v\,du.$$

12. $\displaystyle\frac{d}{d\alpha}\int_{\psi(\alpha)}^{\varphi(\alpha)} f(x, \alpha)\,dx = f[\varphi(\alpha), \alpha]\frac{d\varphi(\alpha)}{d\alpha} - f[\psi(\alpha), \alpha]\frac{d\psi(\alpha)}{d\alpha}$

$$+ \int_{\psi(\alpha)}^{\varphi(\alpha)} \frac{\partial}{\partial\alpha}[f(x, \alpha)]\,dx.$$

(differentiation of a definite integral with respect to a parameter)

13. If $f(x)$ is integrable over the interval $[a, b]$, then

$$\left|\int_a^b f(x)\,dx\right| \le \int_a^b |f(x)|\,dx.$$ **(absolute value integral inequality)**

14. If $f(x)$ and $g(x)$ are integrable over the interval $[a, b]$ and $f(x) \leq g(x)$, then

$$\int_a^b f(x)\, dx \leq \int_a^b g(x)\, dx. \qquad \textbf{(comparison integral inequality)}$$

15. The following are **mean-value theorems for integrals**.

   (i) If $f(x)$ is continuous on the closed interval $[a, b]$, there is a number ξ, with $a < \xi < b$, such that

   $$\int_a^b f(x)\, dx = (b - a)f(\xi).$$

   (ii) If $f(x)$ and $g(x)$ are continuous on the closed interval $[a, b]$ and $g(x)$ is monotonic (either decreasing on increasing) on the open interval $[a, b]$, there is a number ξ, with $a < \xi < b$, such that

   $$\int_a^b f(x)g(x)\, dx = g(a) \int_a^\xi f(x)\, dx + g(b) \int_\xi^b f(x)\, dx.$$

   (iii) If in (ii), $g(x) > 0$ on the open interval $[a, b]$, there is a number ξ, with $a < \xi < b$, such that when $g(x)$ is monotonic decreasing

   $$\int_a^b f(x)g(x)\, dx = g(a) \int_a^\xi f(x)\, dx,$$

   and when $g(x)$ is monotonic increasing

   $$\int_a^b f(x)g(x)\, dx = g(b) \int_\xi^b f(x)\, dx.$$

1.15.3 Reduction formulas

1.15.3.1   When, after integration by parts, an integral containing one or more parameters can be expressed in terms of an integral of similar form involving parameters with reduced values, the result is called a **reduction formula (recursion relation)**. Its importance derives from the fact that it can be used in reverse, because after the simplest form of the integral has been evaluated, the formula can be used to determine more complicated integrals of similar form.

Typical examples of reduction formulas together with an indication of their use follow.

(a) Given $I_n = \int (1 - x^3)^n dx$, $n = 0, 1, 2, \ldots$, integration by parts shows the reduction formula satisfied by I_n to be

$$(3n + 1)I_n = x(1 - x^3)^n + 3nI_{n-1}.$$

Setting $n = 0$ in I_n and omitting the arbitrary constant of integration gives $I_0 = x$. Next, setting $n = 1$ in the reduction formula determines I_1 from

$$4I_1 = x(1 - x^3) + 3I_0 = 4x - x^4, \text{ so}$$

$$I_1 = x - \frac{1}{4}x^4.$$

I_2 follows by using I_1 in the reduction formula with $n = 2$, while $I_3, I_4, \ldots,$ are obtained in similar fashion. Because an indefinite integral is involved, an arbitrary constant must be added to each I_n so obtained to arrive at the most general result.

(b) Given $I_{m,n} = \int \sin^m x \cos^n x \, dx$, repeated integration by parts shows the reduction formula satisfied by $I_{m,n}$ to be

$$(m + n)I_{m,n} = \sin^{m+1} x \cos^{n-1} x + (n - 1)I_{m,n-2}.$$

Given $I_n = \int_0^{\pi/2} \cos^n x \, dx$, repeated integration by parts shows the reduction formula satisfied by I_n to be

(c) $$I_n = \left(\frac{n - 1}{n}\right)I_{n-2}.$$

Since $I_0 = \pi/2$ and $I_1 = 1$, this result implies

$$I_{2n} = \left(\frac{2n - 1}{2n}\right)\left(\frac{2n - 3}{2n - 2}\right)\cdots\frac{1}{2}I_0 = \frac{1 \cdot 3 \cdot 5 \ldots (2n - 1)}{2 \cdot 4 \cdot 6 \ldots 2n} \cdot \frac{\pi}{2},$$

$$I_{2n+1} = \left(\frac{2n}{2n + 1}\right)\left(\frac{2n - 2}{2n - 1}\right)\cdots\frac{2}{3}I_1 = \frac{2 \cdot 4 \cdot 6 \ldots 2n}{3 \cdot 5 \cdot 7 \ldots (2n + 1)}.$$

Combining these results gives

$$\frac{\pi}{2} = \left[\frac{2 \cdot 4 \cdot 6 \ldots 2n}{3 \cdot 5 \cdot 7 \ldots (2n - 1)}\right]^2 \left(\frac{1}{2n + 1}\right)\frac{I_{2n}}{I_{2n+1}},$$

but

$$\lim_{n \to \infty} \frac{I_{2n}}{I_{2n+1}} = 1,$$

so we arrive at the Wallis infinite product (see 1.9.2.15)

$$\frac{\pi}{2} = \prod_{k=1}^{\infty} \left(\frac{2k}{2k - 1}\right)\left(\frac{2k}{2k + 1}\right).$$

1.15.4 Improper integrals

An **improper integral** is a definite integral that possesses one or more of the following properties:

(i)   the interval of integration is either semi-infinite or infinite in length;

(ii) the integrand becomes infinite at an interior point of the interval of integration;

(iii) the integrand becomes infinite at an end point of the interval of integration.

An integral of type (i) is called an **improper integral of the first kind,** while integrals of types (ii) and (iii) are called **improper integrals of the second kind**.

The evaluation of improper integrals.

I. Let $f(x)$ be defined and finite on the semi-infinite interval $[a, \infty)$. Then the improper integral $\int_a^\infty f(x)\,dx$ is defined as

$$\int_a^\infty f(x)\,dx = \lim_{R\to\infty} \int_a^R f(x)\,dx.$$

The improper integral is said to **converge** to the value of this limit when it exists and is finite. If the limit does not exist, or is infinite, the integral is said to **diverge**. Corresponding definitions exist for the improper integrals

$$\int_{-\infty}^a f(x)\,dx = \lim_{R\to\infty} \int_{-R}^a f(x)\,dx,$$

and

$$\int_{-\infty}^\infty f(x)\,dx = \lim_{R_1\to\infty} \int_{-R_1}^a f(x)\,dx + \lim_{R_2\to\infty} \int_a^{R_2} f(x)\,dx, \quad \text{[arbitrary } a\text{]},$$

where R_1 and R_2 tend to infinity *independently* of each other.

II. Let $f(x)$ be defined and finite on the interval $[a, b]$ except at a point ξ interior to $[a, b]$ at which point it becomes infinite. The improper integral $\int_a^b f(x)\,dx$ is then defined as

$$\int_a^b f(x)\,dx = \lim_{\varepsilon\to 0} \int_a^{\xi-\varepsilon} f(x)\,dx + \lim_{\delta\to 0} \int_{\xi+\delta}^b f(x)\,dx,$$

where $\varepsilon > 0, \delta > 0$ tend to zero *independently* of each other. The improper integral is said to **converge** when both limits exist and are finite, and its value is then the sum of the values of the two limits. The integral is said to **diverge** if at least one of the limits is undefined, or is infinite.

III. Let $f(x)$ be defined and finite on the interval $[a, b]$ except at an end point, say, at $x = a$, where it is infinite. The improper integral $\int_a^b f(x)\,dx$ is then defined as

$$\int_a^b f(x)\,dx = \lim_{\varepsilon\to 0} \int_{a+\varepsilon}^b f(x)\,dx,$$

where $\varepsilon > 0$. The improper integral is said to **converge** to the value of this limit when it exists and is finite. If the limit does not exist, or is infinite, the integral is said to **diverge**. A corresponding definition applies when $f(x)$ is infinite at $x = b$.

IV. It may happen that although an improper integral of type (i) is divergent, modifying the limits in I by setting $R_1 = R_2 = R$ gives rise to a finite result. This is said to define the **Cauchy principal** value of the integral, and it is indicated by inserting the letters PV before the integral sign. Thus, when the limit is finite,

$$\text{PV} \int_{-\infty}^{\infty} f(x)\,dx = \lim_{R \to \infty} \int_{-R}^{R} f(x)\,dx.$$

Similarly, it may happen that although an improper integral of type (ii) is divergent, modifying the limits in II by setting $\varepsilon = \delta$, so the limits are evaluated *symmetrically* about $x = \xi$, gives rise to a finite result. This also defines a Cauchy principal value, and when the limits exist and are finite, we have

$$\text{PV} \int_{a}^{b} f(x)\,dx = \lim_{\varepsilon \to 0} \left\{ \int_{a}^{\xi - \varepsilon} f(x)\,dx + \int_{\xi + \varepsilon}^{b} f(x)\,dx \right\}.$$

When an improper integral converges, its value and the Cauchy principal value coincide. Typical examples of improper integrals are:

$$\int_{0}^{\infty} xe^{-x}\,dx = 1, \qquad \int_{1}^{\infty} \frac{dx}{1 + x^2} = \frac{\pi}{4},$$

$$\int_{0}^{2} \frac{dx}{(x - 1)^{2/3}} = 6, \qquad \int_{0}^{1} \frac{dx}{\sqrt{1 - x^2}} = \frac{\pi}{2},$$

$$\int_{0}^{\infty} \sin x\,dx \text{ diverges}, \qquad \text{PV} \int_{-\infty}^{\infty} \frac{\cos x}{a^2 - x^2}\,dx = \frac{\pi}{a} \sin a.$$

Convergence tests for improper integrals.

1. Let $f(x)$ and $g(x)$ be continuous on the semi-infinite interval $[a, \infty)$, and let $g'(x)$ be continuous and monotonic decreasing on this same interval. Suppose

   (i) $\lim_{x \to \infty} g(x) = 0,$

   (ii) $F(x) = \int_{a}^{x} f(x)\,dx$ is bounded on $[a, \infty)$, so that for some $M > 0$,

   $$|F(x)| < M \qquad \text{for } a \le x < \infty,$$

   then the improper integral

   $$\int_{a}^{\infty} f(x)g(x)\,dx$$

   converges. **(Abel's test)**

2. Let $f(x)$ and $g(x)$ be continuous in the semi-infinite interval $[a, \infty)$, and let $g'(x)$ be continuous in this same interval. Suppose also that

   (i) $\lim_{x \to \infty} g(x) = 0,$

   (ii) $\int_{a}^{\infty} |g'(x)|\,dx$ converges,

(iii) $F(x) = \int_a^\infty f(x)\,dx$ is bounded on $[a, \infty)$, so that for some $M > 0$,

$$|F(x)| < M \qquad \text{for all } a \le x < \infty,$$

then

$$\int_a^\infty f(x)g(x)\,dx$$

converges.   **(Dirichlet's test)**

1.15.5 Integration of rational functions

A partial fraction decomposition of a rational function $N(x)/D(x)$ (see 1.72) reduces it to a sum of terms of the type:

(a) a polynomial $P(x)$ when the degree of $N(x)$ equals or exceeds the degree of $D(x)$;

(b) $\dfrac{A}{a + bx}$,

(c) $\dfrac{B}{(a + bx)^m}$ $[m > 1$ an integer$]$,

(d) $\dfrac{Cx + D}{(a + bx + cx^2)^n}$ $[n \ge 1$ an integer$]$.

Integration of these simple rational functions gives:

(a′) If $P(x) = a_0 + a_1 x + \cdots + a_r x^r$, then

$$\int P(x)dx = a_0 x + 1/2 a_1 x^2 + \cdots + \frac{1}{(r+1)} a_r x^{r+1} + \text{const.},$$

(b′) $\displaystyle\int \frac{A}{a + bx}\,dx = \left(\frac{A}{b}\right) \ln|a + bx| + \text{const.},$

(c′) $\displaystyle\int \frac{B}{(a + bx)^m}\,dx = \left(\frac{B}{b}\right)\left(\frac{1}{1 - m}\right)(a + bx)^{1-m} + \text{const.}$ $[m > 1],$

(d′) $\displaystyle\int \frac{Cx + D}{(a + bx + cx^2)^n}\,dx$ should be reexpressed as

$$\int \frac{Cx + D}{(a + bx + cx^2)^n}\,dx = \left(\frac{C}{2c}\right) \int \frac{b + 2cx}{(a + bx + cx^2)^n}\,dx$$

$$+ \left(D - \frac{Cb}{2c}\right) \int \frac{dx}{(a + bx + cx^2)^n}.$$

Then,

(i) $\displaystyle\int \frac{b + 2cx}{(a + bx + cx^2)}\, dx = \ln|a + bx + cx^2| + \text{const.}$ $\qquad [n = 1],$

(ii) $\displaystyle\int \frac{b + 2cx}{(a + bx + cx^2)^n}\, dx = \frac{-1}{(n - 1)}\frac{1}{(a + bx + cx^2)^{n-1}} + \text{const.}$ $\qquad [n > 1],$

(iii) $\displaystyle\int \frac{dx}{a + bx + cx^2} = \frac{1}{c}\int \frac{dx}{[x + b/(2c)]^2 + [a/c - b^2/(4c^2)]}$ $\qquad [n = 1],$

where the integral on the right is a standard integral. When evaluated, the result depends on the sign of $\Delta = 4ac - b^2$. If $\Delta > 0$ integration yields an inverse tangent function (see 4.2.5.1.1), and if $\Delta < 0$ it yields a logarithmic function or, equivalently, an inverse hyperbolic tangent function (see 4.2.5.1.1). If $\Delta = 0$, integration of the right-hand side gives $-2/(b + 2cx) + \text{const.}$

(iv) $\displaystyle\int \frac{dx}{(a + bx + cx^2)^n}$ $\qquad [n > 1 \text{ an integer}].$

This integral may be evaluated by using result (iii) above in conjunction with the reduction formula

$$I_n = \frac{b + 2cx}{(n - 1)(4ac - b^2)D^{n-1}} + \frac{2(2n - 3)c}{(n - 1)(4ac - b^2)}I_{n-1},$$

where

$$D = a + bx + cx^2 \quad \text{and} \quad I_n = \int \frac{dx}{(a + bx + cx^2)^n}.$$

1.15.6 Elementary applications of definite integrals

1.15.6.1　　The elementary applications that follow outline the use of definite integrals for the determination of areas, arc lengths, volumes, centers of mass, and moments of inertia.

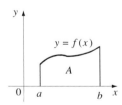

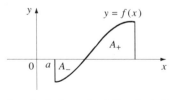

$A = A_- + A_+;\quad A_- < 0,\quad A_+ > 0$

Area under a curve.　　The definite integral

1. $\displaystyle A = \int_a^b f(x)\, dx$

may be interpreted as the algebraic sum of areas above and below the x-axis that are bounded by the curve $y = f(x)$ and the lines $x = a$ and $x = b$, with areas above the x-axis assigned positive values and those below it negative values.

Volume of revolution.

(i)   Let $f(x)$ be a continuous and nonnegative function defined on the interval $a \leq x \leq b$, and A be the area between the curve $y = f(x)$, the x-axis, and the lines $x = a$ and $x = b$. Then the volume of the solid generated by rotating A about the x-axis is

2. $$V = \pi \int_a^b [f(x)]^2 \, dx.$$

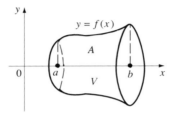

(ii)   Let $g(y)$ be a continuous and nonnegative function defined on the interval $c \leq y \leq d$, and A be the area between the curve $x = g(y)$, the y-axis, and the lines $y = c$ and $y = d$. Then the volume of the solid generated by rotating A about the y-axis is

3. $$V = \pi \int_c^d [g(y)]^2 \, dy.$$

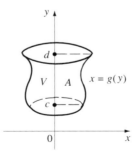

(iii)   Let $g(y)$ be a continuous and nonnegative function defined on the interval $c \leq y \leq d$ with $c \geq 0$, and A be the area between the curve $x = g(y)$, the y-axis, and the lines $y = c$ and $y = d$. Then the volume of the solid generated by rotating A about the x-axis is

4. $V = 2\pi \int_c^d y g(y) \, dy.$

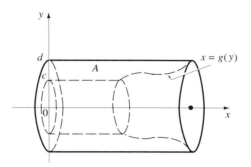

(iv)   Let $f(x)$ be a continuous and nonnegative function defined on the interval
       $a \le x \le b$ with $a \ge 0$, and A be the area between the curve $y = f(x)$, the
       x-axis, and the lines $x = a$ and $x = b$. Then the volume of the solid generated
       by rotating A about the y-axis is

5. $V = 2\pi \int_a^b x f(x) \, dx.$

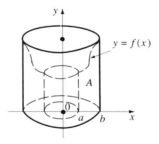

(v)   ***Theorem of Pappus***
       Let a closed curve C in the (x, y)-plane that does not intersect the x-axis
       have a circumference L and area A, and let its centroid be at a perpendicular
       distance $\bar{y}$ from the x-axis. Then the surface area S and volume V of the solid
       generated by rotating the area within the curve C about the x-axis are given by

6. $S = 2\pi \bar{y} L$ and $V = 2\pi \bar{y} A.$

Length of an arc.

(i)   Let $f(x)$ be a function with a continuous derivative that is defined on the
       interval $a \le x \le b$. Then the length of the arc along $y = f(x)$ from $x = a$ to
       $x = b$ is

7. $s = \int_a^b \sqrt{1 + [f'(x)]^2} \, dx,$

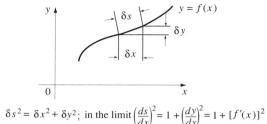

$$\delta s^2 = \delta x^2 + \delta y^2; \text{ in the limit } \left(\frac{ds}{dx}\right)^2 = 1 + \left(\frac{dy}{dx}\right)^2 = 1 + [f'(x)]^2$$

(ii) Let $g(y)$ be a function with a continuous derivative that is defined on the interval $c \le y \le d$. Then the length of the arc along $x = g(y)$ from $y = c$ to $y = d$ is

8. $\quad s = \displaystyle\int_c^d \sqrt{1 + [g'(y)]^2}\, dy.$

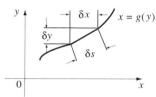

$$\delta s^2 = \delta y^2 + \delta x^2; \text{ in the limit } \left(\frac{ds}{dy}\right)^2 = 1 + \left(\frac{dx}{dy}\right)^2 = 1 + [g'(y)]^2$$

Area of surface of revolution.

(i) Let $f(x)$ be a nonnegative function with a continuous derivative that is defined on the interval $a \le x \le b$. Then the area of the surface of revolution generated by rotating the curve $y = f(x)$ about the x-axis between the planes $x = a$ and $x = b$ is

9. $\quad S = 2\pi \displaystyle\int_a^b f(x)\sqrt{1 + [f'(x)]^2}\, dx.$

(see also Pappus's theorem 1.15.6.1.5)

(ii) Let $g(y)$ be a nonnegative function with a continuous derivative that is defined on the interval $c \le y \le d$. Then the area of the surface generated by rotating the curve $x = g(y)$ about the y-axis between the planes $y = c$ and $y = d$ is

10. $\quad S = 2\pi \displaystyle\int_c^d g(y)\sqrt{1 + [g'(y)]^2}\, dy.$

Center of mass and moment of inertia.

(i) Let a plane lamina in a region R of the (x, y)-plane within a closed plane curve C have the continuous mass density distribution $\rho(x, y)$. Then the **center of mass (gravity)** of the lamina is located at the point G with coordinates $(\bar{x}, \bar{y})$, where

11. $$\bar{x} = \frac{\iint_R x\rho(x, y)\, dA}{M}, \quad \bar{y} = \frac{\iint_R y\rho(x, y)\, dA}{M},$$

with dA the element of area in R and

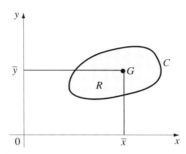

12. $$M = \iint_R \rho(x, y)\, dA. \qquad\qquad \textbf{(mass of lamina)}$$

When this result is applied to the area R within the plane curve C in the (x, y)-plane, which may be regarded as a lamina with a uniform mass density that may be taken to be $\rho(x, y) \equiv 1$, the center of mass is then called the **centroid**.

(ii) The **moments of inertia** of the lamina in (i) about the x-, y-, and z-axes are given, respectively, by

13. $$I_x = \iint_R y^2\rho(x, y)\, dA,$$

$$I_y = \iint_R x^2\rho(x, y)\, dA$$

$$I_2 = \iint_R (x^2 + y^2)\rho(x, y)\, dA.$$

The **radius of gyration** of a body about an axis L denoted by k_L is defined as

14. $$k_L^2 = I_L/M,$$

where I_L is the moment of inertia of the body about the axis L and M is the mass of the body.

2

Functions and Identities

2.1 Complex Numbers and Trigonometric and Hyperbolic Functions

2.1.1 Basic results

2.1.1.1 Modulus-argument representation. In the modulus-argument (r, θ) representation of the complex number $z = x + iy$, located at a point P in the complex plane, r is the radial distance of P from the origin and θ is the angle measured from the positive real axis to the line OP. The number r is called the **modulus** of z (see 1.1.1.1), and θ is called the **argument** of z, written arg z, and it is chosen to lie in the interval

1. $-\pi < \theta \le \pi$.

   By convention, $\theta = \arg z$ is measured *positively* in the counterclockwise sense from the positive real axis, so that $0 \le \theta \le \pi$, and *negatively* in the clockwise sense from the positive real axis, so that $-\pi < \theta \le 0$. Thus,

2. $z = x + iy = r \cos\theta + ir \sin\theta$

   or

3. $z = r(\cos\theta + i\sin\theta)$.

The connection between the **Cartesian representation** $z = x + iy$ and the modulus-argument form is given by

4. $x = r\cos\theta, \quad y = r\sin\theta$

5. $r = (x^2 + y^2)^{1/2}$.

The periodicity of the sine and cosine functions with period 2π means that for given r and θ, the complex number z in 2.1.1.1.3 will be unchanged if $\theta = \arg z$ is replaced by $\theta \pm 2k\pi$, $k = 0, 1, 2, \ldots$. This ambiguity in $\arg z$, which is, in fact, a *set* of values, is removed by constraining θ to satisfy 2.1.1.1.1. When $\arg z$ is chosen in this manner, and z is given by 2.1.1.1.3, θ is called the **principal value** of the argument of z.

Examples of modulus and argument representations follow:

(i) $z = 1 + i$, $r = 2^{1/2}$, $\theta = \arg z = \pi/4$

(ii) $z = 1 - i\sqrt{3}$, $r = 2$, $\theta = \arg z = -\pi/3$

(iii) $z = -2 - 2i$, $r = 2^{3/2}$, $\theta = \arg z = -3\pi/4$

(iv) $z = -2\sqrt{3} + 2i$, $r = 4$, $\theta = \arg z = 5\pi/6$

2.1.1.2 Euler's formula and de Moivre's theorem.

1. $e^{i\theta} = \cos\theta + i\sin\theta$,   **(Euler's formula)**

so an arbitrary complex number can always be written in the form

2. $z = re^{i\theta}$

3. $(\cos\theta + i\sin\theta)^n = \cos n\theta + i\sin n\theta$.   **(de Moivre's theorem)**

Some special complex numbers in modulus-argument form:

$$1 = e^{2k\pi i} \quad [k = 0, 1, 2, \ldots], \qquad -1 = e^{(2k+1)\pi i} \quad [k = 0, 1, \ldots]$$
$$i = e^{\pi i/2}, \qquad -i = e^{3\pi i/2}.$$

Euler's formula and de Moivre's theorem may be used to establish trigonometric identities. For example, from the Euler formula, 2.1.1.2.1,

$$\cos\theta = \frac{1}{2}(e^{i\theta} + e^{-i\theta}),$$

so

$$\begin{aligned}
\cos^5\theta &= \left(\frac{e^{i\theta} + e^{-i\theta}}{2}\right)^5 \\
&= \frac{1}{32}(e^{5i\theta} + 5e^{3i\theta} + 10e^{i\theta} + 10e^{-i\theta} + 5e^{-3i\theta} + e^{-5i\theta}) \\
&= \frac{1}{16}\left[\left(\frac{e^{5i\theta} + e^{-5i\theta}}{2}\right) + 5\left(\frac{e^{3i\theta} + e^{-3i\theta}}{2}\right) + 10\left(\frac{e^{i\theta} + e^{-i\theta}}{2}\right)\right] \\
&= \frac{1}{16}(\cos 5\theta + 5\cos 3\theta + 10\cos\theta),
\end{aligned}$$

which expresses $\cos^5\theta$ in terms of multiple angles.

A similar argument using

$$\sin \theta = \frac{e^{i\theta} - e^{-i\theta}}{2i}$$

shows that, for example,

$$\sin^4 \theta = \frac{1}{8}(\cos 4\theta - 4 \cos 2\theta + 3),$$

which expresses $\sin^4 \theta$ in terms of multiple angles (see 2.4.1.7.3).

Sines and cosines of multiple angles may be expressed in terms of powers of sines and cosines by means of de Moivre's theorem, 2.1.1.2.3. For example, setting $n = 5$ in the theorem gives

$$(\cos \theta + i \sin \theta)^5 = \cos 5\theta + i \sin 5\theta$$

or

$$(\cos^5 \theta - 10 \sin^2 \theta \cos^3 \theta + 5 \sin^4 \theta \cos \theta)$$
$$+ i(5 \sin \theta \cos^4 \theta - 10 \sin^3 \theta \cos^2 \theta + \sin^5 \theta) = \cos 5\theta + i \sin 5\theta.$$

Equating the respective real and imaginary parts of this identity gives

$$\cos 5\theta = \cos^5 \theta - 10 \sin^2 \theta \cos^3 \theta + 5 \sin^4 \theta \cos \theta,$$
$$\sin 5\theta = 5 \sin \theta \cos^4 \theta - 10 \sin^3 \theta \cos^2 \theta + \sin^5 \theta.$$

These identities may be further simplified by using $\cos^2 \theta + \sin^2 \theta = 1$ to obtain

$$\cos 5\theta = 16 \cos^5 \theta - 20 \cos^3 \theta + 5 \cos \theta$$

and

$$\sin 5\theta = 5 \sin \theta - 20 \sin^3 \theta + 16 \sin^5 \theta.$$

2.1.1.3 Roots of a complex number. Let $w^n = z$, with n an integer, so that $w = z^{1/n}$. Then, if $w = \rho e^{i\phi}$ and $z = r e^{i\theta}$,

1. $\rho = r^{1/n}, \quad \phi_k = \dfrac{\theta + 2k\pi}{n}, \qquad [k = 0, 1, \ldots, n - 1],$

   so the n roots of z are

2. $w_k = r^{1/n}\left[\cos\left(\dfrac{\theta + 2k\pi}{n}\right) + i \sin\left(\dfrac{\theta + 2k\pi}{n}\right)\right] \qquad [k = 0, 1, \ldots, n - 1].$

   When $z = 1$, the n roots of $z^{1/n}$ are called the **n'th roots of unity,** and are usually denoted by $\omega_0, \omega_1, \ldots, \omega^{n-1}$, where

3. $\omega_k = e^{2k\pi i/n} \qquad [k = 0, 1, \ldots, n - 1].$

Example of roots of a complex number. If $w^4 = -2\sqrt{3} + 2i$, so $w = (-2\sqrt{3} + 2i)^{1/4}$, if we set $z = -2\sqrt{3} + 2i$ it follows that $r = |z| = 4$ and $\theta = \arg z = 5\pi/6$. So, from 2.1.1.3.2, the required fourth roots of z are

$$w_k = 2^{1/2}\left[\cos\left(\frac{5 + 12k}{24}\right)\pi + i \sin\left(\frac{5 + 12k}{24}\right)\pi\right] \qquad [k = 0, 1, 2, 3].$$

Some special roots:

$$w^2 = i, \text{ or } w = \sqrt{i} \text{ has the two complex roots } (-1+i)/\sqrt{2}$$
$$\text{and } -(1+i)/\sqrt{2},$$
$$w^2 = -i, \text{ or } w = \sqrt{-i} \text{ has the two complex roots } (1-i)/\sqrt{2}$$
$$\text{and } (-1+i)/\sqrt{2},$$
$$w^3 = i, \text{ or } w = i^{1/3} \text{ has the three complex roots } -i, -(\sqrt{3}+i)/2$$
$$\text{and } (\sqrt{3}+i)/2,$$
$$w^3 = -i, \text{ or } w = (-i)^{1/3} \text{ has the three complex roots } i, -(\sqrt{3}+i)/2$$
$$\text{and } (\sqrt{3}-i)/2.$$

2.1.1.4 Relationship between roots. Once a root w_* of $w^n = z$ has been found for a given z and integer n, so that $w_* = z^{1/n}$ is one of the n'th roots of z, the other $n-1$ roots are given by $w_*\omega_1, w_*\omega_2, \ldots, w_*\omega_{n-1}$, where the ω_k are the n'th roots of unity defined in 2.1.1.3.3.

2.1.1.5 Roots of functions, polynomials, and nested multiplication. If $f(x)$ is an arbitrary function of x, a number x_0 such that $f(x_0) = 0$ is said to be a **root** of the function or, equivalently, a **zero** of $f(x)$. The need to determine a root of a function $f(x)$ arises frequently in mathematics, and when it cannot be found analytically it becomes necessary to make use of numerical methods.

In the special case in which the function is the **polynomial of degree n**

1. $P_n(x) \equiv a_0 x^n + a_1 x^{n-1} + \cdots + a_n,$

    with the **coefficients** $a_0, a_1, \ldots, a_n$, it follows from the *fundamental theorem of algebra* that $P_n(x) = 0$ has n roots $x_1, x_2, \ldots, x_n$, although these are not necessarily all real or distinct. If x_i is a root of $P_n(x) = 0$, then $(x - x_i)$ is a factor of the polynomial and it can be expressed as

2. $P_n(x) \equiv a_0(x - x_1)(x - x_2) \cdots (x - x_n).$

    If a root is repeated m times it is said to have **multiplicity** m.

    The important **division algorithm** for polynomials asserts that if $P(x)$ and $Q(x)$ are polynomials with the respective degrees n and m, where $1 \leq m \leq n$, then a polynomial $R(x)$ of degree $n - m$ and a polynomial $S(x)$ of degree $m - 1$ or less, both of which are unique, can be found such that

    $$P(x) = Q(x)R(x) + S(x).$$

Polynomials in which the coefficients $a_0, a_1, \ldots, a_n$ are real numbers have the following special properties, which are often of use:

(i)   If the degree n of $P_n(x)$ is odd it has at least one real root.

(ii) If a root z of $P_n(x)$ is complex, then its complex conjugate $\bar{z}$ is also a root. The quadratic expression $(x - z)(x - \bar{z})$ has *real* coefficients and is a factor of $P_n(x)$.

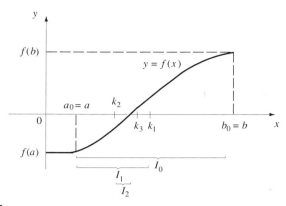

FIGURE 2.1 ∎

(iii) $P_n(x)$ may always be expressed as a product of real linear and quadratic factors with real coefficients, although they may occur with a multiplicity greater than unity.

The following numerical methods for the determination of the roots of a function $f(x)$ are arranged in order of increasing speed of convergence.

The bisection method.   The **bisection method,** which is the most elementary method for the location of a root of $f(x) = 0$, is based on the *intermediate value theorem.* The theorem asserts that if $f(x)$ is continuous on the interval $a \le x \le b$, with $f(a)$ and $f(b)$ of opposite signs, then there will be at least one value $x = c$, strictly intermediate between a and b, such that $f(c) = 0$. The method, which is iterative, proceeds by repeated bisection of intervals containing the required root, where after each bisection the subinterval that contains the root forms the interval to be used in the next bisection. Thus the root is bracketed in a nested set of intervals $I_0, I_1, I_2, \ldots$, where $I_0 = [a, b]$, and after the n'th bisection the interval I_n is of length $(b - a)/2^n$. The method is illustrated in Figure 2.1.

The bisection algorithm.
Step 1.   Find two numbers a_0, b_0 such that $f(a_0)$ and $f(b_0)$ are of opposite sign, so that $f(x)$ has at least one root in the interval $I_0 = [a_0, b_0]$.
Step 2.   Starting from Step 1, construct a new interval $I_{n+1} = [a_{n+1}, b_{n+1}]$ from interval $I_n = [a_n, b_n]$ by setting

3. $k_{n+1} = a_n + \dfrac{1}{2}(b_n - a_n)$

and choosing

4. $a_{n+1} = a_n, \quad b_{n+1} = k_{n+1} \qquad \text{if } f(a_n)f(k_{n+1}) < 0$

and

5. $a_{n+1} = k_{n+1}, \quad b_{n+1} = b_n \qquad$ if $f(a_n)f(k_{n+1}) > 0$.

> *Step 3.*   Terminate the iteration when one of the following conditions is satisfied:

> (i)   For some $n = N$, the number k_N is an exact root of $f(x)$, so $f(k_N) = 0$.

> (ii)   Take k_N as the approximation to the required root if, for some $n = N$ and some preassigned error bound $\varepsilon > 0$, it follows that $|k_N - k_{N-1}| < \varepsilon$.

To avoid excessive iteration caused by round-off error interfering with a small error bound ε it is necessary to place an upper bound M on the total number of iterations to be performed. In the bisection method the number M can be estimated by using $M > \log_2(\frac{b-a}{\varepsilon})$.

The convergence of the bisection method is slow relative to other methods, but it has the advantage that it is *unconditionally convergent*. The bisection method is often used to determine a starting approximation for a more sophisticated and rapidly convergent method, such as Newton's method, which can diverge if a poor approximation is used.

The method of false position (regula falsi).   The **method of false position,** also known as the **regula falsi method,** is a bracketing technique similar to the bisection method, although the nesting of the intervals I_n within which the root of $f(x) = 0$ lies is performed differently. The method starts as in the bisection method with two numbers a_0, b_0 and the interval $I_0 = [a_0, b_0]$ such that $f(a_0)$ and $f(b_0)$ are of opposite signs. The starting approximation to the required root in I_0 is taken to be the point k_0 at which the chord joining the points $(a_0, f(a_0))$ and $(b_0, f(b_0))$ cuts the x-axis. The interval I_0 is then divided into the two subintervals $[a_0, k_0]$ and $[k_0, b_0]$, and the interval I_1 is chosen to be the subinterval at the ends of which $f(x)$ has opposite signs.

Thereafter, the process continues iteratively until, for some $n = N$, $|k_N - k_{N-1}| < \varepsilon$, where $\varepsilon > 0$ is a preassigned error bound. The approximation to the required root is taken to be k_N. The method is illustrated in Figure 2.2.

The false position algorithm.

> *Step 1.*   Find two numbers a_0, b_0 such that $f(a_0)$ and $f(b_0)$ are of opposite signs, so that $f(x)$ has at least one root in the interval $I_0 = [a_0, b_0]$.

> *Step 2.*   Starting from Step 1, construct a new interval $I_{n+1} = [a_{n+1}, b_{n+1}]$ from the interval $I_n = [a_n, b_n]$ by setting

6. $k_{n+1} = a_n - \dfrac{f(a_n)(b_n - a_n)}{f(b_n) - f(a_n)}$

and choosing

7. $a_{n+1} = a_n, \quad b_{n+1} = k_{n+1} \qquad$ if $f(a_n)f(k_{n+1}) < 0$

or

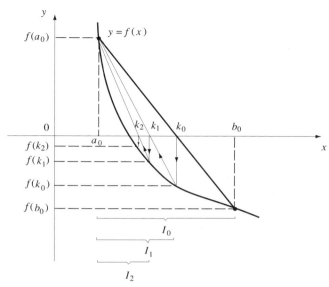

FIGURE 2.2 ■

8. $a_{n+1} = k_{n+1}$, $b_{n+1} = b_n$ if $f(a_n)f(k_{n+1}) > 0$.

    Step 3. Terminate the iterations if either, for some $n = N$, k_N is an exact root so that $f(k_N) = 0$, or for some preassigned error bound $\varepsilon > 0$, $|k_N - k_{N-1}| < \varepsilon$, in which case k_N is taken to be the required approximation to the root. It is necessary to place an upper bound M on the total number of iterations N to prevent excessive iteration that may be caused by round-off errors interfering with a small error bound ε.

The secant method. Unlike the previous methods, the **secant method** does *not* involve bracketing a root of $f(x) = 0$ in a sequence of nested intervals. Thus the convergence of the secant method cannot be guaranteed, although when it does converge the process is usually faster than either of the two previous methods.

    The secant method is started by finding two approximations k_0 and k_1 to the required root of $f(x) = 0$. The next approximation k_2 is taken to be the point at which the secant drawn through the points $(k_0, f(k_0))$ and $(k_1, f(k_1))$ cuts the x-axis. Thereafter, iteration takes place with secants drawn as shown in Figure 2.3.

The secant algorithm. Starting from the approximations k_0, k_1 to the required root of $f(x) = 0$, the approximation k_{n+1} is determined from the approximations k_n and k_{n-1} by using

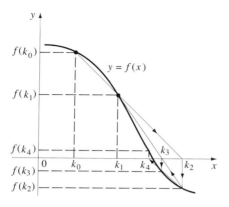

FIGURE 2.3 ∎

9. $k_{n+1} = k_n - \dfrac{f(k_n)(k_n - k_{n-1})}{f(k_n) - f(k_{n-1})}.$

The iteration is terminated when, for some $n = N$ and a preassigned error bound $\varepsilon > 0$, $|k_N - k_{N-1}| < \varepsilon$. An upper bound M must be placed on the number of iterations N in case the method diverges, which occurs when $|k_N|$ increases without bound.

Newton's method.   **Newton's method,** often called the **Newton–Raphson method,** is based on a tangent line approximation to the curve $y = f(x)$ at an approximate root x_0 of $f(x) = 0$. The method may be deduced from a Taylor series approximation to $f(x)$ about the point x_0 as follows. Provided $f(x)$ is differentiable, then if $x_0 + h$ is an exact root,

$$0 = f(x_0 + h) = f(x_0) + hf'(x_0) + \frac{h^2}{2!} f''(x_0) + \cdots .$$

Thus, if h is sufficiently small that higher powers of h may be neglected, an approximate value h_0 to h is given by

$$h_0 = -\frac{f(x_0)}{f'(x_0)},$$

and so a better approximation to x_0 is $x_1 = x_0 + h_0$. This process is iterated until the required accuracy is attained. However, a poor choice of the starting value x_0 may cause the method to diverge. The method is illustrated in Figure 2.4.

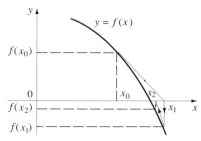

FIGURE 2.4 ▮

The Newton algorithm. Starting from an approximation x_0 to the required root of $f(x) = 0$, the approximation x_{n+1} is determined from the approximation x_n by using

10. $$x_{n+1} = x_n - \frac{f(x_n)}{f'(x_n)}.$$

The iteration is terminated when, for some $n = N$ and a preassigned error bound $\varepsilon > 0$, $|x_N - x_{N-1}| < \varepsilon$. An upper bound M must be placed on the number of iterations N in case the method diverges.

Notice that the secant method is an approximation to Newton's method, as can be seen by replacing the derivative $f'(x_n)$ in the above algorithm with a difference quotient.

Nested multiplication. When working with polynomials in general, or when using them in conjunction with one of the previous root-finding methods, it is desirable to have an efficient method for their evaluation for specific arguments. Such a method is provided by the technique of **nested multiplication.**

Consider, for example, the quartic

$$P_4(x) \equiv 3x^4 + 2x^3 - 4x^2 + 5x - 7.$$

Then, instead of evaluating each term of $P_4(x)$ separately for some $x = c$, say, and summing them to find $P_4(c)$, fewer multiplications are required if $P_4(x)$ is rewritten in the **nested form:**

$$P_4(x) \equiv x\{x[x(3x + 2) - 4] + 5\} - 7.$$

When repeated evaluation of polynomials is required, as with root-finding methods, this economy of multiplication becomes significant.

Nested multiplication is implemented on a computer by means of the simple algorithm given below, which is based on the division algorithm for polynomials, and for this reason the method is sometimes called **synthetic division.**

The nested multiplication algorithm for evaluating $P_n(c)$. Suppose we want to evaluate the polynomial

11. $P_n(x) = a_0 x^n + a_1 x^{n-1} + \cdots + a_n$

for some $x = c$. Set $b_0 = a_0$, and generate the sequence $b_1, b_2, \ldots, b_n$ by means of the algorithm

12. $b_i = c b_{i-1} + a_i$,   for $1 \le i \le n$;

then

13. $P_n(c) = b_n$.

The argument that led to the nested multiplication algorithm for the evaluation of $P_n(c)$ also leads to the following algorithm for the evaluation of $P_n'(c) = [d P_n(x)/dx]_{x=c}$. These algorithms are useful when applying Newton's method to polynomials because all we have to store is the sequence $b_0, b_1, \ldots, b_n$.

Algorithm for evaluating $P_n'(c)$.   Suppose we want to evaluate $P_n'(x)$ when $x = c$, where $P_n(x)$ is the polynomial in the nested multiplication algorithm, and $b_0, b_1, \ldots, b_n$ is the sequence generated by that algorithm. Set $d_0 = b_0$ and generate the sequence $d_1, d_2, \ldots, d_{n-1}$ by means of the algorithm

14. $d_i = c d_{i-1} + b_i$,   for $1 \le i \le n - 1$;

then

15. $P_n'(c) = d_{n-1}$.

EXAMPLE.   Find $P_4(1.4)$ and $P_4'(1.4)$ when

$$P_4(x) = 2.1 x^4 - 3.7 x^3 + 5.4 x^2 - 1.1 x - 7.2.$$

Use the result to perform three iterations of Newton's method to find the root approximated by $x = 1.4$.

Setting $c = 1.4$, $a_0 = 2.1$, $a_1 = -3.7$, $a_2 = 5.4$, $a_3 = -1.1$, and $a_4 = -7.2$ in the nested multiplication algorithm gives

$b_0 = 2.1$, $b_1 = -0.76000$, $b_2 = 4.33600$,

$b_3 = 4.97040$, $b_4 = -0.24144$,

so

$$P_4(1.4) = b_4 = -0.24144.$$

Using these results in the algorithm for $P_4'(1.4)$ with $d_0 = b_0 = 2.1$ gives

$d_1 = 2.18000$, $d_2 = 7.38800$, $d_3 = 15.31360$,

so

$$P_4'(1.4) = d_3 = 15.31360.$$

Because $P_4(1.3) = -1.63508$ and $P_4(1.5) = 1.44375$, a root of $P_4(x)$ must lie in the interval $1.3 < x < 1.5$, so we take as our initial approximation $x_0 = 1.4$. A first application of the Newton algorithm gives

$$x_1 = x_0 - \frac{P_4(1.4)}{P_4'(1.4)}$$

$$= 1.4 - \frac{(-0.24144)}{15.31360} = 1.41576.$$

Repetition of the process gives

$$x_2 = 1.41576 - \frac{P_4(1.41576)}{P_4'(1.41576)}$$

$$= 1.41576 - \frac{0.00355}{15.7784} = 1.41553$$

and

$$x_3 = 1.41553 - \frac{P_4(1.41553)}{P_4'(1.41553)}$$

$$= 1.41553 - \frac{(-0.00008)}{15.7715} = 1.41553.$$

Thus, Newton's method has converged to five decimal places after only three iterations, showing that the required root is $x = 1.41553$.

2.1.1.6 Connection between trigonometric and hyperbolic functions.

1. $\sin x = \dfrac{(e^{ix} - e^{-ix})}{2i}$

2. $\cos x = \dfrac{(e^{ix} + e^{-ix})}{2}$

3. $\tan x = \dfrac{\sin x}{\cos x}$

4. $\sin(ix) = i \sinh x$

5. $\cos(ix) = \cosh x$

6. $\tan(ix) = i \tanh x$

7. $\sinh x = \dfrac{(e^x - e^{-x})}{2}$

8. $\cosh x = \dfrac{(e^x + e^{-x})}{2}$

9. $\tanh x = \dfrac{\sinh x}{\cosh x}$

10. $\sinh(ix) = i \sin x$

11. $\cosh(ix) = \cos x$

12. $\tanh(ix) = i \tan x$

The corresponding results for $\sec x$, $\csc x$, $\cot x$ and $\operatorname{sech} x$, $\operatorname{csch} x$, $\coth x$ follow from the above results and the definitions:

13. $\sec x = \dfrac{1}{\cos x}$

14. $\csc x = \dfrac{1}{\sin x}$

15. $\cot x = \dfrac{\cos x}{\sin x}$

16. $\operatorname{sech} x = \dfrac{1}{\cosh x}$

17. $\operatorname{csch} x = \dfrac{1}{\sinh x}$

18. $\tanh x = \dfrac{\sinh x}{\cosh x}.$

2.2 Logarithms and Exponentials

2.2.1 Basic functional relationships

2.2.1.1 The logarithmic function.   Let $a > 0$, with $a \neq 1$.   Then, if y is the **logarithm** of x to the **base** a,

1. $y = \log_a x$ if and only if $a^y = x$

2. $\log_a 1 = 0$

3. $\log_a a = 1$

For all positive numbers x, y and all real numbers z:

4. $\log_a(xy) = \log_a x + \log_a y$

5. $\log_a(x/y) = \log_a x - \log_a y$

6. $\log_a(x^z) = z \log_a x$

7. $a^{\log_a x} = x$

8. $b^x = a^{x \log_a b}$

9. $y = \ln x$ if and only if $e^y = x$ [$e \approx 2.71828$]
     (**natural logarithm**, or **logarithm to the base e**, or **Naperian logarithm**)

10. $\ln 1 = 0$

11. $\ln e = 1$

12. $\ln(xy) = \ln x + \ln y$

13. $\ln(x/y) = \ln x - \ln y$

14. $\ln(x^z) = z \ln x$

15. $e^{\ln x} = x$

16. $a^x = e^{x \ln a}$

17. $\log_a x = \log_b x / \log_b a$ (**change from base a to base b**)

18. $\log_{10} x = \ln x / \log_{10} e$ or $\ln x = 2.30258509 \log_{10} x$

Graphs of e^x and $\ln x$ are shown in Figure 2.5. The fact that each function is the inverse of the other can be seen by observing that each graph is the reflection of the other in the line $y = x$.

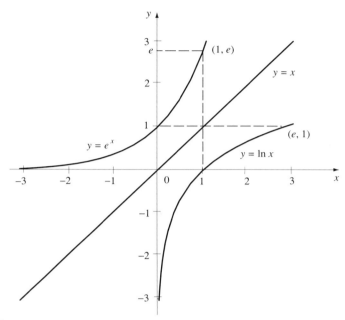

FIGURE 2.5 ■

2.2.2 The number e

2.2.2.1 Definitions.

1. $e = \lim\limits_{n \to \infty} \left(1 + \dfrac{1}{n}\right)^n$

2. $e = \sum\limits_{k=0}^{\infty} \dfrac{1}{k!} = 1 + \dfrac{1}{1!} + \dfrac{1}{2!} + \dfrac{1}{3!} + \cdots$

To fifteen decimal places $e = 2.718\,281\,828\,459\,045$. A direct consequence of 2.2.2.1.1 and 2.2.2.1.2 is that for real a:

3. $e^a = \lim\limits_{n \to \infty} \left(1 + \dfrac{a}{n}\right)^n$

4. $e^a = \sum\limits_{k=0}^{\infty} \dfrac{a^k}{k!} = 1 + a + \dfrac{a^2}{2!} + \dfrac{a^3}{3!} + \cdots.$

2.3 The Exponential Function

2.3.1 Series representations

2.3.1.1

1. $e^x = \sum\limits_{k=0}^{\infty} \dfrac{x^k}{k!} = 1 + \dfrac{x}{1!} + \dfrac{x^2}{2!} + \dfrac{x^3}{3!} + \cdots$

2. $e^{-x} = \sum_{k=0}^{\infty} (-1)^k \frac{x^k}{k!} = 1 - \frac{x}{1!} + \frac{x^2}{2!} - \frac{x^3}{3!} + \cdots$

3. $e^{-x^2} = \sum_{k=0}^{\infty} (-1)^k \frac{x^{2k}}{k!} = 1 - \frac{x^2}{1!} + \frac{x^4}{2!} - \frac{x^6}{3!} + \cdots$

4. $\dfrac{x}{e^x - 1} = 1 - \dfrac{x}{2} + \sum_{k=1}^{\infty} B_{2k} \dfrac{x^{2k}}{(2k)!}$ $[x < 2\pi]$ (see 1.3.1.1.1)

2.3.1.2

1. $e^{\sin x} = 1 + x + \frac{x^2}{2} - \frac{x^4}{8} - \frac{x^5}{15} - \frac{x^6}{240} + \frac{x^7}{90} + \cdots$

2. $e^{\cos x} = e\left(1 - \frac{x^2}{2} + \frac{x^4}{6} - \frac{31x^6}{720} + \frac{379x^8}{40320} - \cdots\right)$

3. $e^{\tan x} = 1 + x + \frac{x^2}{2} + \frac{x^3}{2} + \frac{3x^4}{8} + \frac{37x^5}{120} + \frac{59x^6}{240} + \cdots$

4. $e^{\sec x} = e\left(1 + \frac{x^2}{2} + \frac{x^4}{3} + \frac{151x^6}{720} + \frac{5123x^8}{40320} \cdots\right)$

5. $e^{x \sec x} = 1 + x + \frac{x^2}{2} + \frac{2x^3}{3} + \frac{13x^4}{24} + \frac{7x^5}{15} + \cdots$

2.3.1.3

1. $e^{\arcsin x} = 1 + x + \frac{x^2}{2} + \frac{x^3}{3} + \frac{5x^4}{24} + \frac{x^5}{6} + \cdots$

2. $e^{\arccos x} = e^{\pi/2}\left(1 - x + \frac{x^2}{2} - \frac{x^3}{3} + \frac{5x^4}{24} - \frac{x^5}{6} \cdots\right)$

3. $e^{\arctan x} = 1 + x + \frac{x^2}{2!} - \frac{x^3}{6} - \frac{7x^4}{24} + \frac{x^5}{24} + \cdots$

4. $e^{\sinh x} = 1 + x + \frac{x^2}{2} + \frac{x^3}{3} + \frac{5x^4}{24} + \frac{x^5}{10} + \frac{37x^6}{720} + \cdots$

5. $e^{\cosh x} = e\left(1 + \frac{x^2}{2} + \frac{x^4}{6} + \frac{31x^6}{720} + \frac{379x^8}{40320} + \cdots\right)$

6. $e^{\tanh x} = 1 + x + \frac{x^2}{2} - \frac{x^3}{6} - \frac{7x^4}{24} - \frac{x^5}{40} - \cdots$

7. $e^{\operatorname{arcsinh} x} = x + \sqrt{(x^2 + 1)} = 1 + x + \frac{x^2}{2} - \frac{x^4}{8} + \frac{x^6}{16} - \frac{5x^8}{128} + \cdots$

8. $e^{\operatorname{arctanh} x} = \left(\frac{1 + x}{1 - x}\right)^{1/2} = 1 + x + \frac{x^2}{2} + \frac{x^3}{2} + \frac{3x^4}{8} + \frac{3x^5}{8} + \cdots$

2.4 Trigonometric Identities

2.4.1 Trigonometric functions

2.4.1.1 Basic definitions.

1. $\quad \sin x = \dfrac{1}{2i}(e^{ix} - e^{-ix})$

   2. $\quad \cos x = \dfrac{1}{2}(e^{ix} + e^{-ix})$

3. $\quad \tan x = \dfrac{\sin x}{\cos x}$

   4. $\quad \csc x = \dfrac{1}{\sin x}$

5. $\quad \sec x = \dfrac{1}{\cos x}$

   6. $\quad \cot x = \dfrac{1}{\tan x}$ (also ctn x)

Graphs of these functions are shown in Figure 2.6.

2.4.1.2 Even, odd, and periodic functions.
A function $f(x)$ is said to be an **even** function if it is such that $f(-x) = f(x)$ and to be an **odd** function if it is such that $f(-x) = -f(x)$. In general, an arbitrary function $g(x)$ defined for all x is neither even nor odd, although it can always be represented as the sum of an even function $h(x)$ and an odd function $k(x)$ by writing

$$g(x) = h(x) + k(x),$$

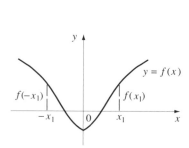

Even function $f(-x) = f(x)$

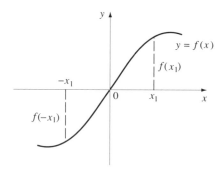

Odd function $f(-x) = -f(x)$

where

$$h(x) = \frac{1}{2}(g(x) + g(-x)), \quad k(x) = \frac{1}{2}(g(x) - g(-x)).$$

The product of two *even* or two *odd* functions is an *even* function, whereas the product of an *even* and an *odd* function is an odd function.

A function $f(x)$ is said to be **periodic** with **period** X if

$$f(x + X) = f(x),$$

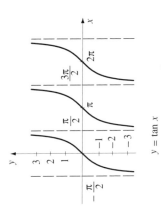

$y = \tan x$

Domain: all $x \neq (2n-1)\dfrac{\pi}{2}$
Range: $(-\infty, \infty)$
Period: π

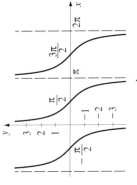

$y = \cot x = \dfrac{1}{\tan x}$

Domain: all $x \neq n\pi$
Range: $(-\infty, \infty)$
Period: π

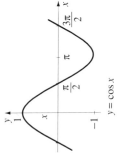

$y = \cos x$

Domain: $-\infty < x < \infty$
Range: $[-1, 1]$
Period: 2π

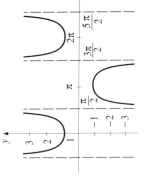

$y = \sec x = \dfrac{1}{\cos x}$

Domain: all $x \neq (2n-1)\dfrac{\pi}{2}$
Range: $(-\infty, -1]$ and $[1, \infty)$
Period: 2π

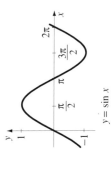

$y = \sin x$

Domain: $-\infty < x < \infty$
Range: $[-1, 1]$
Period: 2π

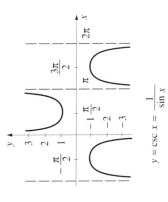

$y = \csc x = \dfrac{1}{\sin x}$

Domain: all $x \neq n\pi$
Range: $(-\infty, -1]$ and $[1, \infty)$
Period: 2π

FIGURE 2.6 ▌ Graphs of trigonometric functions.

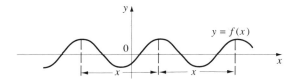

and X is the *smallest* number for which this is true.

2.4.1.3 Basic relationships.

1. $\sin(-x) = -\sin x$ **(odd** function)
2. $\cos(-x) = \cos x$ **(even** function)
3. $\tan(-x) = -\tan x$ (odd function)
4. $\csc(-x) = -\csc x$ (odd function)
5. $\sec(-x) = \sec x$ (even function)
6. $\cot(-x) = -\cot x$ (odd function)
7. $\sin(x + 2\pi) = \sin x$ **(periodic** with **period** 2π)
8. $\cos(x + 2\pi) = \cos x$ (periodic with period 2π)
9. $\tan(x + \pi) = \tan x$ (periodic with period π)
10. $\sin(x + \pi) = -\sin x$
11. $\cos(x + \pi) = -\cos x$
12. $\cot\left(\dfrac{\pi}{2} - x\right) = \tan x$
13. $\csc\left(\dfrac{\pi}{2} - x\right) = \sec x$
14. $\sec\left(\dfrac{\pi}{2} - x\right) = \csc x$
15. $\sin^2 x + \cos^2 x = 1$
16. $\sec^2 x = 1 + \tan^2 x$
17. $\csc^2 x = 1 + \cot^2 x$
18. $\sin x = \pm\sqrt{(1 - \cos^2 x)}$
19. $\cos x = \pm\sqrt{(1 - \sin^2 x)}$
20. $\tan x = \pm\sqrt{(\sec^2 x - 1)}$

The choice of sign in entries 2.4.1.3.18 to 20 is determined by the quadrant in which the argument x is located. For x in the first quadrant the sine, cosine, and tangent functions are all positive, whereas for x in the second, third, and fourth quadrants only the sine, tangent, and cosine functions, respectively, are positive.

2.4.1.4 Sines and cosines of sums and differences.

1. $\sin(x + y) = \sin x \cos y + \cos x \sin y$
2. $\sin(x - y) = \sin x \cos y - \cos x \sin y$
3. $\cos(x + y) = \cos x \cos y - \sin x \sin y$
4. $\cos(x - y) = \cos x \cos y + \sin x \sin y$
5. $\sin x \cos y = \frac{1}{2}\{\sin(x + y) + \sin(x - y)\}$
6. $\cos x \cos y = \frac{1}{2}\{\cos(x + y) + \cos(x - y)\}$
7. $\sin x \sin y = \frac{1}{2}\{\cos(x - y) - \cos(x + y)\}$
8. $\sin^2 x - \sin^2 y = \sin(x + y) \sin(x - y)$
9. $\cos^2 x - \cos^2 y = \sin(x + y) \sin(y - x)$
10. $\cos^2 x - \sin^2 y = \cos(x + y) \cos(x - y)$
11. $\sin x + \sin y = 2 \sin \frac{1}{2}(x + y) \cos \frac{1}{2}(x - y)$
12. $\sin x - \sin y = 2 \sin \frac{1}{2}(x - y) \cos \frac{1}{2}(x + y)$
13. $\cos x + \cos y = 2 \cos \frac{1}{2}(x + y) \cos \frac{1}{2}(x - y)$
14. $\cos x - \cos y = 2 \sin \frac{1}{2}(x + y) \sin \frac{1}{2}(y - x)$
15. $\sin(x + iy) = \sin x \cosh y + i \cos x \sinh y$
16. $\sin(x - iy) = \sin x \cosh y - i \cos x \sinh y$
17. $\cos(x + iy) = \cos x \cosh y - i \sin x \sinh y$
18. $\cos(x - iy) = \cos x \cosh y + i \sin x \sinh y$

2.4.1.5 Tangents and cotangents of sums and differences.

1. $\tan(x + y) = \dfrac{\tan x + \tan y}{1 - \tan x \tan y}$

2. $\tan(x - y) = \dfrac{\tan x - \tan y}{1 + \tan x \tan y}$

3. $\cot(x + y) = \dfrac{\cot x \cot y - 1}{\cot x + \cot y}$

4. $\cot(x - y) = \dfrac{\cot x \cot y + 1}{\cot y - \cot x}$

5. $\tan x + \tan y = \dfrac{\sin(x + y)}{\cos x \cos y}$

6. $\tan x - \tan y = \dfrac{\sin(x - y)}{\cos x \cos y}$

7. $\tan x = \cot x - 2 \cot 2x$ (from 2.4.1.5.3 with $y = x$)

8. $\tan(x + iy) = \dfrac{\sin 2x + i \sinh 2y}{\cos 2x + \cosh 2y}$

9. $\tan(x - iy) = \dfrac{\sin 2x - i \sinh 2y}{\cos 2x + \cosh 2y}$

2.4.1.6 Sines, cosines, and tangents of multiple angles.

1. $\sin 2x = 2 \sin x \cos x$

2. $\sin 3x = 3 \sin x - 4 \sin^3 x$

3. $\sin 4x = \cos x (4 \sin x - 8 \sin^3 x)$

4. $\sin nx = n \cos^{n-1} x \sin x - \dbinom{n}{3} \cos^{n-3} x \sin^3 x + \dbinom{n}{5} \cos^{n-5} x \sin^5 x + \cdots$

$$= \sin x \left[2^{n-1} \cos^{n-1} x - \dbinom{n-2}{1} 2^{n-3} x \cos^{n-3} x \right.$$

$$+ \dbinom{n-3}{2} 2^{n-5} \cos^{n-5} x - \dbinom{n-4}{3} 2^{n-7} \cos^{n-7} x + \cdots \Bigg]$$

$$\left[n = 2, 3, \ldots, \text{ and } \dbinom{m}{k} = 0, k > m \right]$$

5. $\cos 2x = 2 \cos^2 x - 1$

6. $\cos 3x = 4 \cos^3 x - 3 \cos x$

7. $\cos 4x = 8 \cos^4 x - 8 \cos^2 x + 1$

8. $\cos nx = \cos^n x - \dbinom{n}{2} \cos^{n-2} x \sin^2 x + \dbinom{n}{4} \cos^{n-4} x \sin^4 x - \cdots$

$$= 2^{n-1} \cos^n x - \frac{n}{1} 2^{n-3} \cos^{n-2} x + \frac{n}{2} \dbinom{n-3}{1} 2^{n-5} \cos^{n-4} x$$

$$- \frac{n}{3} \dbinom{n-4}{2} 2^{n-7} \cos^{n-6} x + \cdots$$

$$\left[n = 2, 3, \ldots, \text{ and } \dbinom{m}{k} = 0, k > m \right]$$

9. $\tan 2x = \dfrac{2 \tan x}{1 - \tan^2 x}$

10. $\tan 3x = \dfrac{3 \tan x - \tan^3 x}{1 - 3 \tan^2 x}$

11. $\tan 4x = \dfrac{4 \tan x - 4 \tan^3 x}{1 - 6 \tan^2 x + \tan^4 x}$

12. $\tan 5x = \dfrac{5 \tan x - 10 \tan^3 x + \tan^5 x}{1 - 10 \tan^2 x + 5 \tan^4 x}$

2.4.1.7 Powers of sines, cosines, and tangents in terms of multiple angles.

1. $\sin^2 x = \frac{1}{2}(1 - \cos 2x)$

2. $\sin^3 x = \frac{1}{4}(3 \sin x - \sin 3x)$

3. $\sin^4 x = \frac{1}{8}(3 - 4 \cos 2x + \cos 4x)$

4. $\sin^{2n-1} x = \dfrac{1}{2^{2n-2}} \displaystyle\sum_{k=0}^{n-1} (-1)^{n+k-1} \binom{2n-1}{k} \sin(2n - 2k - 1)x$

$$[n = 1, 2, \ldots]$$

5. $\sin^{2n} x = \dfrac{1}{2^{2n}} \left\{ \displaystyle\sum_{k=0}^{n-1} (-1)^{n-k} 2 \binom{2n}{k} \cos 2(n - k)x + \binom{2n}{n} \right\} \qquad [n = 1, 2, \ldots]$

6. $\cos^2 x = \frac{1}{2}(1 + \cos 2x)$

7. $\cos^3 x = \frac{1}{4}(3 \cos x + \cos 3x)$

8. $\cos^4 x = \frac{1}{8}(3 + 4 \cos 2x + \cos 4x)$

9. $\cos^{2n-1} x = \dfrac{1}{2^{2n-2}} \displaystyle\sum_{k=0}^{n-1} \binom{2n-1}{k} \cos(2n - 2k - 1)x \qquad [n = 1, 2, \ldots]$

10. $\cos^{2n} x = \dfrac{1}{2^{2n}} \left\{ \displaystyle\sum_{k=0}^{n-1} 2 \binom{2n}{k} \cos 2(n - k)x + \binom{2n}{n} \right\} \qquad [n = 1, 2, \ldots]$

11. $\tan^2 x = \dfrac{1 - \cos 2x}{1 + \cos 2x}$

12. $\tan^3 x = \dfrac{3 \sin x - \sin 3x}{3 \cos x + \cos 3x}$

13. $\tan^4 x = \dfrac{3 - 4 \cos 2x + \cos 4x}{3 + 4 \cos 2x + \cos 4x}$

14. For $\tan^n x$ use $\tan^n x = \sin^n x / \cos^n x$ with 2.4.1.7.4 and 2.4.1.7.9 or 2.4.1.7.5 and 2.4.1.7.10.

2.4.1.8 Half angle representations of trigonometric functions.

1. $\sin \dfrac{x}{2} = \pm \left(\dfrac{1 - \cos x}{2} \right)^{1/2}$

$$[+ \text{ sign if } 0 < x < 2\pi \text{ and } - \text{ sign if } 2\pi < x < 4\pi]$$

2. $\cos \dfrac{x}{2} = \pm \left(\dfrac{1 + \cos x}{2} \right)^{1/2}$

$$[+ \text{ sign if } -\pi < x < \pi \text{ and } - \text{ sign if } \pi < x < 3\pi]$$

3. $\tan \dfrac{x}{2} = \dfrac{\sin x}{1 + \cos x} = \dfrac{1 - \cos x}{\sin x}$

4. $\cot \dfrac{x}{2} = \dfrac{1 + \cos x}{\sin x} = \dfrac{\sin x}{1 - \cos x}$

5. $\sin x = 2 \sin \dfrac{x}{2} \cos \dfrac{x}{2}$

6. $\cos x = \cos^2 \dfrac{x}{2} - \sin^2 \dfrac{x}{2}$

7. $\tan x = \dfrac{2 \tan \dfrac{x}{2}}{1 - \tan^2 \dfrac{x}{2}} = \dfrac{2 \cot \dfrac{x}{2}}{\csc^2 \dfrac{x}{2} - 2}$

2.5 Hyperbolic Identities

2.5.1 Hyperbolic functions

2.5.1.1 Basic definitions.

1. $\sinh x = \dfrac{1}{2}(e^x - e^{-x})$

2. $\cosh x = \dfrac{1}{2}(e^x + e^{-x})$

3. $\tanh x = \dfrac{\sinh x}{\cosh x}$

4. $\operatorname{csch} x = \dfrac{1}{\sinh x}$

5. $\operatorname{sech} x = \dfrac{1}{\cosh x}$

6. $\coth x = \dfrac{1}{\tanh x}$

Other notations are also in use for these *same* hyperbolic functions. Equivalent notations are sinh, sh; cosh, ch; tanh, th; csch, csh; sech, sch; coth, ctnh, cth. Graphs of these functions are shown in Figure 2.7.

2.5.1.2 Basic relationships.

1. $\sinh(-x) = -\sinh x$ (**odd** function)
2. $\cosh(-x) = \cosh x$ (**even** function)
3. $\tanh(-x) = -\tanh x$ (odd function)
4. $\operatorname{csch}(-x) = -\operatorname{csch}(x)$ (odd function)
5. $\operatorname{sech}(-x) = \operatorname{sech} x$ (even function)
6. $\coth(-x) = -\coth x$ (odd function)
7. $\cosh^2 x - \sinh^2 x = 1$
8. $\operatorname{sech}^2 x = 1 - \tanh^2 x$
9. $\operatorname{csch}^2 x = \coth^2 x - 1$

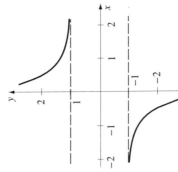

$y = \tanh x$

Domain: $(-\infty, \infty)$
Range: $(-1, 1)$

$y = \coth x = \dfrac{1}{\tanh x}$

Domain: $(-\infty, 0)$ and $(0, \infty)$
Range: $(-\infty, -1)$ and $(1, \infty)$

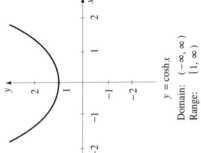

$y = \cosh x$

Domain: $(-\infty, \infty)$
Range: $[1, \infty)$

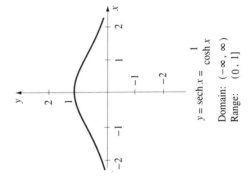

$y = \operatorname{sech} x = \dfrac{1}{\cosh x}$

Domain: $(-\infty, \infty)$
Range: $(0, 1]$

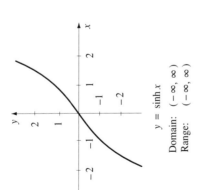

$y = \sinh x$

Domain: $(-\infty, 0)$ and $(0, \infty)$
Range: $(-\infty, \infty)$

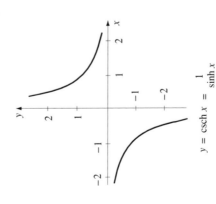

$y = \operatorname{csch} x = \dfrac{1}{\sinh x}$

Domain: $(-\infty, 0)$ and $(0, \infty)$
Range: $(-\infty, 0)$ and $(0, \infty)$

FIGURE 2.7 ▌ Graphs of hyperbolic functions.

10. $\sinh x = \begin{cases} \sqrt{(\cosh^2 x - 1)}, & [x > 0] \\ -\sqrt{(\cosh^2 x - 1)}, & [x < 0] \end{cases}$

11. $\cosh x = \sqrt{(1 + \sinh^2 x)}$

12. $\tanh x = \begin{cases} \sqrt{(1 - \operatorname{sech}^2 x)}, & [x > 0] \\ -\sqrt{(1 - \operatorname{sech}^2 x)}, & [x < 0] \end{cases}$

2.5.1.3 Hyperbolic sines and cosines of sums and differences.

1. $\sinh(x + y) = \sinh x \cosh y + \cosh x \sinh y$
2. $\sinh(x - y) = \sinh x \cosh y - \cosh x \sinh y$
3. $\cosh(x + y) = \cosh x \cosh y + \sinh x \sinh y$
4. $\cosh(x - y) = \cosh x \cosh y - \sinh x \sinh y$
5. $\sinh x \cosh y = \frac{1}{2}\{\sinh(x + y) + \sinh(x - y)\}$
6. $\cosh x \cosh y = \frac{1}{2}\{\cosh(x + y) + \cosh(x - y)\}$
7. $\sinh x \sinh y = \frac{1}{2}\{\cosh(x + y) - \cosh(x - y)\}$
8. $\sinh^2 x - \sinh^2 y = \sinh(x + y) \sinh(x - y)$
9. $\sinh^2 x + \cosh^2 y = \cosh(x + y) \cosh(x - y)$
10. $\sinh x + \sinh y = 2 \sinh \frac{1}{2}(x + y) \cosh \frac{1}{2}(x - y)$
11. $\sinh x - \sinh y = 2 \sinh \frac{1}{2}(x - y) \cosh \frac{1}{2}(x + y)$
12. $\cosh x + \cosh y = 2 \cosh \frac{1}{2}(x + y) \cosh \frac{1}{2}(x - y)$
13. $\cosh x - \cosh y = 2 \sinh \frac{1}{2}(x + y) \sinh \frac{1}{2}(x - y)$
14. $\cosh x - \sinh x = 1/(\sinh x + \cosh x)$
15. $(\sinh x + \cosh x)^n = \sinh nx + \cosh nx$
16. $(\sinh x - \cosh x)^n = \sinh nx - \cosh nx$
17. $\sinh(x + iy) = \sinh x \cos y + i \cosh x \sin y$
18. $\sinh(x - iy) = \sinh x \cos y - i \cosh x \sin y$
19. $\cosh(x + iy) = \cosh x \cos y + i \sinh x \sin y$
20. $\cosh(x - iy) = \cosh x \cos y - i \sinh x \sin y$

2.5.1.4 Hyperbolic tangents and cotangents of sums and differences.

1. $\tanh(x + y) = \dfrac{\tanh x + \tanh y}{1 + \tanh x \tanh y}$

2. $\tanh(x - y) = \dfrac{\tanh x - \tanh y}{1 - \tanh x \tanh y}$

3. $\coth(x + y) = \dfrac{\coth x \coth y + 1}{\coth y + \coth x}$

4. $\coth(x - y) = \dfrac{\coth x \coth y - 1}{\coth y - \coth x}$

5. $\tanh x + \tanh y = \dfrac{\sinh(x + y)}{\cosh x \cosh y}$

6. $\tanh x - \tanh y = \dfrac{\sinh(x - y)}{\cosh x \cosh y}$

7. $\tanh x = 2 \coth 2x - \coth x$ (from 2.5.1.4.3 with $y = x$)

8. $\tanh(x + iy) = \dfrac{\sinh 2x + i \sin 2y}{\cosh 2x + \cos 2y}$

9. $\tanh(x - iy) = \dfrac{\sinh 2x - \sin 2y}{\cosh 2x + \cos 2y}$

2.5.1.5 Hyperbolic sines, cosines, and tangents of multiple arguments.

1. $\sinh 2x = 2 \sinh x \cosh x$
2. $\sinh 3x = 3 \sinh x + 4 \sinh^3 x$
3. $\sinh 4x = \cosh x (4 \sinh x + 8 \sinh^3 x)$
4. $\sinh nx = \displaystyle\sum_{k=1}^{[(n+1)/2]} \binom{n}{2k - 1} \sinh^{2k-1} x \cosh^{n-2k+1} x$

   $[n = 2, 3, \ldots$ with $[(n + 1)/2]$ denoting the *integral part* of $(n + 1)/2]$
5. $\cosh 2x = 2 \cosh^2 x - 1$
6. $\cosh 3x = 4 \cosh^3 x - 3 \cosh x$
7. $\cosh 4x = 8 \cosh^4 x - 8 \cosh^2 x + 1$
8. $\cosh nx = 2^{n-1} \cosh^n x + n \displaystyle\sum_{k=1}^{[n/2]} (-1)^k \dfrac{1}{k} \binom{n - k - 1}{k - 1} 2^{n-2k-1} \cosh^{n-2k} x$

   $\left[n = 2, 3, \ldots \text{ with } [n/2] \text{ denoting the } \textit{integral part} \text{ of } n/2 \right]$

9. $\tanh 2x = \dfrac{2 \tanh x}{1 + \tanh^2 x}$

10. $\tanh 3x = \dfrac{\tanh^3 x + 3 \tanh x}{1 + 3 \tanh^2 x}$

11. $\tanh 4x = \dfrac{4 \tanh^3 x + 4 \tanh x}{1 + 6 \tanh^2 x + \tanh^4 x}$

2.5.1.6 Powers of hyperbolic sines, cosines, and tangents in terms of multiple arguments.

1. $\sinh^2 x = \frac{1}{2}(\cosh 2x - 1)$

2. $\sinh^3 x = \frac{1}{4}(\sinh 3x - 3\sinh x)$

3. $\sinh^4 x = \frac{1}{8}(3 - 4\cosh 2x + \cosh 4x)$

4. $\sinh^{2n-1} x = \dfrac{(-1)^{n-1}}{2^{2n-2}} \displaystyle\sum_{k=0}^{n-1} (-1)^{n+k-1} \binom{2n-1}{k} \sinh(2n - 2k - 1)x$

   $[n = 1, 2, \ldots]$

5. $\sinh^{2n} x = \dfrac{(-1)^n}{2^{2n}} \left\{ \displaystyle\sum_{k=0}^{n-1} (-1)^{n-k} 2\binom{2n}{k} \cosh 2(n - k)x + \binom{2n}{n} \right\}$

   $[n = 1, 2, \ldots]$

6. $\cosh^2 x = \frac{1}{2}(1 + \cosh 2x)$

7. $\cosh^3 x = \frac{1}{4}(3\cosh x + \cosh 3x)$

8. $\cosh^4 x = \frac{1}{8}(3 + 4\cosh 2x + \cosh 4x)$

9. $\cosh^{2n-1} x = \dfrac{1}{2^{2n-2}} \displaystyle\sum_{k=0}^{n-1} \binom{2n-1}{k} \cosh(2n - 2k - 1)x \qquad [n = 1, 2, \ldots]$

10. $\cosh^{2n} x = \dfrac{1}{2^{2n}} \left\{ \displaystyle\sum_{k=0}^{n-1} 2\binom{2n}{k} \cosh 2(n - k)x + \binom{2n}{n} \right\} \qquad [n = 1, 2, \ldots]$

11. $\tanh^2 x = \dfrac{\cosh 2x - 1}{1 + \cosh 2x}$

12. $\tanh^3 x = \dfrac{\sinh 3x - 3\sinh x}{3\cosh x + \cosh 3x}$

13. $\tanh^4 x = \dfrac{3 - 4\cosh 2x + \cosh 4x}{3 + 4\cosh 2x + \cosh 4x}$

14. For $\tanh^n x$ use $\tanh^n x = \sinh^n x / \cosh^n x$ with 2.5.1.6.4 and 2.5.1.6.9, or 2.5.1.6.5 and 2.5.1.6.10.

2.5.1.7 Half-argument representations of hyperbolic functions.

1. $\sinh \dfrac{x}{2} = \pm\left(\dfrac{\cosh x - 1}{2}\right)^{1/2} \qquad [+ \text{ sign if } x > 0 \text{ and } - \text{ sign if } x < 0]$

2. $\cosh \dfrac{x}{2} = \left(\dfrac{1 + \cosh x}{2}\right)^{1/2}$

3. $\tanh \dfrac{x}{2} = \dfrac{\sinh x}{1 + \cosh x} = \dfrac{\cosh x - 1}{\sinh x}$

4. $\coth \dfrac{x}{2} = \dfrac{\sinh x}{\cosh x - 1} = \dfrac{\cosh x + 1}{\sinh x}$

5. $\sinh x = 2 \sinh \dfrac{x}{2} \cosh \dfrac{x}{2}$

6. $\cosh x = \cosh^2 \dfrac{x}{2} + \sinh^2 \dfrac{x}{2}$

7. $\tanh x = \dfrac{2 \tanh \frac{x}{2}}{1 + \tanh^2 \frac{x}{2}} = \dfrac{2 \coth \frac{x}{2}}{\operatorname{csch}^2 \frac{x}{2} + 2}$

2.6 The Logarithm

2.6.1 Series representations

2.6.1.1

1. $\ln(1 + x) = x - \dfrac{1}{2}x^2 + \dfrac{1}{3}x^3 - \dfrac{1}{4}x^4 + \cdots = \displaystyle\sum_{k=1}^{\infty} (-1)^{k+1} \dfrac{x^k}{k}$ $[-1 < x \le 1]$

2. $\ln(1 - x) = -\left[x + \dfrac{x^2}{2} + \dfrac{x^3}{3} + \dfrac{x^4}{4} + \cdots \right] = -\displaystyle\sum_{k=1}^{\infty} \dfrac{x^k}{k}$ $[-1 \le x < 1]$

3. $\ln 2 = 1 - \dfrac{1}{2} + \dfrac{1}{3} - \dfrac{1}{4} + \dfrac{1}{5} - \cdots$

2.6.1.2

1. $\ln x = (x - 1) - \dfrac{1}{2}(x - 1)^2 + \dfrac{1}{3}(x - 1)^3 - \cdots = \displaystyle\sum_{k=1}^{\infty} (-1)^{k+1} \dfrac{(x - 1)^k}{k}$

    $[0 < x \le 2]$

2. $\ln x = 2\left[\dfrac{x - 1}{x + 1} + \dfrac{1}{3}\left(\dfrac{x - 1}{x + 1}\right)^3 + \dfrac{1}{5}\left(\dfrac{x - 1}{x + 1}\right)^5 + \cdots \right]$

    $= 2 \displaystyle\sum_{k=1}^{\infty} \dfrac{1}{2k - 1}\left(\dfrac{x - 1}{x + 1}\right)^{2k-1}$ $[0 < x]$

3. $\ln x = \dfrac{x - 1}{x} + \dfrac{1}{2}\left(\dfrac{x - 1}{x}\right)^2 + \dfrac{1}{3}\left(\dfrac{x - 1}{x}\right)^3 + \cdots = \displaystyle\sum_{k=1}^{\infty} \dfrac{1}{k}\left(\dfrac{x - 1}{x}\right)^k$

    $\left[x \ge \dfrac{1}{2} \right]$

2.6.1.3

1. $\ln\left(\dfrac{1 + x}{1 - x}\right) = 2 \displaystyle\sum_{k=1}^{\infty} \dfrac{1}{2k - 1} x^{2k-1} = 2\operatorname{arctanh} x$ $[x^2 < 1]$

2. $\ln\left(\dfrac{x+1}{x-1}\right) = 2\displaystyle\sum_{k=1}^{\infty}\dfrac{1}{(2k-1)x^{2k-1}} = 2\,\mathrm{arccoth}\,x \qquad [x^2 > 1]$

3. $\ln\left(\dfrac{x}{x-1}\right) = \displaystyle\sum_{k=1}^{\infty}\dfrac{1}{kx^k} \qquad [x \le -1 \text{ or } x > 1]$

4. $\ln\left(\dfrac{1}{1-x}\right) = \displaystyle\sum_{k=1}^{\infty}\dfrac{x^k}{k} \qquad [-1 \le x < 1]$

5. $\left(\dfrac{1-x}{x}\right)\ln\left(\dfrac{1}{1-x}\right) = 1 - \displaystyle\sum_{k=1}^{\infty}\dfrac{x^k}{k(k+1)} \qquad [-1 \le x < 1]$

6. $\left(\dfrac{1}{1-x}\right)\ln\left(\dfrac{1}{1-x}\right) = \displaystyle\sum_{k=1}^{\infty}\left\{x^k\sum_{n=1}^{k}\dfrac{1}{n}\right\} \qquad [x^2 < 1]$

2.6.1.4

1. $\ln\left(1+\sqrt{1+x^2}\right) = \ln 2 + \dfrac{1\cdot 1}{2\cdot 2}x^2 + \dfrac{1\cdot 1\cdot 3}{2\cdot 4\cdot 4}x^4 + \dfrac{1\cdot 1\cdot 3\cdot 5}{2\cdot 4\cdot 6\cdot 6}x^6 - \cdots$

$$= \ln 2 - \sum_{k=1}^{\infty}(-1)^k\dfrac{(2k-1)!}{2^{2k}(k!)^2}x^{2k} \qquad [x^2 \le 1]$$

2. $\ln\left(1+\sqrt{1+x^2}\right) = \ln x + \dfrac{1}{x} - \dfrac{1}{2\cdot 3x^3} + \dfrac{1\cdot 3}{2\cdot 4\cdot 5x^5} - \cdots$

$$= \ln x + \dfrac{1}{x} + \sum_{k=1}^{\infty}(-1)^k\dfrac{(2k-1)!}{2^{2k-1}\cdot k!\,(k-1)!\,(2k+1)x^{2k+1}}$$

$$[x^2 \ge 1]$$

2.6.1.5

1. $\ln\sin x = \ln x - \dfrac{x^2}{6} - \dfrac{x^4}{180} - \dfrac{x^6}{2835} - \cdots$

$$= \ln x + \sum_{k=1}^{\infty}\dfrac{(-1)^k 2^{2k-1}B_{2k}x^{2k}}{k(2k)!} \qquad [0 < x < \pi]$$

2. $\ln\cos x = -\dfrac{x^2}{2} - \dfrac{x^4}{12} - \dfrac{x^6}{45} - \dfrac{17x^8}{2520} - \cdots$

$$= -\sum_{k=1}^{\infty}\dfrac{2^{2k-1}(2^{2k}-1)|B_{2k}|}{k(2k)!}x^{2k} = -\dfrac{1}{2}\sum_{k=1}^{\infty}\dfrac{\sin^{2k}x}{k} \qquad \left[x^2 < \dfrac{\pi^2}{4}\right]$$

3. $\ln\tan x = \ln x + \dfrac{x^2}{3} + \dfrac{7}{90}x^4 + \dfrac{62}{2835}x^6 + \dfrac{127}{18,900}x^8 + \cdots$

$$= \ln x + \sum_{k=1}^{\infty}(-1)^{k+1}\dfrac{(2^{2k-1}-1)2^{2k}B_{2k}x^{2k}}{k(2k)!} \qquad \left[0 < x < \dfrac{\pi}{2}\right]$$

2.7 Inverse Trigonometric and Hyperbolic Functions

2.7.1 Domains of definition and principal values

Other notations are also in use for the *same* inverse trigonometric and hyperbolic functions. Equivalent notations are arcsin, $\sin^{-1}$; arccos, $\cos^{-1}$; arctan, $\tan^{-1}$, arctg; arccot, $\cot^{-1}$, arctg; arcsinh, $\sinh^{-1}$, arsh; arccosh, $\cosh^{-1}$, arch; arctanh, $\tanh^{-1}$, arth; arccoth, $\coth^{-1}$, arcth. Graphs of these functions are shown in Figures 2.8 and 2.9.

Function	Inverse function	Principal value	Domain
$x = \sin y$	$y = \arcsin x$	$-\frac{\pi}{2} \leq \arcsin x \leq \frac{\pi}{2}$	$-1 \leq x \leq 1$
$x = \cos y$	$y = \arccos x$	$0 \leq \arccos x \leq \pi$	$-1 \leq x \leq 1$
$x = \tan y$	$y = \arctan x$	$-\frac{\pi}{2} < \arctan x < \frac{\pi}{2}$	$-\infty < x < \infty$
$x = \cot y$	$y = \text{arccot } x$	$0 < \text{arccot } x < \pi$	$-\infty < x < \infty$
$x = \sinh y$	$y = \text{arcsinh } x$	$-\infty < \text{arcsinh } x < \infty$	$-\infty < x < \infty$
$x = \cosh y$	$y = \text{arccosh } x$	$0 \leq \text{arccosh } x < \infty$	$1 \leq x < \infty$
$x = \tanh y$	$y = \text{arctanh } x$	$-\infty < \text{arctanh } x < \infty$	$-1 < x < 1$
$x = \coth y$	$y = \text{arccoth } x$	$\begin{cases} -\infty < \text{arccoth } x \leq 0 \\ 0 \leq \text{arccoth } x < \infty \end{cases}$	$\begin{matrix} -\infty < x < -1 \\ 1 < x < \infty \end{matrix}$

2.7.2 Functional relations

2.7.2.1 Relationship between trigonometric and inverse trigonometric functions.

1. $\arcsin(\sin x) = x - 2n\pi \qquad \left[2n\pi - \frac{1}{2}\pi \leq x \leq 2n\pi + \frac{1}{2}\pi \right]$

   $= -x + (2n + 1)\pi$

   $$\left[(2n + 1)\pi - \frac{1}{2}\pi \leq x \leq (2n + 1)\pi + \frac{1}{2}\pi \right]$$

2. $\arccos(\cos x) = x - 2n\pi \qquad [2n\pi \leq x \leq (2n + 1)\pi]$

   $= -x + 2(n + 1)\pi \qquad [(2n + 1)\pi \leq x \leq 2(n + 1)\pi]$

3. $\arctan(\tan x) = x - n\pi \qquad \left[n\pi - \frac{1}{2}\pi < x < n\pi + \frac{1}{2}\pi \right]$

4. $\text{arccot}(\cot x) = x - n\pi \qquad [n\pi < x < (n + 1)\pi]$

2.7.2.2 Relationships between inverse trigonometric functions, inverse hyperbolic functions, and the logarithm.

1. $\arcsin z = \frac{1}{i} \ln(iz + \sqrt{1 - z^2}) = \frac{1}{i} \text{arcsinh}(iz)$

2. $\arccos z = \frac{1}{i} \ln(z + \sqrt{z^2 - 1}) = \frac{1}{i} \text{arccosh } z$

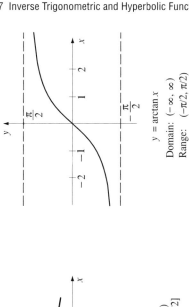

$y = \arctan x$
Domain: $(-\infty, \infty)$
Range: $(-\pi/2, \pi/2)$

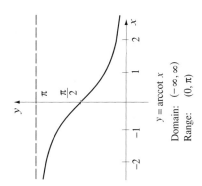

$y = \text{arccot } x$
Domain: $(-\infty, \infty)$
Range: $(0, \pi)$

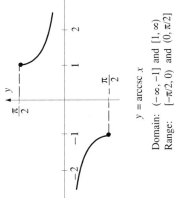

$y = \text{arccsc } x$
Domain: $(-\infty, -1]$ and $[1, \infty)$
Range: $[-\pi/2, 0)$ and $(0, \pi/2]$

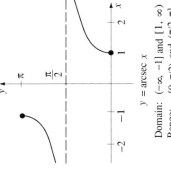

$y = \text{arcsec } x$
Domain: $(-\infty, -1]$ and $[1, \infty)$
Range: $[0, \pi/2)$ and $(\pi/2, \pi]$

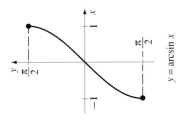

$y = \arcsin x$
Domain: $[-1, 1]$
Range: $[-\pi/2, \pi/2]$

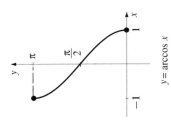

$y = \arccos x$
Domain: $[-1, 1]$
Range: $(0, \pi)$

FIGURE 2.8 ▌ Graphs of inverse trigonometric functions.

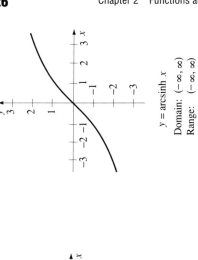

$y = \operatorname{arcsinh} x$
Domain: $(-\infty, \infty)$
Range: $(-\infty, \infty)$

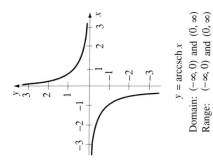

$y = \operatorname{arccsch} x$
Domain: $(-\infty, 0)$ and $(0, \infty)$
Range: $(-\infty, 0)$ and $(0, \infty)$

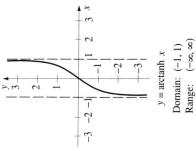

$y = \operatorname{arctanh} x$
Domain: $(-1, 1)$
Range: $(-\infty, \infty)$

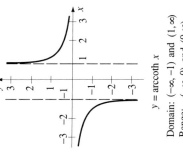

$y = \operatorname{arccoth} x$
Domain: $(-\infty, -1)$ and $(1, \infty)$
Range: $(-\infty, 0)$ and $(0, \infty)$

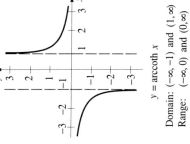

$y = \operatorname{arccosh} x$
Domain: $[1, \infty)$
Range: $[0, \infty)$

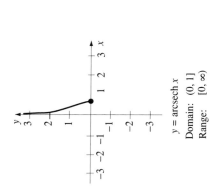

$y = \operatorname{arcsech} x$
Domain: $(0, 1]$
Range: $[0, \infty)$

FIGURE 2.9 ▮ Graphs of inverse hyperbolic functions.

3. $\arctan z = \dfrac{1}{2i} \ln\left(\dfrac{1 + iz}{1 - iz}\right) = \dfrac{1}{i} \operatorname{arctanh}(iz)$

4. $\operatorname{arccot} z = \dfrac{1}{2i} \ln\left(\dfrac{iz - 1}{iz + 1}\right) = i \operatorname{arccoth}(iz)$

5. $\operatorname{arcsinh} z = \ln(z + \sqrt{z^2 + 1}) = \dfrac{1}{i} \arcsin(iz)$

6. $\operatorname{arccosh} z = \ln(z + \sqrt{z^2 - 1}) = i \arccos z$

7. $\operatorname{arctanh} z = \dfrac{1}{2} \ln\left(\dfrac{1 + z}{1 - z}\right) = \dfrac{1}{i} \arctan(iz)$

8. $\operatorname{arccoth} z = \dfrac{1}{2} \ln\left(\dfrac{z + 1}{z - 1}\right) = \dfrac{1}{i} \operatorname{arccot}(-iz)$

2.7.2.3 Relationships between different inverse trigonometric functions.

1. $\arcsin x + \arccos x = \dfrac{\pi}{2}$

2. $\operatorname{arctg} x + \operatorname{arcctg} x = \dfrac{\pi}{2}$

2.7.2.4

1. $\arcsin x = \arccos \sqrt{1 - x^2} \qquad [0 \le x \le 1]$
   $\qquad\quad = -\arccos \sqrt{1 - x^2} \qquad [-1 \le x \le 0]$

2. $\arcsin x = \arctan \dfrac{x}{\sqrt{1 - x^2}} \qquad [x^2 < 1]$

3. $\arcsin x = \operatorname{arccot} \dfrac{\sqrt{1 - x^2}}{x} \qquad [0 < x \le 1]$
   $\qquad\quad = \operatorname{arccot} \dfrac{\sqrt{1 - x^2}}{x} - \pi \qquad [-1 \le x < 0]$

4. $\arccos x = \arcsin \sqrt{1 - x^2} \qquad [0 \le x \le 1]$
   $\qquad\quad = \pi - \arcsin \sqrt{1 - x^2} \qquad [-1 \le x \le 0]$

5. $\arccos x = \arctan \dfrac{\sqrt{1 - x^2}}{x} \qquad [0 < x \le 1]$
   $\qquad\quad = \pi + \operatorname{arccot} \dfrac{\sqrt{1 - x^2}}{x} \qquad [-1 \le x < 0]$

6. $\arccos x = \operatorname{arccot} \dfrac{x}{\sqrt{1 - x^2}} \qquad [-1 \le x < 1]$

7. $\arctan x = \arcsin \dfrac{x}{\sqrt{1+x^2}}$ $[-\infty < x < \infty]$

8. $\arctan x = \arccos \dfrac{1}{\sqrt{1+x^2}}$ $[x \geq 0]$

$\qquad\qquad = -\arccos \dfrac{1}{\sqrt{1+x^2}}$ $[x \leq 0]$

9. $\arctan x = \operatorname{arccot} \dfrac{1}{x}$ $[x > 0]$

$\qquad\qquad = -\operatorname{arccot} \dfrac{1}{x} - \pi$ $[x < 0]$

10. $\arctan x = \arcsin \dfrac{1}{\sqrt{1+x^2}}$ $[x > 0]$

$\qquad\qquad = \pi - \arcsin \dfrac{1}{\sqrt{1+x^2}}$ $[x < 0]$

11. $\operatorname{arccot} x = \arccos \dfrac{x}{\sqrt{1+x^2}}$

12. $\operatorname{arccot} x = \arctan \dfrac{1}{x}$ $[x > 0]$

$\qquad\qquad = \pi + \arctan \dfrac{1}{x}$ $[x < 0]$

2.7.2.5 Relationships between the inverse hyperbolic functions.

1. $\operatorname{arcsinh} x = \operatorname{arccosh} \sqrt{x^2+1} = \operatorname{arctanh} \dfrac{x}{\sqrt{x^2+1}}$

2. $\operatorname{arccosh} x = \operatorname{arcsinh} \sqrt{x^2-1} = \operatorname{arctanh} \dfrac{\sqrt{x^2-1}}{x}$

3. $\operatorname{arctanh} x = \operatorname{arcsinh} \dfrac{x}{\sqrt{1-x^2}} = \operatorname{arccosh} \dfrac{1}{\sqrt{1-x^2}} = \operatorname{arccoth} \dfrac{1}{x}$

4. $\operatorname{arcsinh} x \pm \operatorname{arcsinh} y = \operatorname{arcsinh}(x\sqrt{1+y^2} \pm y\sqrt{1+x^2})$

5. $\operatorname{arccosh} x \pm \operatorname{arccosh} y = \operatorname{arccosh}(xy \pm \sqrt{(x^2-1)(y^2-1)})$

6. $\operatorname{arctanh} x \pm \operatorname{arctanh} y = \operatorname{arctanh} \dfrac{x \pm y}{1 \pm xy}$

2.8 Series Representations of Trigonometric and Hyperbolic Functions

2.8.1 Trigonometric functions

2.8.1.1

1. $\sin x = x - \dfrac{x^3}{3!} + \dfrac{x^5}{5!} - \dfrac{x^7}{7!} + \cdots = \displaystyle\sum_{k=0}^{\infty} (-1)^k \dfrac{x^{2k+1}}{(2k+1)!}$

2. $\cos x = 1 - \dfrac{x^2}{2!} + \dfrac{x^4}{4!} - \dfrac{x^6}{6!} + \cdots = \displaystyle\sum_{k=0}^{\infty} (-1)^k \dfrac{x^{2k}}{(2k)!}$

3. $\tan x = x + \dfrac{x^3}{3} + \dfrac{2}{15}x^5 + \dfrac{17}{315}x^7 \cdots = \displaystyle\sum_{k=1}^{\infty} (-1)^{k+1} \dfrac{2^{2k}(2^{2k}-1)}{(2k)!} B_{2k} x^{2k-1}$

$$\left[|x| < \dfrac{\pi}{2}; \text{ see } 1.3.1.4.3 \right]$$

4. $\csc x = \dfrac{1}{x} + \dfrac{x}{6} + \dfrac{7}{360}x^3 + \dfrac{31}{51120}x^5 + \cdots$

$$= \dfrac{1}{x} + \sum_{k=1}^{\infty} \dfrac{2(2^{2k-1}-1)|B_{2k}|}{(2k)!} x^{2k-1} \qquad [|x| < \pi]$$

5. $\sec x = 1 + \dfrac{x^2}{2} + \dfrac{5}{24}x^4 + \dfrac{61}{720}x^6 + \cdots$

$$= 1 + \sum_{k=1}^{\infty} \dfrac{E_{2k}}{(2k)!} x^{2k} \qquad \left[|x| < \dfrac{\pi}{2}; \text{ see } 1.3.1.1.5 \right]$$

6. $\cot x = \dfrac{1}{x} - \dfrac{x}{3} - \dfrac{x^3}{45} - \dfrac{2x^5}{945} - \cdots$

$$= \dfrac{1}{x} - \sum_{k=1}^{\infty} -\dfrac{2^{2k}|B_{2k}|}{(2k)!} x^{2k-1} \qquad [|x| < \pi]$$

7. $\sin^2 x = x^2 - \dfrac{x^4}{3} + \dfrac{2x^6}{45} - \dfrac{x^8}{315} + \cdots = \displaystyle\sum_{k=1}^{\infty} (-1)^{k+1} \dfrac{2^{2k-1}x^{2k}}{(2k)!}$

8. $\cos^2 x = 1 - x^2 + \dfrac{x^4}{3} - \dfrac{2x^6}{45} + \dfrac{x^8}{315} - \cdots = 1 - \displaystyle\sum_{k=1}^{\infty} \dfrac{2^{2k-1}x^{2k}}{(2k)!}$

9. $\tan^2 x = x^2 + \dfrac{2x^4}{3} + \dfrac{17}{45}x^6 + \dfrac{62}{315}x^8 + \cdots \qquad \left[|x| < \dfrac{\pi}{2} \right]$

2.8.2 Hyperbolic functions

2.8.2.1

1. $\sinh x = x + \dfrac{x^3}{3!} + \dfrac{x^5}{5!} + \dfrac{x^7}{7!} + \cdots = \displaystyle\sum_{k=0}^{\infty} \dfrac{x^{2k+1}}{(2k+1)!}$

2. $\cosh x = 1 + \dfrac{x^2}{2!} + \dfrac{x^4}{4!} + \dfrac{x^6}{6!} + \cdots = \displaystyle\sum_{k=0}^{\infty} \dfrac{x^{2k}}{(2k)!}$

3. $\tanh x = x - \dfrac{x^3}{3} + \dfrac{2}{15}x^5 - \dfrac{17}{315}x^7 + \cdots = \displaystyle\sum_{k=1}^{\infty} \dfrac{2^{2k}(2^{2k}-1)}{(2k)!} B_{2k} x^{2k-1}$

$$\left[|x| < \frac{\pi}{2} \right]$$

4. $\operatorname{csch} x = \dfrac{1}{x} - \dfrac{x}{6} + \dfrac{7}{360}x^3 - \dfrac{31}{15120}x^5 + \cdots = \dfrac{1}{x} - \displaystyle\sum_{k=1}^{\infty} \dfrac{2(2^{2k-1}-1)B_{2k}}{(2k)!} x^{2k-1}$

$$[|x| < \pi]$$

5. $\operatorname{sech} x = 1 - \dfrac{x^2}{2!} + \dfrac{5}{4!}x^4 - \dfrac{61}{6!}x^6 + \cdots = 1 + \displaystyle\sum_{k=1}^{\infty} \dfrac{E_{2k}}{(2k)!} x^{2k}$ $\left[|x| < \dfrac{\pi}{2} \right]$

6. $\coth x = \dfrac{1}{x} + \dfrac{x}{3} - \dfrac{x^3}{45} + \dfrac{2}{945}x^5 - \cdots = \dfrac{1}{x} + \displaystyle\sum_{k=1}^{\infty} \dfrac{2^{2k} B_{2k}}{(2k)!} x^{2k-1}$ $[|x| < \pi]$

7. $\sinh^2 x = x^2 + \dfrac{x^4}{3} + \dfrac{2x^6}{45} + \dfrac{x^8}{315} + \cdots = \displaystyle\sum_{k=1}^{\infty} \dfrac{2^{2k-1} x^{2k}}{(2k)!}$

8. $\cosh^2 x = 1 + x^2 + \dfrac{x^4}{3} + \dfrac{2x^6}{45} + \dfrac{x^8}{315} + \cdots = 1 + \displaystyle\sum_{k=1}^{\infty} \dfrac{2^{2k-1} x^{2k}}{(2k)!}$

9. $\tanh^2 x = x^2 - \dfrac{2x^4}{3} + \dfrac{17}{45}x^6 - \dfrac{62}{315}x^8 + \cdots$

2.8.3 Inverse trigonometric functions

2.8.3.1

1. $\arcsin x = x + \dfrac{1}{2 \cdot 3}x^3 + \dfrac{1 \cdot 3}{2 \cdot 4 \cdot 5}x^5 + \dfrac{1 \cdot 3 \cdot 5}{2 \cdot 4 \cdot 6 \cdot 7}x^7 + \cdots$

$$= \sum_{k=0}^{\infty} \dfrac{(2k)! x^{2k+1}}{2^{2k}(k!)^2(2k+1)} \qquad [|x| < 1]$$

2. $\arccos x = \dfrac{\pi}{2} - \left(x + \dfrac{x^3}{2 \cdot 3} + \dfrac{1 \cdot 3}{2 \cdot 4 \cdot 5}x^5 + \dfrac{1 \cdot 3 \cdot 5}{2 \cdot 4 \cdot 6 \cdot 7}x^7 + \cdots \right)$

$$= \dfrac{\pi}{2} - \sum_{k=0}^{\infty} \dfrac{(2k)! x^{2k+1}}{2^{2k}(k!)^2(2k+1)} \qquad [|x| < 1]$$

3. $\arctan x = x - \dfrac{x^3}{3} + \dfrac{x^5}{5} - \dfrac{x^7}{7} + \cdots = \displaystyle\sum_{k=0}^{\infty} \dfrac{(-1)^k x^{2k+1}}{2k+1}$ $[|x| < 1]$

    [obtained by integrating $1/(1 + x^2)$; see 0.7.3]

4. $\arctan x = \dfrac{\pi}{2} - \dfrac{1}{x} + \dfrac{1}{3x^3} - \dfrac{1}{5x^5} + \dfrac{1}{7x^7} - \cdots = \dfrac{\pi}{2} - \displaystyle\sum_{k=0}^{\infty}(-1)^k \dfrac{1}{(2k+1)x^{2k+1}}$

    $[x^2 \geq 1]$

5. $\arctan x = \dfrac{x}{\sqrt{1+x^2}} \left\{ 1 + \dfrac{1}{6}\left(\dfrac{x^2}{1+x^2}\right) + \dfrac{3}{40}\left(\dfrac{x^2}{1+x^2}\right)^2 \right.$

    $\left. + \dfrac{5}{112}\left(\dfrac{x^2}{1+x^2}\right)^3 + \cdots \right\}$

    $= \dfrac{x}{\sqrt{1+x^2}} \displaystyle\sum_{k=0}^{\infty} \dfrac{(2k)!}{2^{2k}(k!)^2(2k+1)} \left(\dfrac{x^2}{1+x^2}\right)^k$ $[x^2 < \infty]$

6. $\text{arcsec } x = \dfrac{\pi}{2} - \dfrac{1}{x} - \dfrac{1}{2 \cdot 3x^3} - \dfrac{1 \cdot 3}{2 \cdot 4 \cdot 5x^5} - \cdots = \dfrac{\pi}{2} - \displaystyle\sum_{k=0}^{\infty} \dfrac{(2k)! \, x^{-(2k+1)}}{(k!)^2 2^{2k}(2k+1)}$

    $[x^2 > 1]$

7. $\text{arccsc } x = \dfrac{1}{x} + \dfrac{1}{2 \cdot 3x^3} + \dfrac{1 \cdot 3}{2 \cdot 4 \cdot 5x^5} + \cdots = \displaystyle\sum_{k=0}^{\infty} \dfrac{(2k)! \, x^{-(2k+1)}}{(k!)^2 2^{2k}(2k+1)}$ $[x^2 > 1]$

2.8.4 Inverse hyperbolic functions

2.8.4.1

1. $\text{arcsinh } x = x - \dfrac{1}{2 \cdot 3}x^3 + \dfrac{1 \cdot 3}{2 \cdot 4 \cdot 5}x^5 - \cdots = \displaystyle\sum_{k=0}^{\infty}(-1)^k \dfrac{(2k)! \, x^{2k+1}}{2^{2k}(k!)^2(2k+1)}$

    $= \ln[x + \sqrt{(x^2+1)}]$ $[|x| \leq 1]$

2. $\text{arcsinh } x = \ln 2x + \dfrac{1}{2}\dfrac{1}{2x^2} - \dfrac{1 \cdot 3}{2 \cdot 4}\dfrac{1}{4x^4} + \cdots$

    $= \ln 2x + \displaystyle\sum_{k=1}^{\infty}(-1)^{k+1} \dfrac{(2k)! \, x^{-2k}}{2^{2k}(k!)^2 2k}$ $[x \geq 1]$

3. $\text{arcsinh } x = -\ln|2x| - \dfrac{1}{2} \cdot \dfrac{1}{2x^2} + \dfrac{1 \cdot 3}{2 \cdot 4}\dfrac{1}{4x^4} - \cdots$

    $= -\ln|2x| - \displaystyle\sum_{k=1}^{\infty}(-1)^{k+1} \dfrac{(2k)! \, x^{-2k}}{2^{2k}(k!)^2 2k}$ $[x < -1]$

4. $\arccosh x = \ln 2x - \dfrac{1}{2}\dfrac{1}{2x^2} - \dfrac{1\cdot 3}{2\cdot 4}\dfrac{1}{4x^4} - \cdots = \ln 2x - \displaystyle\sum_{k=1}^{\infty}\dfrac{(2k)!\,x^{-2k}}{2^{2k}(k!)^2 2k}$

$= \ln[x + \sqrt{(x^2-1)}]$ $[x \geq 1]$

5. $\arctanh x = x + \dfrac{x^3}{3} + \dfrac{x^5}{5} + \dfrac{x^7}{7} + \cdots = \displaystyle\sum_{k=0}^{\infty}\dfrac{x^{2k+1}}{2k+1}$

$= \dfrac{1}{2}\ln\left(\dfrac{1+x}{1-x}\right)$ $[|x| < 1]$

6. $\arccoth x = \dfrac{1}{x} + \dfrac{1}{3x^3} + \dfrac{1}{5x^5} + \dfrac{1}{7x^7} + \cdots = \displaystyle\sum_{k=0}^{\infty}\dfrac{1}{(2k+1)x^{2k+1}}$

$= \dfrac{1}{2}\ln\left(\dfrac{x+1}{x-1}\right)$ $[|x| > 1]$

2.9 Useful Limiting Values and Inequalities Involving Elementary Functions

2.9.1 Logarithmic functions

1. $\dfrac{x}{1+x} < \ln(1+x) < x$ $[x \neq 0, -1 < x]$

2. $x < -\ln(1-x) < \dfrac{x}{1-x}$ $[x \neq 0, x < 1]$

3. $\displaystyle\lim_{x\to\infty}[x^{-\alpha}\ln x] = 0$ $[\alpha > 0]$

4. $\displaystyle\lim_{x\to 0}[x^{\alpha}\ln x] = 0$ $[\alpha > 0]$

2.9.2 Exponential functions

1. $1 + x < e^x$

2. $e^x < \dfrac{1}{1-x}$ $[x < 1]$

3. $\left(1 + \dfrac{x}{n}\right)^n < e^x$ $[n > 0]$

4. $\dfrac{x}{1+x} < 1 - e^{-x} < x$ $[-1 < x]$

5. $x < e^x - 1 < \dfrac{x}{1-x}$ $[x < 1]$

6. $\lim\limits_{x\to\infty} (x^{\alpha} e^{-x}) = 0$

7. $\lim\limits_{n\to\infty} \left(1 + \dfrac{x}{n}\right)^{n} = e^{x}$

2.9.3 Trigonometric and hyperbolic functions

1. $\dfrac{\sin x}{x} > \dfrac{2}{\pi}$ $[|x| < \pi/2]$

2. $\sin x \leq x \leq \tan x$ $[0 \leq x < \pi/2]$

3. $\cos x \leq \dfrac{\sin x}{x} \leq 1$ $[0 \leq x \leq \pi]$

4. $\cot x \leq \dfrac{1}{x} \leq \operatorname{cosec} x$ $[0 \leq x \leq \pi/2]$

If $z = x + iy$, then:

5. $|\sinh y| \leq |\sin z| \leq \cosh y$

6. $|\sinh y| \leq |\cos z| \leq \cosh y$

7. $|\sin z| \leq \sinh |z|$

8. $|\cos z| \leq \cosh |z|$

9. $\sin|x| \leq |\cosh z| \leq \cosh x$

10. $\sinh|x| \leq |\sinh z| \leq \cosh x$

11. $\lim\limits_{x\to 0} \left(\dfrac{\sin kx}{x}\right) = k$

12. $\lim\limits_{x\to 0} \left(\dfrac{\tan kx}{x}\right) = k$

13. $\lim\limits_{n\to\infty} \left(n \sin \dfrac{x}{n}\right) = x$

14. $\lim\limits_{n\to\infty} \left(n \tan \dfrac{x}{n}\right) = x$

3

Derivatives of Elementary Functions

3.1 Derivatives of Algebraic, Logarithmic, and Exponential Functions

3.1.1

Let $u(x)$ be a differentiable function with respect to x, and α, a, and k be constants.

1. $\dfrac{d}{dx}[x^\alpha] = \alpha x^{\alpha-1}$

2. $\dfrac{d}{dx}[u^\alpha] = \alpha u^{\alpha-1}\dfrac{du}{dx}$

3. $\dfrac{d}{dx}[x^{1/2}] = \dfrac{1}{2x^{1/2}}$

4. $\dfrac{d}{dx}[u^{1/2}] = \dfrac{1}{2u^{1/2}}\dfrac{du}{dx}$

5. $\dfrac{d}{dx}[x^{-\alpha}] = -\dfrac{\alpha}{x^{\alpha+1}}$

6. $\dfrac{d}{dx}[u^{-\alpha}] = -\dfrac{\alpha}{u^{\alpha+1}}\dfrac{du}{dx}$

7. $\dfrac{d}{dx}[\ln x] = \dfrac{1}{x}$

8. $\dfrac{d}{dx}[\ln u] = \dfrac{1}{u}\dfrac{du}{dx}$

9. $\dfrac{d^n}{dx^n}[\ln x] = \dfrac{(-1)^{n-1}(n-1)!}{x^n}$

10. $\dfrac{d^2}{dx^2}[\ln u] = -\dfrac{1}{u^2}\left(\dfrac{du}{dx}\right)^2 + \dfrac{1}{u}\dfrac{d^2u}{dx^2}$

11. $\dfrac{d}{dx}[x \ln x] = \ln x + 1$

12. $\dfrac{d}{dx}[x \ln u] = \ln u + \dfrac{x}{u}\dfrac{du}{dx}$

13. $\dfrac{d}{dx}[x^n \ln x] = (n \ln x + 1)x^{n-1}$

14. $\dfrac{d}{dx}[u \ln u] = (\ln u + 1)\dfrac{du}{dx}$

15. $\dfrac{d}{dx}[e^x] = e^x$

16. $\dfrac{d}{dx}[e^{kx}] = ke^{kx}$

17. $\dfrac{d}{dx}[e^u] = e^u \dfrac{du}{dx}$

18. $\dfrac{d}{dx}[a^x] = a^x \ln a$

19. $\dfrac{d}{dx}[a^u] = a^u (\ln a)\dfrac{du}{dx}$

20. $\dfrac{d}{dx}[x^x] = (1 + \ln x)x^x$

3.2 Derivatives of Trigonometric Functions

3.2.1

Let $u(x)$ be a differentiable function with respect to x.

1. $\dfrac{d}{dx}[\sin x] = \cos x$

2. $\dfrac{d}{dx}[\sin u] = \cos u \dfrac{du}{dx}$

3. $\dfrac{d}{dx}[\cos x] = -\sin x$

4. $\dfrac{d}{dx}[\cos u] = -\sin u \dfrac{du}{dx}$

5. $\dfrac{d}{dx}[\tan x] = \sec^2 x$

6. $\dfrac{d}{dx}[\tan u] = \sec^2 u \dfrac{du}{dx}$

7. $\dfrac{d}{dx}[\csc x] = -\csc x \cot x$

8. $\dfrac{d}{dx}[\csc u] = -\csc u \cot u \dfrac{du}{dx}$

9. $\dfrac{d}{dx}[\sec x] = \sec x \tan x$

10. $\dfrac{d}{dx}[\sec u] = \sec u \tan u \dfrac{du}{dx}$

11. $\dfrac{d}{dx}[\cot x] = -\csc^2 x$

12. $\dfrac{d}{dx}[\cot u] = -\csc^2 u \dfrac{du}{dx}$

13. $\dfrac{d^n}{dx^n}[\sin x] = \sin\left(x + \dfrac{1}{2}n\pi\right)$

14. $\dfrac{d^n}{dx^n}[\cos x] = \cos\left(x + \dfrac{1}{2}n\pi\right)$

3.3 Derivatives of Inverse Trigonometric Functions

3.3.1

Let $u(x)$ be a differentiable function with respect to x.

1. $\dfrac{d}{dx}\left[\arcsin \dfrac{x}{a}\right] = \dfrac{1}{(a^2 - x^2)^{1/2}}$ $\qquad \left[-\dfrac{\pi}{2} < \arcsin \dfrac{x}{a} < \dfrac{\pi}{2}\right]$

2. $\dfrac{d}{dx}\left[\arcsin\dfrac{u}{a}\right] = \dfrac{1}{(a^2-u^2)^{1/2}}\dfrac{du}{dx}$ $\qquad \left[-\dfrac{\pi}{2} < \arcsin\dfrac{u}{a} < \dfrac{\pi}{2}\right]$

3. $\dfrac{d}{dx}\left[\arccos\dfrac{x}{a}\right] = \dfrac{-1}{(a^2-x^2)^{1/2}}$ $\qquad \left[0 < \arccos\dfrac{x}{a} < \pi\right]$

4. $\dfrac{d}{dx}\left[\arccos\dfrac{u}{a}\right] = \dfrac{-1}{(a^2-u^2)^{1/2}}\dfrac{du}{dx}$ $\qquad \left[0 < \arccos\dfrac{u}{a} < \pi\right]$

5. $\dfrac{d}{dx}\left[\arctan\dfrac{x}{a}\right] = \dfrac{a}{a^2+x^2}$

6. $\dfrac{d}{dx}\left[\arctan\dfrac{u}{a}\right] = \dfrac{a}{a^2+u^2}\dfrac{du}{dx}$

7. $\dfrac{d}{dx}\left[\operatorname{arccsc}\dfrac{x}{a}\right] = \dfrac{-a}{x(x^2-a^2)^{1/2}}$ $\qquad \left[0 < \operatorname{arccsc}\dfrac{x}{a} < \dfrac{\pi}{2}\right]$

8. $\dfrac{d}{dx}\left[\operatorname{arccsc}\dfrac{x}{a}\right] = \dfrac{a}{x(x^2-a^2)^{1/2}}$ $\qquad \left[-\dfrac{\pi}{2} < \operatorname{arccsc}\dfrac{x}{a} < 0\right]$

9. $\dfrac{d}{dx}\left[\operatorname{arccsc}\dfrac{u}{a}\right] = \dfrac{-a}{u(u^2-a^2)^{1/2}}\dfrac{du}{dx}$ $\qquad \left[0 < \operatorname{arccsc}\dfrac{u}{a} < \dfrac{\pi}{2}\right]$

10. $\dfrac{d}{dx}\left[\operatorname{arccsc}\dfrac{u}{a}\right] = \dfrac{a}{u(u^2-a^2)^{1/2}}\dfrac{du}{dx}$ $\qquad \left[-\dfrac{\pi}{2} < \operatorname{arccsc}\dfrac{u}{a} < 0\right]$

11. $\dfrac{d}{dx}\left[\operatorname{arcsec}\dfrac{x}{a}\right] = \dfrac{a}{x(x^2-a^2)^{1/2}}$ $\qquad \left[0 < \operatorname{arcsec}\dfrac{x}{a} < \dfrac{\pi}{2}\right]$

12. $\dfrac{d}{dx}\left[\operatorname{arcsec}\dfrac{x}{a}\right] = \dfrac{-a}{x(x^2-a^2)^{1/2}}$ $\qquad \left[\dfrac{\pi}{2} < \operatorname{arcsec}\dfrac{x}{a} < \pi\right]$

13. $\dfrac{d}{dx}\left[\operatorname{arcsec}\dfrac{u}{a}\right] = \dfrac{a}{u(u^2-a^2)^{1/2}}\dfrac{du}{dx}$ $\qquad \left[0 < \operatorname{arcsec}\dfrac{u}{a} < \dfrac{\pi}{2}\right]$

14. $\dfrac{d}{dx}\left[\operatorname{arcsec}\dfrac{u}{a}\right] = \dfrac{-a}{u(u^2-a^2)^{1/2}}\dfrac{du}{dx}$ $\qquad \left[\dfrac{\pi}{2} < \operatorname{arcsec}\dfrac{u}{a} < \pi\right]$

3.4 Derivatives of Hyperbolic Functions

3.4.1

Let $u(x)$ be a differentiable function with respect to x.

1. $\dfrac{d}{dx}[\sinh x] = \cosh x$ $\qquad\qquad$ 2. $\dfrac{d}{dx}[\sinh u] = \cosh u \dfrac{du}{dx}$

3. $\dfrac{d}{dx}[\cosh x] = \sinh x$

4. $\dfrac{d}{dx}[\cosh u] = \sinh u \dfrac{du}{dx}$

5. $\dfrac{d}{dx}[\tanh x] = \operatorname{sech}^2 x$

6. $\dfrac{d}{dx}[\tanh u] = \operatorname{sech}^2 u \dfrac{du}{dx}$

7. $\dfrac{d}{dx}[\operatorname{csch} x] = -\operatorname{csch} x \coth x$

8. $\dfrac{d}{dx}[\operatorname{csch} u] = -\operatorname{csch} u \coth u \dfrac{du}{dx}$

9. $\dfrac{d}{dx}[\operatorname{sech} x] = -\operatorname{sech} x \tanh x$

10. $\dfrac{d}{dx}[\operatorname{sech} u] = -\operatorname{sech} u \tanh u \dfrac{du}{dx}$

11. $\dfrac{d}{dx}[\coth x] = -\operatorname{csch}^2 x$

12. $\dfrac{d}{dx}[\coth u] = \operatorname{csch}^2 u \dfrac{du}{dx}$

3.5 Derivatives of Inverse Hyperbolic Functions

3.5.1

Let $u(x)$ be a differentiable function with respect to x

1. $\dfrac{d}{dx}\left[\operatorname{arcsinh}\dfrac{x}{a}\right] = \dfrac{1}{(x^2 + a^2)^{1/2}}$

2. $\dfrac{d}{dx}\left[\operatorname{arcsinh}\dfrac{u}{a}\right] = \dfrac{1}{(u^2 + a^2)^{1/2}}\dfrac{du}{dx}$

3. $\dfrac{d}{dx}\left[\operatorname{arccosh}\dfrac{x}{a}\right] = \dfrac{1}{(x^2 - a^2)^{1/2}}$ $\left[\dfrac{x}{a} > 1, \operatorname{arccosh}\dfrac{x}{a} > 0\right]$

4. $\dfrac{d}{dx}\left[\operatorname{arccosh}\dfrac{x}{a}\right] = \dfrac{-1}{(x^2 - a^2)^{1/2}}$ $\left[\dfrac{x}{a} > 1, \operatorname{arccosh}\dfrac{x}{a} < 0\right]$

5. $\dfrac{d}{dx}\left[\operatorname{arccosh}\dfrac{u}{a}\right] = \dfrac{1}{(u^2 - a^2)^{1/2}}\dfrac{du}{dx}$ $\left[\dfrac{u}{a} > 1, \operatorname{arccosh}\dfrac{u}{a} > 0\right]$

6. $\dfrac{d}{dx}\left[\operatorname{arccosh}\dfrac{u}{a}\right] = \dfrac{-1}{(u^2 - a^2)^{1/2}}\dfrac{du}{dx}$ $\left[\dfrac{u}{a} > 1, \operatorname{arccosh}\dfrac{u}{a} < 0\right]$

7. $\dfrac{d}{dx}\left[\operatorname{arctanh}\dfrac{x}{a}\right] = \dfrac{a}{a^2 - x^2}$ $[x^2 < a^2]$

8. $\dfrac{d}{dx}\left[\operatorname{arctanh}\dfrac{u}{a}\right] = \dfrac{a}{a^2 - u^2}\dfrac{du}{dx}$ $[u^2 < a^2]$

9. $\dfrac{d}{dx}\left[\operatorname{arccsch}\dfrac{x}{a}\right] = \dfrac{-a}{|x|(x^2 + a^2)^{1/2}}$ $[x \neq 0]$

10. $\dfrac{d}{dx}\left[\operatorname{arccsch}\dfrac{u}{a}\right] = \dfrac{-a}{|u|(u^2 + a^2)^{1/2}}\dfrac{du}{dx}$ $[u \neq 0]$

11. $\dfrac{d}{dx}\left[\text{arcsech } \dfrac{x}{a}\right] = \dfrac{-a}{x(a^2 - x^2)^{1/2}}$ $\quad \left[0 < \dfrac{x}{a} < 1, \text{arcsech } \dfrac{x}{a} > 0\right]$

12. $\dfrac{d}{dx}\left[\text{arcsech } \dfrac{x}{a}\right] = \dfrac{a}{x(a^2 - x^2)^{1/2}}$ $\quad \left[0 < \dfrac{x}{a} < 1, \text{arcsech } \dfrac{x}{a} < 0\right]$

13. $\dfrac{d}{dx}\left[\text{arcsech } \dfrac{u}{a}\right] = \dfrac{-a}{u(a^2 - u^2)^{1/2}}\dfrac{du}{dx}$ $\quad \left[0 < \dfrac{u}{a} < 1, \text{arcsech } \dfrac{u}{a} > 0\right]$

14. $\dfrac{d}{dx}\left[\text{arcsech } \dfrac{u}{a}\right] = \dfrac{a}{u(a^2 - u^2)^{1/2}}\dfrac{du}{dx}$ $\quad \left[0 < \dfrac{u}{a} < 1, \text{arcsech } \dfrac{u}{a} < 0\right]$

15. $\dfrac{d}{dx}\left[\text{arccoth } \dfrac{x}{a}\right] = \dfrac{a}{a^2 - x^2}$ $\quad [x^2 > a^2]$

16. $\dfrac{d}{dx}\left[\text{arccoth } \dfrac{u}{a}\right] = \dfrac{a}{a^2 - u^2}\dfrac{du}{dx}$ $\quad [u^2 > a^2]$

CHAPTER 4

Indefinite Integrals of Algebraic Functions

4.1 Algebraic and Transcendental Functions

4.1.1 Definitions

4.1.1.1 A function $f(x)$ is said to be **algebraic** if a polynomial $P(x, y)$ in the two variables x, y can be found with the property that $P(x, f(x)) = 0$, for all x for which $f(x)$ is defined. Thus, the function

$$f(x) = x^2 - (1 - x^4)^{1/2}$$

is an *algebraic function*, because the polynomial

$$P(x, y) = y^2 - 2x^2 y + 2x^4 - 1$$

has the necessary property.

Functions that are not algebraic are called **transcendental functions**. Examples of transcendental functions are

$$f(x) = \sin x, \quad f(x) = e^x + \ln x, \quad \text{and} \quad f(x) = \tan x + (1 - x^2)^{1/2}.$$

141

A fundamental difference between algebraic and transcendental functions is that whereas algebraic functions can only have a finite number of zeros, a transcendental function can have an infinite number. Thus, for example, $\sin x$ has zeros at $x = \pm n\pi$, $n = 0, 1, 2, \ldots$. Transcendental functions arise in many different ways, one of which is as a result of integrating algebraic functions in, for example, the following cases:

$$\int \frac{dx}{x} = \ln x, \quad \int \frac{dx}{(x^2 - a^2)^{1/2}} = \ln|x + (x^2 - a^2)^{1/2}|, \quad \text{and}$$

$$\int \frac{dx}{x^2 + a^2} = \frac{1}{a} \arctan\left(\frac{x}{a}\right).$$

Within the class of algebraic functions there is an important and much simpler subclass called **rational functions**. A function $f(x)$ is said to be a **rational function** if

$$f(x) = \frac{P(x)}{Q(x)},$$

where $P(x)$ and $Q(x)$ are both polynomials (see 1.7.1.1). These simple algebraic functions can be integrated by first expressing their integrands in terms of partial fractions (see 1.7), and then integrating the result term by term. Hereafter, for convenience of reference, indefinite integrals will be classified according to their integrands as rational functions, nonrational (irrational) algebraic functions, or transcendental functions.

4.2 Indefinite Integrals of Rational Functions

4.2.1 Integrands involving x^n

4.2.1.1

1. $\displaystyle\int dx = x$

2. $\displaystyle\int x\, dx = \frac{1}{2}x^2$

3. $\displaystyle\int x^2\, dx = \frac{1}{3}x^3$

4. $\displaystyle\int x^n\, dx = \frac{x^{n+1}}{n+1}$ $\qquad [n \neq -1]$

5. $\displaystyle\int \frac{dx}{x} = \ln|x|$

6. $\displaystyle\int \frac{dx}{x^2} = -\frac{1}{x}$

7. $\displaystyle\int \frac{dx}{x^3} = -\frac{1}{2x^2}$

8. $\displaystyle\int \frac{dx}{x^n} = -\frac{1}{(n-1)x^{n-1}}$ $\qquad [n \neq 1]$

4.2.2 Integrands involving $a + bx$

4.2.2.1

1. $\displaystyle\int (a + bx)^n\, dx = \frac{(a + bx)^{n+1}}{b(n+1)}$ $\qquad [n \neq -1]$

2. $\displaystyle\int \frac{dx}{a + bx} = \frac{1}{b} \ln|a + bx|$

3. $\displaystyle\int \frac{dx}{(a+bx)^2} = -\frac{1}{b(a+bx)}$

4. $\displaystyle\int \frac{dx}{(a+bx)^3} = -\frac{1}{2b(a+bx)^2}$

5. $\displaystyle\int \frac{dx}{(a+bx)^n} = -\frac{1}{b(n-1)(a+bx)^{n-1}}$ $[n \neq 1]$

4.2.2.2

1. $\displaystyle\int \frac{x\,dx}{a+bx} = \frac{x}{b} - \frac{a}{b^2}\ln|a+bx|$ (see 4.2.2.6.1)

2. $\displaystyle\int \frac{x\,dx}{(a+bx)^2} = -\frac{x}{b(a+bx)} + \frac{1}{b^2}\ln|a+bx|$ (see 4.2.2.6.2)

3. $\displaystyle\int \frac{x\,dx}{(a+bx)^3} = -\left(\frac{x}{b} + \frac{a}{2b^2}\right)\frac{1}{(a+bx)^2}$

4. $\displaystyle\int \frac{x\,dx}{(a+bx)^n} = \frac{x}{b(2-n)(a+bx)^{n-1}} - \frac{a}{b(2-n)}\int \frac{dx}{(a+bx)^n}$

                [reduction formula for $n \neq 2$]

4.2.2.3

1. $\displaystyle\int \frac{x^2\,dx}{a+bx} = \frac{x^2}{2b} - \frac{ax}{b^2} + \frac{a^2}{b^3}\ln|a+bx|$ (see 4.2.2.6.1)

2. $\displaystyle\int \frac{x^2\,dx}{(a+bx)^2} = \frac{x}{b^2} - \frac{a^2}{b^3(a+bx)} - \frac{2a}{b^3}\ln|a+bx|$ (see 4.2.2.6.2)

3. $\displaystyle\int \frac{x^2\,dx}{(a+bx)^3} = \left(\frac{2ax}{b^2} + \frac{3a^2}{2b^3}\right)\frac{1}{(a+bx)^2} + \frac{1}{b^3}\ln|a+bx|$

4. $\displaystyle\int \frac{x^2\,dx}{(a+bx)^n} = \frac{x^2}{b(3-n)(a+bx)^{n-1}} - \frac{2a}{b(3-n)}\int \frac{x\,dx}{(a+bx)^n}$

                [reduction formula for $n \neq 3$]

4.2.2.4

1. $\displaystyle\int \frac{x^3\,dx}{(a+bx)} = \frac{x^3}{3b} - \frac{ax^2}{2b^2} + \frac{a^2x}{b^3} - \frac{a^3}{b^4}\ln|a+bx|$ (see 4.2.2.6.1)

2. $\displaystyle\int \frac{x^3\,dx}{(a+bx)^2} = \frac{x^2}{2b^2} - \frac{2ax}{b^3} + \frac{a^3}{b^4(a+bx)} + \frac{3a^2}{b^4}\ln|a+bx|$ (see 4.2.2.6.2)

3. $\displaystyle\int \frac{x^3\,dx}{(a+bx)^3} = \left(\frac{x^3}{b} + \frac{2ax^2}{b^2} - \frac{2a^2x}{b^3} - \frac{5a^3}{2b^4}\right)\frac{1}{(a+bx)^2} - \frac{3a}{b^4}\ln|a+bx|$

4. $\displaystyle\int \frac{x^3 dx}{(a+bx)^n} = \frac{x^3}{b(4-n)(a+bx)^{n-1}} - \frac{3a}{b(4-n)}\int \frac{x^2 dx}{(a+bx)^n}$

[reduction formula for $n \neq 4$]

4.2.2.5   For arbitrary positive integers m, n the following reduction formula applies:

1. $\displaystyle\int \frac{x^m\, dx}{(a+bx)^n} = \frac{-x^m}{b(m+1-n)(a+bx)^{n-1}} - \frac{ma}{b(m+1-n)}\int \frac{x^{m-1} dx}{(a+bx)^n}.$

For $m = n - 1$ the above reduction formula can be replaced with

2. $\displaystyle\int \frac{x^{n-1}\, dx}{(a+bx)^n} = \frac{x^{n-1}}{b(n-1)(a+bx)^{n-1}} + \frac{1}{b}\int \frac{x^{n-2}\, dx}{(a+bx)^{n-1}}.$

4.2.2.6

1. $\displaystyle\int \frac{x^n dx}{a+bx} = \frac{x^n}{nb} - \frac{ax^{n-1}}{(n-1)b^2} + \frac{a^2 x^{n-2}}{(n-2)b^3} - \cdots + (-1)^{n-1}\frac{a^{n-1}x}{1 \cdot b^n}$

$\displaystyle + \frac{(-1)^n a^n}{b^{n+1}}\ln|a+bx|$

2. $\displaystyle\int \frac{x^n dx}{(a+bx)^2} = \sum_{k=1}^{n-1}(-1)^{k-1}\frac{ka^{k-1}x^{n-k}}{(n-k)b^{k+1}} + (-1)^{n-1}\frac{a^n}{b^{n+1}(a+bx)}$

$\displaystyle + (-1)^{n+1}\frac{na^{n-1}}{b^{n+1}}\ln|a+bx|$

4.2.2.7

1. $\displaystyle\int \frac{dx}{x(a+bx)} = -\frac{1}{a}\ln\left|\frac{a+bx}{x}\right|$

2. $\displaystyle\int \frac{dx}{x(a+bx)^2} = \frac{1}{a(a+bx)} - \frac{1}{a^2}\ln\left|\frac{a+bx}{x}\right|$

3. $\displaystyle\int \frac{dx}{x(a+bx)^3} = \left(\frac{3}{2a} + \frac{bx}{a^2}\right)\frac{1}{(a+bx)^2} - \frac{1}{a^3}\ln\left|\frac{a+bx}{x}\right|$

4. $\displaystyle\int \frac{dx}{x(a+bx)^n} = \frac{1}{a(n-1)(a+bx)^{n-1}} + \frac{1}{a}\int \frac{dx}{x(a+bx)^{n-1}}$

[reduction formula for $n \neq 1$]

4.2.2.8

1. $\displaystyle\int \frac{dx}{x^2(a+bx)} = -\frac{1}{ax} + \frac{b}{a^2}\ln\left|\frac{a+bx}{x}\right|$

2. $\displaystyle\int \frac{dx}{x^2(a+bx)^2} = -\left(\frac{1}{ax} + \frac{2b}{a^2}\right)\frac{1}{(a+bx)} + \frac{2b}{a^3}\ln\left|\frac{a+bx}{x}\right|$

3. $\int \dfrac{dx}{x^2(a+bx)^3} = -\left(\dfrac{1}{ax} + \dfrac{9b}{2a^2} + \dfrac{3b^2x}{a^3}\right)\dfrac{1}{(a+bx)^2} + \dfrac{3b}{a^4}\ln\left|\dfrac{a+bx}{x}\right|$

4. $\int \dfrac{dx}{x^2(a+bx)^n} = \dfrac{-1}{ax(a+bx)^{n-1}} - \dfrac{nb}{a}\int \dfrac{dx}{x(a+bx)^n}$

[reduction formula]

4.2.3 Integrands involving linear factors

4.2.3.1 Integrals of the form

$$\int \frac{x^m\,dx}{(x-a)^n(x-b)^r \cdots (x-q)^s}$$

can be integrated by first expressing the integrand in terms of partial fractions (see 1.7), and then integrating the result term by term.

4.2.3.2

1. $\int \left(\dfrac{a+bx}{c+dx}\right)dx = \dfrac{bx}{d} + \left(\dfrac{ad-bc}{d^2}\right)\ln|c+dx|$

2. $\int \dfrac{dx}{(x-a)(x-b)} = \dfrac{1}{(a-b)}\ln\left|\dfrac{x-a}{x-b}\right| \qquad [a \neq b]$

3. $\int \dfrac{x\,dx}{(x-a)(x-b)} = \dfrac{a}{(a-b)}\ln|x-a| - \dfrac{b}{(a-b)}\ln|x-b| \qquad [a \neq b]$

4. $\int \dfrac{dx}{(x-a)^2(x-b)} = \dfrac{-1}{(x-a)(a-b)} - \dfrac{1}{(a-b)^2}\ln\left|\dfrac{x-a}{x-b}\right| \qquad [a \neq b]$

5. $\int \dfrac{dx}{(x-a)^2(x-b)^2} = \dfrac{a+b-2x}{(x-a)(x-b)(a-b)^2} - \dfrac{2}{(a-b)^3}\ln\left|\dfrac{x-a}{x-b}\right|$

$[a \neq b]$

6. $\int \dfrac{x\,dx}{(x-a)^2(x-b)} = \dfrac{-a}{(x-a)(a-b)} + \dfrac{b}{(a-b)^2}\ln\left|\dfrac{x-b}{x-a}\right| \qquad [a \neq b]$

7. $\int \dfrac{x\,dx}{(x-a)^2(x-b)^2} = \dfrac{2ab-(a+b)x}{(x-a)(x-b)(a-b)^2} + \dfrac{(a+b)}{(a-b)^3}\ln\left|\dfrac{x-b}{x-a}\right|$

$[a \neq b]$

8. $\int \dfrac{x^2\,dx}{(x-a)^2(x-b)^2} = \dfrac{ab(a+b)-(a^2+b^2)x}{(x-a)(x-b)(a-b)^2} + \dfrac{2ab}{(a-b)^3}\ln\left|\dfrac{x-b}{x-a}\right|$

$[a \neq b]$

4.2.4 Integrands involving $a^2 \pm b^2 x^2$

4.2.4.1

1. $\displaystyle \int \frac{dx}{a^2 + b^2 x^2} = \frac{1}{ab} \arctan\left(\frac{bx}{a}\right)$ $\left[-\frac{\pi}{2} < \arctan\left(\frac{bx}{a}\right) < \frac{\pi}{2} \right]$

2. $\displaystyle \int \frac{dx}{(a^2 + b^2 x^2)^2} = \frac{x}{2a^2(a^2 + b^2 x^2)} + \frac{1}{2a^3 b} \arctan\left(\frac{bx}{a}\right)$

$$\left[-\frac{\pi}{2} < \arctan\left(\frac{bx}{a}\right) < \frac{\pi}{2} \right]$$

3. $\displaystyle \int \frac{dx}{(a^2 + b^2 x^2)^3} = \frac{x(5a^2 + 3b^2 x^2)}{8a^4(a^2 + b^2 x^2)^2} + \frac{3}{8a^5 b} \arctan\left(\frac{bx}{a}\right)$

$$\left[-\frac{\pi}{2} < \arctan\left(\frac{bx}{a}\right) < \frac{\pi}{2} \right]$$

4. $\displaystyle \int \frac{dx}{(a^2 + b^2 x^2)^n} = \frac{x}{2(n-1)a^2(a^2 + b^2 x^2)^{n-1}} + \frac{(2n-3)}{2(n-1)a^2}$

$$\times \int \frac{dx}{(a^2 + b^2 x^2)^{n-1}} \qquad \text{[reduction formula } n > 1\text{]}$$

4.2.4.2

1. $\displaystyle \int \frac{x\, dx}{a^2 + b^2 x^2} = \frac{1}{2b^2} \ln(a^2 + b^2 x^2)$

2. $\displaystyle \int \frac{x\, dx}{(a^2 + b^2 x^2)^2} = \frac{-1}{2b^2(a^2 + b^2 x^2)}$

3. $\displaystyle \int \frac{x\, dx}{(a^2 + b^2 x^2)^3} = \frac{-1}{4b^2(a^2 + b^2 x^2)^2}$

4. $\displaystyle \int \frac{x\, dx}{(a^2 + b^2 x^2)^n} = \frac{-1}{2(n-1)b^2(a^2 + b^2 x^2)^{n-1}}$ $[n \neq 1]$

4.2.4.3

1. $\displaystyle \int \frac{x^2\, dx}{a^2 + b^2 x^2} = \frac{x}{b^2} - \frac{a}{b^3} \arctan\left(\frac{bx}{a}\right)$ $\left[-\frac{\pi}{2} < \arctan\left(\frac{bx}{a}\right) < \frac{\pi}{2} \right]$

2. $\displaystyle \int \frac{x^2\, dx}{(a^2 + b^2 x^2)^2} = \frac{-x}{2b^2(a^2 + b^2 x^2)} + \frac{1}{2ab^3} \arctan\left(\frac{bx}{a}\right)$

$$\left[-\frac{\pi}{2} < \arctan\left(\frac{bx}{a}\right) < \frac{\pi}{2} \right]$$

3. $\displaystyle\int \frac{x^2\,dx}{(a^2+b^2x^2)^3} = \frac{x(b^2x^2-a^2)}{8a^2b^2(a^2+b^2x^2)^2} + \frac{1}{8a^3b^3}\arctan\left(\frac{bx}{a}\right)$

$$\left[-\frac{\pi}{2} < \arctan\left(\frac{bx}{a}\right) < \frac{\pi}{2}\right]$$

4.2.4.4

1. $\displaystyle\int \frac{x^3\,dx}{a^2+b^2x^2} = \frac{x^2}{2b^2} - \frac{a^2}{2b^4}\ln(a^2+b^2x^2)$

2. $\displaystyle\int \frac{x^3\,dx}{(a^2+b^2x^2)^2} = \frac{a^2}{2b^4(a^2+b^2x^2)} + \frac{1}{2b^4}\ln(a^2+b^2x^2)$

3. $\displaystyle\int \frac{x^3\,dx}{(a^2+b^2x^2)^3} = \frac{-(a^2+2b^2x^2)}{4b^3(a^2+b^2x^2)^2}$

4. $\displaystyle\int \frac{x^3\,dx}{(a^2+b^2x^2)^n} = \frac{-[a^2+(n-1)b^2x^2]}{2(n-1)(n-2)b^4(a^2+b^2x^2)^{n-1}}$ $\qquad [n>2]$

4.2.4.5

1. $\displaystyle\int \frac{dx}{a^2-b^2x^2} = -\frac{1}{2ab}\ln\left|\frac{bx-a}{bx+a}\right|$

2. $\displaystyle\int \frac{dx}{(a^2-b^2x^2)^2} = \frac{x}{2a^2(a^2-b^2x^2)} - \frac{1}{4a^3b}\ln\left|\frac{bx-a}{bx+a}\right|$

3. $\displaystyle\int \frac{dx}{(a^2-b^2x^2)^3} = \frac{x(5a^2-3b^2x^2)}{8a^4(a^2-b^2x^2)^2} - \frac{3}{16a^5b}\ln\left|\frac{bx-a}{bx+a}\right|$

4. $\displaystyle\int \frac{dx}{(a^2-b^2x^2)^n} = \frac{x}{2(n-1)a^2(a^2-b^2x^2)^{n-1}} + \frac{(2n-3)}{2(n-1)a^2}$

$$\times \int \frac{dx}{(a^2-b^2x^2)^{n-1}} \qquad \text{[reduction formula for } n>1\text{]}$$

4.2.4.6

1. $\displaystyle\int \frac{x\,dx}{(a^2-b^2x^2)} = -\frac{1}{2b^2}\ln|a^2-b^2x^2|$

2. $\displaystyle\int \frac{x\,dx}{(a^2-b^2x^2)^2} = \frac{1}{2b^2(a^2-b^2x^2)}$

3. $\displaystyle\int \frac{x\,dx}{(a^2-b^2x^2)^3} = \frac{1}{4b^2(a^2-b^2x^2)^2}$

4. $\displaystyle\int \frac{x\,dx}{(a^2-b^2x^2)^n} = \frac{1}{2(n-1)b^2(a^2-b^2x^2)^{n-1}}$

4.2.4.7

1. $\displaystyle \int \frac{x^2\,dx}{a^2 - b^2x^2} = -\frac{x^2}{b^2} - \frac{a}{2b^3}\ln\left|\frac{bx - a}{bx + a}\right|$

2. $\displaystyle \int \frac{x^2\,dx}{(a^2 - b^2x^2)^2} = \frac{x}{2b^2(a^2 - b^2x^2)} + \frac{1}{4ab^3} + \ln\left|\frac{bx - a}{bx + a}\right|$

3. $\displaystyle \int \frac{x^2\,dx}{(a^2 - b^2x^2)^3} = \frac{x(a^2 + b^2x^2)}{8a^2b^2(a^2 - b^2x^2)^2} + \frac{1}{16a^3b^3}\ln\left|\frac{bx - a}{bx + a}\right|$

4.2.4.8

1. $\displaystyle \int \frac{x^3\,dx}{a^2 - b^2x^2} = -\frac{x^2}{2b^2} - \frac{a^2}{2b^4}\ln|a^2 - b^2x^2|$

2. $\displaystyle \int \frac{x^3\,dx}{(a^2 - b^2x^2)^2} = \frac{a^2}{2b^4(a^2 - b^2x^2)} + \frac{1}{2b^4}\ln|a^2 - b^2x^2|$

3. $\displaystyle \int \frac{x^3\,dx}{(a^2 - b^2x^2)^3} = \frac{2b^2x^2 - a^2}{4b^4(a^2 - b^2x^2)^2}$

4.2.4.9

1. $\displaystyle \int \frac{dx}{a^2 + x^2} = \frac{1}{a}\arctan\frac{x}{a}$

2. $\displaystyle \int \frac{dx}{x(a^2 + x^2)} = \frac{1}{2a^2}\ln\left(\frac{x^2}{a^2 + x^2}\right)$

3. $\displaystyle \int \frac{dx}{x^2(a^2 + x^2)} = -\frac{1}{a^2x} - \frac{1}{a^3}\arctan\frac{x}{a}$

4. $\displaystyle \int \frac{dx}{x^3(a^2 + x^2)} = -\frac{1}{2a^2x^2} - \frac{1}{2a^4}\ln\left(\frac{x^2}{a^2 + x^2}\right)$

5. $\displaystyle \int \frac{dx}{x^4(a^2 + x^2)} = -\frac{1}{3a^2x^3} + \frac{1}{a^4x} + \frac{1}{a^5}\arctan\frac{x}{a}$

6. $\displaystyle \int \frac{dx}{x(a^2 + x^2)^2} = \frac{1}{2a^2(a^2 + x^2)} + \frac{1}{2a^4}\ln\left(\frac{x^2}{a^2 + x^2}\right)$

7. $\displaystyle \int \frac{dx}{x^2(a^2 + x^2)^2} = -\frac{1}{a^4x} - \frac{x}{2a^4(a^2 + x^2)} - \frac{3}{2a^5}\arctan\frac{x}{a}$

8. $\displaystyle \int \frac{dx}{(a^2 - x^2)} = \frac{1}{2a}\ln\left|\frac{a + x}{a - x}\right|$

9. $\displaystyle \int \frac{dx}{x(a^2 - x^2)} = \frac{1}{2a^2}\ln\left|\frac{x^2}{a^2 - x^2}\right|$

10. $$\int \frac{dx}{x^2(a^2 - x^2)} = -\frac{1}{a^2x} + \frac{1}{2a^3} \ln\left|\frac{a+x}{a-x}\right|$$

11. $$\int \frac{dx}{x^3(a^2 - x^2)} = -\frac{1}{2a^2x^2} + \frac{1}{2a^4} \ln\left|\frac{x^2}{a^2 - x^2}\right|$$

12. $$\int \frac{dx}{x^4(a^2 - x^2)} = -\frac{1}{3a^2x^3} - \frac{1}{a^4x} + \frac{1}{2a^5} \ln\left|\frac{a+x}{a-x}\right|$$

13. $$\int \frac{dx}{x(a^2 - x^2)^2} = \frac{1}{2a^2(a^2 - x^2)} + \frac{1}{2a^4} \ln\left|\frac{x^2}{a^2 - x^2}\right|$$

14. $$\int \frac{dx}{x^2(a^2 - x^2)^2} = -\frac{1}{a^4x} + \frac{x}{2a^4(a^2 - x^2)} + \frac{3}{4a^3} \ln\left|\frac{a+x}{a-x}\right|$$

4.2.4.10   The change of variable $x^2 = u$ reduces integrals of the form

$$\int \frac{x^{2m+1} dx}{(a^2 \pm b^2 x^2)^n}$$

to the simpler form

$$\frac{1}{2} \int \frac{u^m du}{(a^2 \pm b^2 u)^n},$$

as listed in 4.2.2.

4.2.4.11   The change of variable $x^2 = u$ reduces integrals of the form

$$\int \frac{x^{2m+1} dx}{(a^4 \pm b^4 x^4)^n}$$

to the simpler form

$$\frac{1}{2} \int \frac{u^m du}{(a^4 \pm b^4 u^2)^n},$$

as listed in 4.2.4.1–4.2.4.8.

4.2.5 Integrands involving $a + bx + cx^2$

Notation: $R = a + bx + cx^2$, $\Delta = 4ac - b^2$

4.2.5.1

1. $$\int \frac{dx}{R} = \frac{1}{\sqrt{-\Delta}} \ln\left|\frac{\sqrt{-\Delta} - (b + 2cx)}{(b + 2cx) + \sqrt{-\Delta}}\right|$$

$$= \frac{-2}{\sqrt{-\Delta}} \operatorname{arctanh}\left(\frac{b + 2cx}{\sqrt{-\Delta}}\right) \qquad [\Delta < 0]$$

$$= \frac{-2}{b + 2cx} \qquad [\Delta = 0]$$

$$= \frac{2}{\sqrt{\Delta}} \arctan\left(\frac{b + 2cx}{\sqrt{\Delta}}\right) \qquad [\Delta > 0]$$

2. $\displaystyle\int \frac{dx}{R^2} = \frac{b+2cx}{\Delta R} + \frac{2c}{\Delta}\int \frac{dx}{R}$

3. $\displaystyle\int \frac{dx}{R^3} = \left(\frac{b+2cx}{\Delta}\right)\left(\frac{1}{2R^2} + \frac{3c}{\Delta R}\right) + \frac{6c^2}{\Delta^2}\int \frac{dx}{R}$

4. $\displaystyle\int \frac{dx}{R^{n+1}} = \frac{b+2cx}{n\Delta R^n} + \frac{(4n-2)c}{n\Delta}\int \frac{dx}{R^n}$

$$= \frac{(b+2cx)}{2n+1}\sum_{k=0}^{n-1} \frac{2k(2n+1)(2n-1)(2n-3)\cdots(2n-2k+1)c^k}{n(n-1)\cdots(n-k)\Delta^{k+1}R^{n-k}}$$

$$+ \frac{2^n(2n-1)!!c^n}{n!\Delta^n}\int \frac{dx}{R}$$

[general reduction formula: **double factorial** $(2n-1)!! = 1\cdot 3\cdot 5\cdots(2n-1)$]

4.2.5.2

1. $\displaystyle\int \frac{x\,dx}{R} = \frac{1}{2c}\ln R - \frac{b}{2c}\int \frac{dx}{R}$

2. $\displaystyle\int \frac{x\,dx}{R^2} = -\left(\frac{2a+bx}{\Delta R}\right) - \frac{b}{\Delta}\int \frac{dx}{R}$

3. $\displaystyle\int \frac{x\,dx}{R^3} = -\left(\frac{2a+bx}{2\Delta R^2}\right) - \frac{3b(b+2cx)}{2\Delta^2 R} - \frac{3bc}{\Delta^2}\int \frac{dx}{R}$

4. $\displaystyle\int \frac{x^2\,dx}{R^2} = \frac{x}{c} - \frac{b}{2c^2}\ln R + \left(\frac{b^2-2ac}{2c^2}\right)\int \frac{dx}{R}$

5. $\displaystyle\int \frac{x^2\,dx}{R^2} = \frac{ab+(b^2-2ac)x}{c\Delta R} + \frac{2a}{\Delta}\int \frac{dx}{R}$

6. $\displaystyle\int \frac{x^2\,dx}{R^3} = \frac{ab+(b^2-2ac)x}{2c\Delta R^2} + \frac{(2ac+b^2)(b+2cx)}{2c\Delta^2 R} + \frac{(2ac+b^2)}{\Delta^2}\int \frac{dx}{R}$

7. $\displaystyle\int \frac{x^m\,dx}{R^n} = \frac{-x^{m-1}}{(2n-m-1)cR^{n-1}} - \frac{(n-m)b}{(2n-m-1)c}\int \frac{x^{m-1}\,dx}{R^n}$

$$+ \frac{(m-1)a}{(2n-m-1)c}\int \frac{x^{m-2}\,dx}{R^n}$$

[general reduction formula: $m \neq 2n-1$]

8. $\displaystyle\int \frac{x^{2n-1}\,dx}{R^n} = \frac{1}{c}\int \frac{x^{2n-3}\,dx}{R^{n-1}} - \frac{a}{c}\int \frac{x^{2n-3}\,dx}{R^n} - \frac{b}{c}\int \frac{x^{2n-2}\,dx}{R^n}$

[general reduction formula: case $m = 2n-1$ in 4.2.5.2.7]

4.2.5.3

1. $\displaystyle \int \frac{dx}{xR} = \frac{1}{2a}\ln\frac{x^2}{R} - \frac{b}{2a}\int\frac{dx}{R}$

2. $\displaystyle \int \frac{dx}{xR^2} = \frac{1}{2a^2}\ln\frac{x^2}{R} + \frac{1}{2aR}\left[1 - \frac{b(b+2cx)}{\Delta}\right] - \frac{b}{2a^2}\left(1 + \frac{2ac}{\Delta}\right)\int\frac{dx}{R}$

3. $\displaystyle \int \frac{dx}{x^2R} = -\frac{b}{2a^2}\ln\frac{x^2}{R} - \frac{1}{ax} + \frac{(b^2-2ac)}{2a^2}\int\frac{dx}{R}$

4. $\displaystyle \int \frac{dx}{x^mR^n} = \frac{-1}{(m-1)ax^{m-1}R^{n-1}} - \frac{b(m+n-2)}{a(m-1)}\int\frac{dx}{x^{m-1}R^n}$
$\displaystyle \qquad\qquad - \frac{c(m+2n-3)}{a(m-1)}\int\frac{dx}{x^{m-2}R^n}$ [general reduction formula]

4.2.6 Integrands involving $a + bx^3$

Notation: $\alpha = (a/b)^{1/3}$

4.2.6.1

1. $\displaystyle \int \frac{dx}{a+bx^3} = \frac{\alpha}{3a}\left\{\frac{1}{2}\ln\left|\frac{(x+\alpha)^2}{x^2-\alpha x+\alpha^2}\right| + \sqrt{3}\arctan\left(\frac{x\sqrt{3}}{2\alpha-x}\right)\right\}$
$\displaystyle \qquad\qquad = \frac{\alpha}{3a}\left\{\frac{1}{2}\ln\left|\frac{(x+\alpha)^2}{x^2-\alpha x+\alpha^2}\right| + \sqrt{3}\arctan\left(\frac{2x-\alpha}{\alpha\sqrt{3}}\right)\right\}$

2. $\displaystyle \int \frac{x\,dx}{a+bx^3} = -\frac{1}{3b\alpha}\left\{\frac{1}{2}\ln\left|\frac{(x+\alpha)^2}{x^2-\alpha x+\alpha^2}\right| - \sqrt{3}\arctan\left(\frac{2x-\alpha}{\alpha\sqrt{3}}\right)\right\}$

3. $\displaystyle \int \frac{x^2dx}{a+bx^3} = \frac{1}{3b}\ln\left|1+(x/\alpha)^3\right| = \frac{1}{3b}\ln\left|a+bx^3\right|$

4. $\displaystyle \int \frac{x^3dx}{a+bx^3} = \frac{x}{b} - \frac{a}{b}\int\frac{dx}{a+bx^3}$

5. $\displaystyle \int \frac{dx}{(a+bx^3)^2} = \frac{x}{3a(a+bx^3)} + \frac{2}{3a}\int\frac{dx}{a+bx^3}$

6. $\displaystyle \int \frac{x\,dx}{(a+bx^3)^2} = \frac{x^2}{3a(a+bx^3)} + \frac{1}{3a}\int\frac{x\,dx}{a+bx^3}$

7. $\displaystyle \int \frac{x^n\,dx}{(a+bx^3)^m} = \frac{x^{n-2}}{(n+1-3m)b(a+bx^3)^{m-1}}$
$\displaystyle \qquad\qquad - \frac{a(n-2)}{b(n+1-3m)}\int\frac{x^{n-3}\,dx}{(a+bx^3)^m}$
$\displaystyle \qquad\qquad = \frac{x^{n+1}}{3a(m-1)(a+bx^3)^{m-1}} - \frac{(n+4-3m)}{3a(m-1)}\int\frac{x^n\,dx}{(a+bx^3)^{m-1}}$
$\displaystyle \qquad\qquad\qquad\qquad\qquad\qquad\qquad\qquad\qquad\qquad\quad$ [general reduction formula]

8. $\int \dfrac{dx}{x^n (a + bx^3)^m} = -\dfrac{1}{(n-1)ax^{n-1}(a+bx^3)^{m-1}}$

$$-\dfrac{b(3m+n-4)}{a(n-1)} \int \dfrac{dx}{x^{n-3}(a+bx^3)^m}$$

$$= \dfrac{1}{3a(m-1)x^{n-1}(a+bx^3)^{m-1}} + \dfrac{(n+3m-4)}{3a(m-1)}$$

$$\times \int \dfrac{dx}{x^n(a+bx^3)^{m-1}} \qquad \text{[general reduction formula]}$$

4.2.7 Integrands involving $a + bx^4$

Notation: $\alpha = (a/b)^{1/4}$, $\alpha' = (-a/b)^{1/4}$

4.2.7.1

1. $\int \dfrac{dx}{a + bx^4} = \dfrac{\alpha}{4a\sqrt{2}} \left\{ \ln\left| \dfrac{x^2 + \alpha x\sqrt{2} + \alpha^2}{x^2 - \alpha x\sqrt{2} + \alpha^2} \right| + 2\arctan\left(\dfrac{\alpha x\sqrt{2}}{\alpha^2 - x^2} \right) \right\}$

$$[ab > 0] \qquad \text{[see 4.2.7.1.3]}$$

$$= \dfrac{\alpha'}{4a} \left\{ \ln\left| \dfrac{x + \alpha'}{x - \alpha'} \right| + 2\arctan\left(\dfrac{x}{\alpha'} \right) \right\} \qquad [ab < 0] \text{ [see 4.2.7.1.4]}$$

2. $\int \dfrac{x\,dx}{a + bx^4} = \dfrac{1}{2\sqrt{ab}} \arctan\left(x^2 \sqrt{\dfrac{b}{a}} \right) \qquad [ab > 0] \qquad \text{[see 4.2.7.1.5]}$

$$= \dfrac{1}{4i\sqrt{ab}} \ln\left| \dfrac{a + x^2 i\sqrt{ab}}{a - x^2 i\sqrt{ab}} \right| \qquad [ab < 0] \qquad \text{[see 4.2.7.1.6]}$$

3. $\int \dfrac{dx}{a^4 + x^4} = \dfrac{1}{4a^3\sqrt{2}} \ln\left| \dfrac{x^2 + ax\sqrt{2} + a^2}{x^2 - ax\sqrt{2} + a^2} \right| + \dfrac{1}{2a^3\sqrt{2}} \arctan\left(\dfrac{ax\sqrt{2}}{a^2 - x^2} \right)$

$$\text{[special case of 4.2.7.1.1]}$$

4. $\int \dfrac{dx}{a^4 - x^4} = -\dfrac{1}{4a^3} \ln\left| \dfrac{a+x}{a-x} \right| - \dfrac{1}{2a^3} \arctan\left(\dfrac{x}{a} \right) \quad \text{[special case of 4.2.7.1.1]}$

5. $\int \dfrac{x\,dx}{a^4 + x^4} = \dfrac{1}{2a^2} \arctan\left(\dfrac{x^2}{a^2} \right) \qquad \text{[special case of 4.2.7.1.2]}$

6. $\int \dfrac{x\,dx}{a^4 - x^4} = \dfrac{1}{4a^2} \ln\left| \dfrac{a^2 + x^2}{a^2 - x^2} \right| \qquad \text{[special case of 4.2.7.1.2]}$

7. $\int \dfrac{x^2\,dx}{a+bx^4} = \dfrac{1}{4b\alpha\sqrt{2}}\left\{\ln\left|\dfrac{x^2-\alpha x\sqrt{2}+\alpha^2}{x^2+\alpha x\sqrt{2}+\alpha^2}\right| + 2\arctan\left(\dfrac{\alpha x\sqrt{2}}{\alpha^2-x^2}\right)\right\}$

$$[ab>0]$$

$$= -\dfrac{1}{4b\alpha'}\left\{\ln\left|\dfrac{(x+\alpha')}{x-\alpha'}\right| - 2\arctan\left(\dfrac{x}{\alpha'}\right)\right\} \qquad [ab<0]$$

8. $\int \dfrac{x^2\,dx}{a^4+x^4} = -\dfrac{1}{4a\sqrt{2}}\ln\left|\dfrac{x^2+ax\sqrt{2}+a^2}{x^2-ax\sqrt{2}+a^2}\right| + \dfrac{1}{2a\sqrt{2}}\arctan\left(\dfrac{ax\sqrt{2}}{a^2-x^2}\right)$

[special case of 4.2.7.1.7]

9. $\int \dfrac{x^2\,dx}{a^4-x^4} = \dfrac{1}{4a}\ln\left|\dfrac{a+x}{a-x}\right| - \dfrac{1}{2a}\arctan\left(\dfrac{x}{a}\right)$ [special case of 4.2.7.1.7]

10. $\int \dfrac{dx}{x(a+bx^4)} = \dfrac{1}{a}\ln|x| - \dfrac{1}{4a}\ln|a+bx^4|$

11. $\int \dfrac{dx}{x^2(a+bx^4)} = -\dfrac{1}{ax} - \dfrac{b}{a}\int \dfrac{x^2\,dx}{a+bx^4}$

12. $\int \dfrac{x^n\,dx}{(a+bx^4)^m} = \dfrac{x^{n+1}}{4a(m-1)(a+bx^4)^{m-1}}$

$$+ \dfrac{(4m-n-5)}{4a(m-1)}\int \dfrac{x^n\,dx}{(a+bx^4)^{(m-1)}}$$

$$= \dfrac{x^{n-3}}{(n+1-4m)b(a+bx^4)^{m-1}}$$

$$- \dfrac{(n-3)a}{b(n+1-4m)}\int \dfrac{x^{n-4}\,dx}{(a+bx^4)^m}$$

[general reduction formula]

13. $\int \dfrac{dx}{x^n(a+bx^4)^m} = -\dfrac{1}{(n-1)ax^{n-1}(a+bx^4)^{m-1}}$

$$- \dfrac{b(4m+n-5)}{(n-1)a}\int \dfrac{dx}{x^{n-4}(a+bx^4)^m}$$

[general reduction formula: $n \neq 1$]

14. $\int \dfrac{dx}{x(a+bx^4)^m} = \dfrac{1}{a}\int \dfrac{dx}{x(a+bx^4)^{m-1}} - \dfrac{b}{a}\int \dfrac{x^3\,dx}{(a+bx^4)^m}$

[general reduction formula: case $n=1$ in 4.2.7.1.12]

4.3 Nonrational Algebraic Functions

4.3.1 Integrands containing $a + bx^k$ and $\sqrt{x}$

4.3.1.1

1. $\displaystyle\int x^{n/2}\,dx = \frac{2}{n+2}\,x^{(n+2)/2}$ $[n = 0, 1, \ldots]$

2. $\displaystyle\int \frac{dx}{x^{n/2}} = -\frac{2}{(n-2)x^{(n-2)/2}}$ $[n = 0, 1, \ldots]$

4.3.1.2

1. $\displaystyle\int \frac{dx}{x^{1/2}(a+bx)} = \frac{2}{\sqrt{ab}}\arctan\left(\left(\frac{bx}{a}\right)^{1/2}\right)$ $[ab > 0]$

   $\displaystyle\qquad\qquad\qquad\quad = \frac{1}{i\sqrt{ab}}\ln\left|\frac{a - bx + 2i\sqrt{abx}}{a+bx}\right|,$ $[ab < 0]$

2. $\displaystyle\int \frac{dx}{x^{1/2}(a+bx)^2} = \frac{x^{1/2}}{a(a+bx)} + \frac{1}{2a}\int \frac{dx}{x^{1/2}(a+bx)}$

3. $\displaystyle\int \frac{dx}{x^{1/2}(a+bx)^3} = x^{1/2}\left\{\frac{1}{2a(a+bx)^2} + \frac{3}{4a^2(a+bx)}\right\}$

   $\displaystyle\qquad\qquad\qquad\quad + \frac{3}{8a^2}\int \frac{dx}{x^{1/2}(a+bx)}$

4. $\displaystyle\int \frac{x^{1/2}\,dx}{a+bx} = \frac{2x^{1/2}}{b} - \frac{a}{b}\int \frac{dx}{x^{1/2}(a+bx)}$

5. $\displaystyle\int \frac{x^{1/2}\,dx}{(a+bx)^2} = -\frac{x^{1/2}}{b(a+bx)} + \frac{1}{2b}\int \frac{dx}{x^{1/2}(a+bx)}$

6. $\displaystyle\int \frac{x^{1/2}\,dx}{(a+bx)^3} = x^{1/2}\left\{-\frac{1}{2b(a+bx)^2} + \frac{1}{4ab(a+bx)}\right\} + \frac{1}{8ab}\int \frac{dx}{x^{1/2}(a+bx)}$

7. $\displaystyle\int \frac{x^{3/2}\,dx}{a+bx} = 2x^{1/2}\left\{\frac{x}{3b} - \frac{a}{b^2}\right\} + \frac{a^2}{b^2}\int \frac{dx}{x^{1/2}(a+bx)}$

8. $\displaystyle\int \frac{x^{(2m+1)/2}\,dx}{(a+bx)} = 2x^{1/2}\sum_{k=0}^{m}\frac{(-1)^k a^k x^{m-k}}{(2m-2k+1)b^{k+1}}$

   $\displaystyle\qquad\qquad\qquad\quad + (-1)^{m+1}\frac{a^{m+1}}{b^{m+1}}\int \frac{dx}{x^{1/2}(a+bx)}$

9. $\displaystyle\int \frac{x^{3/2}\,dx}{(a+bx)^2} = \frac{2x^{3/2}}{b(a+bx)} - \frac{3a}{b}\int \frac{x^{1/2}\,dx}{(a+bx)^2}$

10. $\displaystyle\int \frac{x^{3/2}\,dx}{(a+bx)^3} = -\frac{2x^{3/2}}{b(a+bx)^2} + \frac{3a}{b}\int \frac{x^{1/2}\,dx}{(a+bx)^3}$

11. $\int \dfrac{x^{5/2}\,dx}{a+bx} = 2x^{1/2}\left(\dfrac{x^2}{5b} - \dfrac{ax}{3b^2} + \dfrac{a^2}{b^3}\right) - \dfrac{a^3}{b^3}\int \dfrac{dx}{x^{1/2}(a+bx)}$

12. $\int \dfrac{x^{5/2}\,dx}{(a+bx)^2} = \dfrac{2x^{1/2}}{(a+bx)}\left(\dfrac{x^2}{3b} - \dfrac{5ax}{3b^2}\right) + \dfrac{5a^2}{b^2}\int \dfrac{x^{1/2}\,dx}{(a+bx)^2}$

13. $\int \dfrac{x^{5/2}\,dx}{(a+bx)^3} = \dfrac{2x^{1/2}}{(a+bx)^2}\left(\dfrac{x^2}{b} + \dfrac{5ax}{b^2}\right) - \dfrac{15a^2}{b^2}\int \dfrac{x^{1/2}\,dx}{(a+bx)^3}$

14. $\int \dfrac{dx}{x^{1/2}(a+bx)^2} = \dfrac{x^{1/2}}{a(a+bx)} + \dfrac{1}{2a}\int \dfrac{dx}{x^{1/2}(a+bx)}$

15. $\int \dfrac{dx}{x^{1/2}(a+bx)^3} = x^{1/2}\left\{\dfrac{1}{2a(a+bx)^2} + \dfrac{3}{4a^2(a+bx)}\right\}$
$\qquad\qquad + \dfrac{3}{8a^2}\int \dfrac{dx}{x^{1/2}(a+bx)}$

16. $\int \dfrac{dx}{x^{3/2}(a+bx)} = -\dfrac{2}{ax^{1/2}} - \dfrac{2}{a^{1/2}}\left(\dfrac{b}{a}\right)^{1/2}\arctan\left[\left(\dfrac{bx}{a}\right)^{1/2}\right]$

17. $\int \dfrac{dx}{x^{3/2}(a+bx)^2} = -\dfrac{(2a+3bx)}{a^2x^{1/2}(a+bx)} - \dfrac{3}{a^2}\left(\dfrac{b}{a}\right)^{1/2}\arctan\left[\left(\dfrac{bx}{a}\right)^{1/2}\right]$

18. $\int \dfrac{dx}{x^{3/2}(a+bx)^3} = -\dfrac{(8a^2 + 25abx + 15b^2x^2)}{4a^3x^{1/2}(a+bx)^2}$
$\qquad\qquad - \dfrac{15}{4a^3}\left(\dfrac{b}{a}\right)^{1/2}\arctan\left(\dfrac{bx}{a}\right)^{1/2}$

4.3.1.3 Notation: $\alpha = (a/b)^{1/4}, \alpha' = (-a/b)^{1/4}$

1. $\int \dfrac{dx}{x^{1/2}(a+bx^2)} = \dfrac{1}{b\alpha^3\sqrt{2}}\left\{\ln\left|\dfrac{x+\alpha\sqrt{2x}+\alpha^2}{(a+bx^2)^{1/2}}\right| + \arctan\left(\dfrac{\alpha\sqrt{2x}}{\alpha^2-x}\right)\right\}$
$$\left[\tfrac{a}{b} > x^2 > 0\right]$$

$\qquad = \dfrac{1}{2b\alpha'^3}\left\{\ln\left|\dfrac{\alpha'-x^{1/2}}{\alpha'+x^{1/2}}\right| - 2\arctan\left(\dfrac{x^{1/2}}{\alpha'}\right)\right\}$
$$\left[-\tfrac{a}{b} > x^2 < 0\right]$$

2. $\int \dfrac{x^{1/2}\,dx}{a+bx^2} = \dfrac{1}{b\alpha\sqrt{2}}\left\{-\ln\left|\dfrac{x+\alpha\sqrt{2x}+\alpha^2}{(a+bx^2)^{1/2}}\right| + \arctan\left(\dfrac{\alpha\sqrt{2x}}{\alpha^2-x}\right)\right\}$
$$\left[\tfrac{a}{b} > x^2 > 0\right]$$

$\qquad = \dfrac{1}{2b\alpha'}\left\{\ln\left|\dfrac{\alpha'-x^{1/2}}{\alpha'+x^{1/2}}\right| + 2\arctan\left(\dfrac{x^{1/2}}{\alpha'}\right)\right\} \qquad \left[-\tfrac{a}{b} > x^2 < 0\right]$

3. $\displaystyle\int \frac{x^{3/2}\,dx}{a+bx^2} = \frac{2x^{1/2}}{b} - \frac{a}{b}\int \frac{dx}{x^{1/2}(a+bx^2)}$

4. $\displaystyle\int \frac{x^{5/2}\,dx}{a+bx^2} = \frac{2x^{3/2}}{3b} - \frac{a}{b}\int \frac{x^{1/2}\,dx}{a+bx^2}$

5. $\displaystyle\int \frac{x^{1/2}\,dx}{(a+bx^2)^2} = \frac{x^{3/2}}{2a(a+bx^2)} + \frac{1}{4a}\int \frac{x^{1/2}\,dx}{(a+bx^2)}$

6. $\displaystyle\int \frac{dx}{x^{1/2}(a+bx^2)^2} = \frac{x^{1/2}}{2a(a+bx^2)} + \frac{3}{4a}\int \frac{dx}{x^{1/2}(a+bx^2)}$

7. $\displaystyle\int \frac{dx}{x^{1/2}(a+bx^2)^3} = x^{1/2}\left(\frac{1}{4a(a+bx^2)^2} + \frac{7}{16a^2(a+bx^2)}\right)$

$$+ \frac{21}{32a^2}\int \frac{dx}{x^{1/2}(a+bx^2)}$$

4.3.2 Integrands containing $(a+bx)^{1/2}$

4.3.2.1

1. $\displaystyle\int \frac{dx}{(a+bx)^{1/2}} = \frac{2}{b}(a+bx)^{1/2}$

2. $\displaystyle\int \frac{x\,dx}{(a+bx)^{1/2}} = \frac{2}{b^2}(a+bx)^{1/2}\left\{\frac{1}{3}(a+bx) - a\right\}$

3. $\displaystyle\int \frac{x^2\,dx}{(a+bx)^{1/2}} = \frac{2}{b^3}(a+bx)^{1/2}\left\{\frac{1}{5}(a+bx)^2 - \frac{2}{3}a(a+bx) + a^2\right\}$

4. $\displaystyle\int \frac{dx}{(a+bx)^{3/2}} = -\frac{2}{b(a+bx)^{1/2}}$

5. $\displaystyle\int \frac{x\,dx}{(a+bx)^{3/2}} = \frac{2(2a+bx)}{b^2(a+bx)^{1/2}}$

6. $\displaystyle\int \frac{x^2\,dx}{(a+bx)^{3/2}} = \frac{2}{b^3(a+bx)^{1/2}}\left(\frac{1}{3}(a+bx)^2 - 2a(a+bx) - a^2\right)$

7. $\displaystyle\int \frac{x^3\,dx}{(a+bx)^{3/2}} = \frac{2(b^3x^3 - 2ab^2x^2 + 8a^2bx + 16a^3)}{5b^4(a+bx)^{1/2}}$

8. $\displaystyle\int \frac{dx}{x(a+bx)^{1/2}} = \frac{1}{\sqrt{a}}\ln\left|\frac{(a+bx)^{1/2} - \sqrt{a}}{(a+bx)^{1/2} + \sqrt{a}}\right| \qquad [a>0]$

$$= \frac{2}{\sqrt{-a}}\arctan\frac{(a+bx)^{1/2}}{\sqrt{-a}} \qquad [a<0]$$

9. $\displaystyle\int \frac{dx}{x(a+bx)^{3/2}} = \frac{2}{a(a+bx)^{1/2}} + \frac{1}{a}\int \frac{dx}{x(a+bx)^{1/2}}$

10. $\displaystyle\int \frac{dx}{x(a+bx)^{3/2}} = \frac{2}{a(a+bx)^{1/2}} + \frac{1}{a^2}\int \frac{dx}{x(a+bx)^{1/2}}$

11. $\displaystyle\int \frac{dx}{x^n(a+bx)^{1/2}} = -\frac{(a+bx)^{1/2}}{(n-1)ax^{n-1}} - \frac{(2n-3)}{(2n-2)}\left(\frac{b}{a}\right)\int \frac{dx}{x^{n-1}(a+bx)^{1/2}}$

12. $\displaystyle\int (a+bx)^{1/2}\,dx = \frac{2}{3b}(a+bx)^{3/2}$

13. $\displaystyle\int (a+bx)^{3/2}\,dx = \frac{2}{5b}(a+bx)^{5/2}$

14. $\displaystyle\int (a+bx)^{n/2}\,dx = \frac{2(a+bx)^{(n+2)/2}}{b(n+2)}$

15. $\displaystyle\int x(a+bx)^{1/2}\,dx = \frac{2}{b^2}\left\{\frac{1}{5}(a+bx)^{5/2} - \frac{a}{3}(a+bx)^{3/2}\right\}$

16. $\displaystyle\int x(a+bx)^{3/2}\,dx = \frac{2}{b^2}\left\{\frac{1}{7}(a+bx)^{7/2} - \frac{a}{5}(a+bx)^{5/2}\right\}$

17. $\displaystyle\int \frac{(a+bx)^{1/2}}{x}\,dx = 2(a+bx)^{1/2} + a\int \frac{dx}{x(a+bx)^{1/2}}$

18. $\displaystyle\int \frac{(a+bx)^{3/2}}{x}\,dx = \frac{2}{3}(a+bx)^{3/2} + 2a(a+bx)^{1/2} + a^2\int \frac{dx}{x(a+bx)^{1/2}}$

19. $\displaystyle\int \frac{(a+bx)^{1/2}}{x^2}\,dx = -\frac{(a+bx)^{1/2}}{x} + \frac{b}{2}\int \frac{dx}{x(a+bx)^{1/2}}$

20. $\displaystyle\int \frac{(a+bx)^{1/2}}{x^3}\,dx = -\frac{(2a+bx)(a+bx)^{1/2}}{4ax^2} - \frac{b^2}{8a}\int \frac{dx}{x(a+bx)^{1/2}}$

21. $\displaystyle\int \frac{(a+bx)^{(2m-1)/2}}{x^n}\,dx = -\frac{(a+bx)^{(2m+1)/2}}{(n-1)ax^{n-1}} + \frac{(2m-2n+3)}{2(n-1)}\left(\frac{b}{a}\right)$
$$\times \int \frac{(a+bx)^{(2m-1)/2}}{x^{n-1}}\,dx \qquad [n \neq 1]$$

4.3.3 Integrands containing $(a+cx^2)^{1/2}$

$$I_1 = \int \frac{dx}{(a+cx^2)^{1/2}},$$

where

$$I_1 = \frac{1}{\sqrt{c}} \ln|x\sqrt{c} + (a + cx^2)^{1/2}| \qquad [c > 0]$$

$$= \frac{1}{\sqrt{-c}} \arcsin\left(x\sqrt{\frac{-c}{a}}\right) \qquad [c < 0, a > 0]$$

$$I_2 = \int \frac{dx}{x(a + cx^2)^{1/2}},$$

where

$$I_2 = \frac{1}{2\sqrt{a}} \ln\left|\frac{(a + cx^2)^{1/2} - \sqrt{a}}{(a + cx^2)^{1/2} + \sqrt{a}}\right| \qquad [a > 0, c > 0]$$

$$= \frac{1}{2\sqrt{a}} \ln\left|\frac{\sqrt{a} - (a + cx^2)^{1/2}}{\sqrt{a} + (a + cx^2)^{1/2}}\right| \qquad [a > 0, c < 0]$$

$$= \frac{1}{\sqrt{-a}} \operatorname{arcsec}\left(x\sqrt{\frac{-c}{a}}\right) = \frac{1}{\sqrt{-a}} \arccos\left(\frac{1}{x}\sqrt{\frac{-a}{c}}\right) \qquad [a < 0, c > 0]$$

4.3.3.1

1. $\int (a + cx^2)^{3/2}\, dx = \frac{1}{4}x(a + cx^2)^{3/2} + \frac{3}{8}ax(a + cx^2)^{1/2} + \frac{3}{8}a^2 I_1$

2. $\int (a + cx^2)^{1/2}\, dx = \frac{1}{2}x(a + cx^2)^{1/2} + \frac{1}{2}a I_1$

3. $\int \frac{dx}{(a + cx^2)^{1/2}} = I_1$

4. $\int \frac{dx}{(a + cx^2)^{3/2}} = \frac{1}{a}\frac{x}{(a + cx^2)^{1/2}}$

5. $\int \frac{dx}{(a + cx^2)^{(2n+1)/2}} = \frac{1}{a^n}\sum_{k=0}^{n-1} \frac{(-1)^k}{2k+1}\binom{n-1}{k}\frac{c^k x^{2k+1}}{(a + cx^2)^{(2k+1)/2}}$

6. $\int \frac{x\, dx}{(a + cx^2)^{(2n+1)/2}} = -\frac{1}{(2n-1)c(a + cx^2)^{(2n-1)/2}}$

7. $\int \frac{x^2\, dx}{(a + cx^2)^{(2n+1)/2}} = \frac{1}{a^{n-1}}\sum_{k=0}^{n-2} \frac{(-1)^k}{2k+3}\binom{n-2}{k}\frac{c^k x^{2k+3}}{(a + cx^2)^{(2k+3)/2}}$

8. $\displaystyle\int \frac{x^3\,dx}{(a+cx^2)^{(2n+1)/2}} = -\frac{1}{(2n-3)c^2(a+cx^2)^{(2n-3)/2}}$
$$+ \frac{a}{(2n-1)c^2(a+cx^2)^{(2n-1)/2}}$$

9. $\displaystyle\int x^2(a+cx^2)^{1/2}\,dx = \frac{1}{4c}x(a+cx^2)^{3/2} - \frac{1}{8c}ax(a+cx^2)^{1/2} - \frac{1}{8}\frac{a^2}{c}I_1$

10. $\displaystyle\int \frac{x^2}{(a+cx^2)^{1/2}}\,dx = \frac{1}{2c}x(a+cx^2)^{1/2} - \frac{1}{2}\frac{a}{c}I_1$

11. $\displaystyle\int \frac{x^2}{(a+cx^2)^{3/2}}\,dx = \frac{-x}{c(a+cx^2)^{1/2}} + \frac{1}{c}I_1$

12. $\displaystyle\int \frac{x^2}{(a+cx^2)^{5/2}}\,dx = \frac{x^3}{3a(a+cx^2)^{3/2}}$

13. $\displaystyle\int \frac{(a+cx^2)^{3/2}}{x}\,dx = \frac{1}{3}(a+cx^2)^{3/2} + a(a+cx^2)^{1/2} + a^2 I_2$

14. $\displaystyle\int \frac{(a+cx^2)^{1/2}}{x}\,dx = (a+cx^2)^{1/2} + aI_2$

15. $\displaystyle\int \frac{dx}{x(a+cx^2)^{1/2}} = I_2$

16. $\displaystyle\int \frac{dx}{x(a+cx^2)^{(2n+1)/2}} = \frac{1}{a^n}I_2 + \sum_{k=0}^{n-1}\frac{1}{(2k+1)a^{n-k}(a+cx^2)^{(2k+1)/2}}$

17. $\displaystyle\int \frac{(a+cx^2)^{1/2}}{x^2}\,dx = -\frac{(a+cx^2)^{1/2}}{x} + cI_1$

18. $\displaystyle\int \frac{(a+cx^2)^{3/2}}{x^2}\,dx = -\frac{(a+cx^2)^{3/2}}{x} + \frac{3}{2}cx(a+cx^2)^{1/2} + \frac{3}{2}aI_1$

19. $\displaystyle\int \frac{(a+cx^2)^{1/2}}{x^3}\,dx = -\frac{(a+cx^2)^{1/2}}{2x^2} + \frac{c}{2}I_2$

20. $\displaystyle\int \frac{(a+cx^2)^{3/2}}{x^3}\,dx = -\frac{(a+cx^2)^{3/2}}{2x^2} + \frac{3}{2}c(a+cx^2)^{1/2} + \frac{3}{2}acI_2$

21. $\displaystyle\int \frac{dx}{x^3(a+cx^2)^{1/2}} = -\frac{(a+cx^2)^{1/2}}{2ax^2} - \frac{c}{2a}I_2$

22. $\displaystyle\int \frac{dx}{x^3(a+cx^2)^{3/2}} = -\frac{1}{2ax^2(a+cx^2)^{1/2}} - \frac{3c}{2a^2(a+cx^2)^{1/2}} - \frac{3c}{2a^2}I_2$

4.3.4 Integrands containing $(a + bx + cx^2)^{1/2}$

Notation: $R = a + bx + cx^2, \Delta = 4ac - b^2$

4.3.4.1

1. $\displaystyle \int \frac{dx}{R^{1/2}} = \frac{1}{\sqrt{c}} \ln|2(cR)^{1/2} + 2cx + b|$ $[c > 0]$

$$= \frac{1}{\sqrt{c}} \operatorname{arcsinh}\left(\frac{2cx + b}{\Delta^{1/2}}\right) \quad [c > 0, \Delta > 0]$$

$$= \frac{-1}{\sqrt{-c}} \arcsin\left(\frac{2cx + b}{(-\Delta)^{1/2}}\right) \quad [c < 0, \Delta < 0]$$

$$= \frac{1}{\sqrt{c}} \ln|2cx + b| \quad [c > 0, \Delta = 0]$$

2. $\displaystyle \int \frac{x\,dx}{R^{1/2}} = \frac{R^{1/2}}{c} - \frac{b}{2c} \int \frac{dx}{R^{1/2}}$

3. $\displaystyle \int \frac{x^2\,dx}{R^{1/2}} = \left(\frac{x}{2c} - \frac{3b}{4c^2}\right) R^{1/2} + \left(\frac{3b^2}{8c^2} - \frac{a}{2c}\right) \int \frac{dx}{R^{1/2}}$

4. $\displaystyle \int \frac{x^3\,dx}{R^{1/2}} = \left(\frac{x^2}{3c} - \frac{5bx}{12c^2} + \frac{5b^2}{8c^3} - \frac{2a}{3c^2}\right) R^{1/2} - \left(\frac{5b^3}{16c^3} - \frac{3ab}{4c^2}\right) \int \frac{dx}{R^{1/2}}$

5. $\displaystyle \int R^{1/2}\,dx = \frac{(2cx + b)}{4c} R^{1/2} + \frac{\Delta}{8c} \int \frac{dx}{R^{1/2}}$

6. $\displaystyle \int x R^{1/2}\,dx = \frac{R^{3/2}}{3c} - \frac{(2cx + b)b}{8c^2} R^{1/2} - \frac{b\Delta}{16c^2} \int \frac{dx}{R^{1/2}}$

7. $\displaystyle \int x^2 R^{1/2}\,dx = \left(\frac{x}{4c} - \frac{5b}{24c^2}\right) R^{3/2} + \left(\frac{5b^2}{16c^2} - \frac{a}{4c}\right) \frac{(2cx + b)R^{1/2}}{4c}$

$$+ \left(\frac{5b^2}{16c^2} - \frac{a}{4c}\right) \frac{\Delta}{8c} \int \frac{dx}{R^{1/2}}$$

8. $\displaystyle \int R^{3/2}\,dx = \left(\frac{R}{8c} + \frac{3\Delta}{64c^2}\right)(2cx + b)R^{1/2} + \frac{3\Delta^2}{128c^2} \int \frac{dx}{R^{1/2}}$

9. $\displaystyle \int x R^{3/2}\,dx = \frac{R^{5/2}}{5c} - (2cx + b)\left(\frac{b}{16c^2} R^{3/2} + \frac{3\Delta b}{128c^3} R^{1/2}\right) - \frac{3\Delta^2 b}{256c^3} \int \frac{dx}{R^{1/2}}$

10. $\displaystyle \int \frac{dx}{x R^{1/2}} = -\frac{1}{\sqrt{a}} \ln\left|\frac{2a + bx + 2(aR)^{1/2}}{x}\right|$ $[a > 0]$

$$= \frac{1}{\sqrt{-a}} \arcsin\left(\frac{2a + bx}{x(b^2 - 4ac)^{1/2}}\right) \quad [a < 0, \Delta < 0]$$

$$= \frac{1}{\sqrt{-a}} \arctan\left(\frac{2a + bx}{2\sqrt{-a}R^{1/2}}\right) \qquad [a < 0]$$

$$= -\frac{1}{\sqrt{-a}} \operatorname{arcsinh}\left(\frac{2a + bx}{x\Delta^{1/2}}\right) \qquad [a > 0, \Delta > 0]$$

$$= -\frac{1}{\sqrt{a}} \operatorname{arctanh}\left(\frac{2a + bx}{2\sqrt{a}R^{1/2}}\right) \qquad [a > 0]$$

$$= \frac{1}{\sqrt{a}} \ln\left|\frac{x}{2a + bx}\right| \qquad [a > 0, \Delta = 0]$$

$$= -\frac{2(bx + cx^2)^{1/2}}{bx} \qquad [a = 0, b \neq 0]$$

11. $\displaystyle\int \frac{dx}{xR^{(2n+1)/2}} = \frac{1}{(2n-1)aR^{(2n-1)/2}} - \frac{b}{2a}\int \frac{dx}{R^{(2n+1)/2}} + \frac{1}{a}\int \frac{dx}{xR^{(2n-1)/2}}$

12. $\displaystyle\int \frac{dx}{x^m R^{(2n+1)/2}} = -\frac{1}{(m-1)ax^{m-1}R^{(2n-1)/2}} - \frac{(2n+2m-3)b}{2(m-1)a}$

$$\times \int \frac{dx}{x^{m-1}R^{(2n+1)/2}} - \frac{(2n+m-2)c}{(m-1)a}\int \frac{dx}{x^{m-2}R^{(2n+1)/2}}$$

13. $\displaystyle\int \frac{dx}{x^2 R^{1/2}} = -\frac{R^{1/2}}{ax} - \frac{b}{2a}\int \frac{dx}{xR^{1/2}}$

14. $\displaystyle\int \frac{dx}{x^2(bx + cx^2)^{1/2}} = \frac{2}{3}\left(-\frac{1}{bx^2} + \frac{2c}{b^2x}\right)(bx + cx^2)^{1/2}$

15. $\displaystyle\int \frac{dx}{x^3 R^{1/2}} = \left(-\frac{1}{2ax^2} + \frac{3b}{4a^2x}\right)R^{1/2} + \left(\frac{3b^2}{8a^2} - \frac{c}{2a}\right)\int \frac{dx}{xR^{1/2}}$

16. $\displaystyle\int \frac{dx}{x^3(bx + cx^2)^{1/2}} = \frac{2}{5}\left(-\frac{1}{bx^3} + \frac{4c}{3b^2x^2} - \frac{8c^2}{3b^3x}\right)(bx + cx^2)^{1/2}$

5

Indefinite Integrals of Exponential Functions

5.1 Basic Results

5.1.1 Indefinite integrals involving e^{ax}

5.1.1.1

1. $\displaystyle\int e^x \, dx = e^x$

2. $\displaystyle\int e^{-x} \, dx = -e^{-x}$

3. $\displaystyle\int e^{ax} \, dx = \frac{1}{a}e^{ax}$

4. $\displaystyle\int a^x \, dx = \int e^{x \ln a} \, dx = \frac{a^x}{\ln a}$

5.1.1.2   As in 5.1.1.1.4, when e^{ax} occurs in an integrand it should be replaced by $a^x = e^{x \ln a}$ (see 2.2.1).

5.1.2 Integrands involving the exponential functions combined with rational functions of x

5.1.2.1 Positive powers of x.

1. $\displaystyle \int x e^{ax}\, dx = e^{ax}\left(\frac{x}{a} - \frac{1}{a^2}\right)$

2. $\displaystyle \int x^2 e^{ax}\, dx = e^{ax}\left(\frac{x^2}{a} - \frac{2x}{a^2} + \frac{2}{a^3}\right)$

3. $\displaystyle \int x^3 e^{ax}\, dx = e^{ax}\left(\frac{x^3}{a} - \frac{3x^2}{a^2} + \frac{6x}{a^3} - \frac{6}{a^4}\right)$

4. $\displaystyle \int x^4 e^{ax}\, dx = e^{ax}\left(\frac{x^4}{a} - \frac{4x^3}{a^2} + \frac{12x^2}{a^3} - \frac{24x}{a^4} + \frac{24}{a^5}\right)$

5. $\displaystyle \int x^m e^{ax}\, dx = \frac{x^m e^{ax}}{a} - \frac{m}{a}\int x^{m-1} e^{ax}\, dx$

6. $\displaystyle \int P_m(x) e^{ax}\, dx = \frac{e^{ax}}{a}\sum_{k=0}^{m}(-1)^k \frac{P^{(k)}(x)}{a^k},$

where $P_m(x)$ is a polynomial in x of degree m and $P^{(k)}(x)$ is the k'th derivative of $P_m(x)$ with respect to x.

5.1.2.2   The **exponential integral** function $Ei(x)$, defined by

1. $\displaystyle Ei(x) = \int \frac{e^x}{x}\, dx,$

is a transcendental function with the series representation

2. $\displaystyle Ei(x) = \ln|x| + \frac{x}{1!} + \frac{x^2}{2\cdot 2!} + \frac{x^3}{3\cdot 3!} + \cdots + \frac{x^k}{k\cdot k!} + \cdots \qquad [x^2 < \infty].$

When $Ei(x)$ is tabulated, 5.1.2.2.1, with x replaced by the dummy variable t, is integrated over $-\infty < t \leq x$. The principal value of the integral is used and to the series 5.1.2.2.2 must then be added the Euler constant γ.

5.1.2.3 Negative powers of x.

1. $\displaystyle \int \frac{e^{ax}}{x}\, dx = Ei(ax)$

2. $\displaystyle \int \frac{e^{ax}}{x^2}\, dx = -\frac{e^{ax}}{x} + a\,Ei(ax)$

3. $\displaystyle \int \frac{e^{ax}}{x^3}\, dx = -\frac{e^{ax}}{2x^2} - \frac{a e^{ax}}{2x} + \frac{a^2}{2}Ei(ax)$

4. $\displaystyle\int \frac{e^{ax}}{x^n}\,dx = -\frac{e^{ax}}{(n-1)x^{n-1}} - \frac{ae^{ax}}{(n-1)(n-2)x^{n-2}} - \cdots$

$$-\frac{a^{n-2}e^{ax}}{(n-1)!x} + \frac{a^{n-1}}{(n-1)!}\,Ei(ax)$$

5.1.2.4 Integrands involving $(a + be^{mx})^{-1}$.

1. $\displaystyle\int \frac{dx}{1+e^x} = \ln\!\left(\frac{e^x}{1+e^x}\right) = x - \ln(1+e^x)$

2. $\displaystyle\int \frac{dx}{a+be^{mx}} = \frac{1}{am}[mx - \ln(a+be^{mx})]$

3. $\displaystyle\int \frac{e^{mx}}{a+be^{mx}}\,dx = \frac{1}{mb}\ln(a+be^{mx})$

4. $\displaystyle\int \frac{e^{2mx}}{a+be^{mx}}\,dx = \frac{1}{mb}e^{mx} - \frac{a}{mb^2}\ln(a+be^{mx})$

5. $\displaystyle\int \frac{e^{3mx}}{a+be^{mx}}\,dx = \frac{1}{2mb}e^{2mx} - \frac{a}{mb^2}e^{mx} + \frac{a^2}{mb^3}\ln(a+be^{mx})$

5.1.2.5 Integrands involving $xe^{ax}(1+ax)^{-m}$.

1. $\displaystyle\int \frac{xe^x\,dx}{(1+x)^2} = \frac{e^x}{1+x}$

2. $\displaystyle\int \frac{xe^{ax}\,dx}{(1+ax)^2} = \frac{e^{ax}}{a^2(1+ax)}$

5.1.3 Integrands involving the exponential functions combined with trigonometric functions

5.1.3.1

1. $\displaystyle\int e^{ax}\sin bx\,dx = \frac{e^{ax}}{a^2+b^2}(a\sin bx - b\cos bx)$

2. $\displaystyle\int e^{ax}\sin^2 bx\,dx = \frac{e^{ax}}{a(a^2+4b^2)}(a^2\sin^2 bx - 2ab\sin bx\cos bx + 2b^2)$

3. $\displaystyle\int e^{ax}\sin^3 bx\,dx = \frac{e^{ax}}{(a^2+b^2)(a^2+9b^2)}$

$$\times\{a\sin bx[(a^2+b^2)\sin^2 bx + 6b^2]$$

$$-3b\cos bx[(a^2+b^2)\sin^2 bx + 2b^2]\}$$

4. $\displaystyle \int e^{ax} \cos bx \, dx = \frac{e^{ax}}{a^2 + b^2} (a \cos bx + b \sin bx)$

5. $\displaystyle \int e^{ax} \cos^2 bx \, dx = \frac{e^{ax}}{a(a^2 + 4b^2)} (a^2 \cos^2 bx + 2ab \sin bx \cos bx + 2b^2)$

6. $\displaystyle \int e^{ax} \cos^3 bx \, dx = \frac{e^{ax}}{(a^2 + b^2)(a^2 + 9b^2)}$

$$\times [a(a^2 + b^2) \cos^3 bx + 3b(a^2 + b^2) \sin bx \cos^2 bx$$

$$+ 6ab^2 \cos bx + 6b^3 \sin bx]$$

7. $\displaystyle \int e^{ax} \sin^n bx \, dx = \frac{1}{a^2 + n^2 b^2} [(a \sin bx - nb \cos bx) e^{ax} \sin^{n-1} bx$

$$+ n(n-1) b^2 \int e^{ax} \sin^{n-2} bx \, dx]$$

8. $\displaystyle \int e^{ax} \cos^n bx \, dx = \frac{1}{a^2 + n^2 b^2} \bigg[(a \cos bx + nb \sin bx) e^{ax} \cos^{n-1} bx$

$$+ n(n-1) b \int e^{ax} \cos^{n-2} bx \, dx \bigg]$$

As an alternative to using the recurrence relations 5.1.3.1.7 and 8, integrals of this type may be evaluated entirely in terms of exponential functions by substituting either $\cos bx = (e^{ibx} + e^{-ibx})/2$, or $\sin bx = (e^{ibx} - e^{-ibx})/(2i)$.

For example,

$$\int e^{ax} \cos^2 bx \, dx = \int e^{ax} [(e^{ibx} + e^{-ibx})/2]^2 \, dx$$

$$= \frac{1}{4} \int \left[e^{(a+i2b)x} + 2e^{ax} + e^{(a-i2b)x} \right] dx$$

$$= \frac{e^{ax}}{2a} + \frac{e^{ax}}{4} \left[\frac{e^{i2bx}}{(a + i2b)} + \frac{e^{-i2bx}}{(a - i2b)} \right]$$

$$= \frac{1}{2a} e^{ax} + \frac{e^{ax}}{2(a^2 + 4b^2)} [a \cos 2bx + 2b \sin 2bx],$$

where the arbitrary additive constant of integration has been omitted.

Notice that this method of integration produces a different form of result compared to the one given in 5.1.3.1.5. That the two results are identical can be seen by substituting $\cos 2bx = 2\cos^2 bx - 1$, $\sin 2bx = 2 \sin bx \cos bx$ and then combining the terms.

We remark that 5.1.3.1.2 can be deduced directly from 5.1.3.1.5 and, conversely, by using the fact that

$$\int e^{ax} [\cos^2 bx + \sin^2 bx] \, dx = \int e^{ax} dx = \frac{1}{a} e^{ax},$$

so

$$\int e^{ax} \sin^2 bx \, dx = \frac{1}{a} e^{ax} - \int e^{ax} \cos^2 bx \, dx,$$

where again the arbitrary additive constant of integration has been omitted.

To evaluate integrals of the form $\int e^{ax} \sin(bx \pm c) \, dx$ and $\int e^{ax} \cos(bx \pm c) \, dx$, use the identities

$$\sin(bx \pm c) = \cos c \sin bx \pm \sin c \cos bx$$

$$\cos(bx \pm c) = \cos c \cos bx \mp \sin c \sin bx,$$

to simplify the integrands and then use 5.1.3.1.1 and 4.

To evaluate integrals of the form $\int e^{ax} \sin bx \sin cx \, dx$, $\int e^{ax} \sin bx \cos cx \, dx$, and $\int e^{ax} \cos bx \cos cx \, dx$, use the identities

$$\sin bx \sin cx = \tfrac{1}{2}[\cos(b - c)x - \cos(b + c)x],$$

$$\sin bx \cos cx = \tfrac{1}{2}[\sin(b + c)x + \sin(b - c)x],$$

$$\cos bx \cos cx = \tfrac{1}{2}[\cos(b + c)x + \cos(b - c)x],$$

to simplify the integrands and then use 5.1.3.1.1 and 4.

9. $\displaystyle \int x e^{ax} \sin bx \, dx$

$$= \frac{e^{ax}}{a^2 + b^2} \left[\left(ax - \frac{a^2 - b^2}{a^2 + b^2} \right) \sin bx - \left(bx - \frac{2ab}{a^2 + b^2} \right) \cos bx \right]$$

10. $\displaystyle \int x e^{ax} \cos bx \, dx$

$$= \frac{e^{ax}}{a^2 + b^2} \left[\left(ax - \frac{a^2 - b^2}{a^2 + b^2} \right) \cos bx + \left(bx - \frac{2ab}{a^2 + b^2} \right) \sin bx \right]$$

11. $\displaystyle \int x^n e^{ax} \sin bx \, dx = e^{ax} \sum_{k=1}^{n+1} \frac{(-1)^{k+1} n! x^{n-k+1}}{(n - k + 1)!(a^2 + b^2)^{k/2}} \sin(bx + kt)$

12. $\displaystyle \int x^n e^{ax} \cos bx \, dx = e^{ax} \sum_{k=1}^{n+1} \frac{(-1)^{k+1} n! x^{n-k+1}}{(n - k + 1)!(a^2 + b^2)^{k/2}} \cos(bx + kt)$

where

$$\sin t = \frac{-b}{(a^2 + b^2)^{1/2}} \quad \text{and} \quad \cos t = \frac{a}{(a^2 + b^2)^{1/2}}.$$

6

Indefinite Integrals of Logarithmic Functions

6.1 Combinations of Logarithms and Polynomials

6.1.1 The logarithm

6.1.1.1 Integrands involving $\ln^m(ax)$.

1. $\displaystyle\int \ln x \, dx = x \ln x - x$

2. $\displaystyle\int \ln(ax) \, dx = x \ln(ax) - x$

3. $\displaystyle\int \ln^2(ax) \, dx = x \ln^2(ax) - 2x \ln(ax) + 2x$

4. $\displaystyle\int \ln^3(ax) \, dx = x \ln^3(ax) - 3x \ln^2(ax) + 6x \ln(ax) - 6x$

5. $\displaystyle\int \ln^4(ax) \, dx = x \ln^4(ax) - 4x \ln^3(ax) + 12x \ln^2(ax) - 24x \ln(ax) + 24x$

6. $\displaystyle\int \ln^m(ax)\,dx = x\ln^m(ax) - m\int \ln^{m-1}(ax)\,dx$

$$= \frac{x}{m+1}\sum_{k=0}^{m}(-1)^k(m+1)m(m-1)\cdots(m-k+1)\ln^{m-k}(ax)$$

$$[m > 0]$$

6.1.2 Integrands involving combinations of ln(ax) and powers of x

6.1.2.1 Integrands involving $x^m\ln^n(ax)$.

1. $\displaystyle\int x\ln(ax)\,dx = \frac{x^2}{2}\ln(ax) - \frac{x^2}{4}$

2. $\displaystyle\int x^2\ln(ax)\,dx = \frac{x^3}{3}\ln(ax) - \frac{x^3}{9}$

3. $\displaystyle\int x^3\ln(ax)\,dx = \frac{x^4}{4}\ln(ax) - \frac{x^4}{16}$

4. $\displaystyle\int x^n\ln(ax)\,dx = x^{n+1}\left[\frac{\ln(ax)}{n+1} - \frac{1}{(n+1)^2}\right]$ $\quad[n > 0]$

5. $\displaystyle\int x^n\ln^2(ax)\,dx = x^{n+1}\left[\frac{\ln^2(ax)}{n+1} - \frac{2\ln(ax)}{(n+1)^2} + \frac{2}{(n+1)^3}\right]$

6. $\displaystyle\int x^n\ln^3(ax)\,dx = x^{n+1}\left[\frac{\ln^3(ax)}{n+1} - \frac{3\ln^2(ax)}{(n+1)^2} + \frac{6\ln(ax)}{(n+1)^3} - \frac{6}{(n+1)^4}\right]$

7. $\displaystyle\int x^n\ln^m(ax)\,dx$

$$= \frac{x^{n+1}}{m+1}\sum_{k=0}^{m}(-1)^k\frac{(m+1)m(m-1)\cdots(m-k+1)\ln^{m-k}(ax)}{(n+1)^{k+1}}$$

$$[n > 0, m > 0]$$

6.1.2.2 Integrands involving $\ln^m(ax)/x^n$.

1. $\displaystyle\int \frac{\ln(ax)}{x}\,dx = \frac{1}{2}\{\ln(ax)\}^2$

2. $\displaystyle\int \frac{\ln(ax)}{x^2}\,dx = -\frac{1}{x}[\ln(ax) + 1]$

3. $\displaystyle\int \frac{\ln(ax)}{x^3}\,dx = -\frac{1}{2x^2}\left[\ln(ax) + \frac{1}{2}\right]$

4. $\displaystyle\int \frac{\ln^2(ax)}{x}\,dx = \frac{\ln(a^3x^3)}{3}$

5. $\displaystyle\int \frac{\ln^2(ax)}{x^2}\,dx = -\frac{1}{x}[\ln(a^2x^2) + 2\ln(ax) + 2]$

6. $\displaystyle\int \frac{\ln^2(ax)}{x^3}\,dx = -\frac{1}{2x^2}\left[\ln^2(ax) + \ln(ax) + \frac{1}{2}\right]$

7. $\displaystyle\int \frac{\ln^n(ax)}{x^m}\,dx = \frac{-1}{(n+1)x^{m-1}}\sum_{k=0}^{n}\frac{(n+1)n(n-1)\cdots(n-k+1)\ln^{n-k}(ax)}{(m-1)^{k+1}}$

$$[m > 1]$$

8. $\displaystyle\int \frac{\ln^n(ax)}{x}\,dx = \frac{\ln^{n+1}(ax)}{n+1}$

6.1.2.3 Integrands involving $[x\ln^m(ax)]^{-1}$.

1. $\displaystyle\int \frac{dx}{x\ln(ax)} = \ln|\ln(ax)|$

2. $\displaystyle\int \frac{dx}{x\ln^m(ax)} = \frac{-1}{(m-1)\ln^{m-1}(ax)} \qquad [m > 1]$

6.1.3 Integrands involving $(a + bx)^m\ln^n x$

6.1.3.1 Integrands involving $(a + bx)^m \ln x$.

1. $\displaystyle\int (a + bx)\ln x\,dx = \left[\frac{(a+bx)^2}{2b} - \frac{a^2}{2b}\right]\ln x - \left(ax + \frac{1}{4}bx^2\right)$

2. $\displaystyle\int (a + bx)^2\ln x\,dx = \frac{1}{3b}[(a+bx)^3 - a^3]\ln x - \left(a^2x + \frac{abx^2}{2} + \frac{b^2x^3}{9}\right)$

3. $\displaystyle\int (a + bx)^3\ln x\,dx = \frac{1}{4b}[(a+bx)^4 - a^4]\ln x$

$$- \left(a^3x + \frac{3}{4}a^2bx^2 + \frac{1}{3}ab^2x^3 + \frac{1}{16}b^3x^4\right)$$

4. $\displaystyle\int (a + bx)^m \ln x\,dx$

$$= \frac{1}{(m+1)b}[(a+bx)^{m+1} - a^{m+1}]\ln x - \sum_{k=0}^{m}\frac{\binom{m}{k}a^{m-k}b^kx^{k+1}}{(k+1)^2}$$

To integrate expressions of the form $(a + bx)^m \ln(cx)$, use the fact that

$$(a + bx)^m \ln(cx) = (a + bx)^m \ln c + (a + bx)^m \ln x$$

to reexpress the integral in terms of 6.1.3.1.1–4.

6.1.3.2 Integrands involving $\ln x/(a + bx)^m$.

1. $\displaystyle\int \frac{\ln x\,dx}{(a + bx)^2} = \frac{-\ln x}{b(a+bx)} + \frac{1}{ab}\ln\left(\frac{x}{a+bx}\right)$

2. $\displaystyle\int \frac{\ln x\, dx}{(a+bx)^3} = \frac{-\ln x}{2b(a+bx)^2} + \frac{1}{2ab(a+bx)} + \frac{1}{2a^2 b}\ln\left(\frac{x}{a+bx}\right)$

3. $\displaystyle\int \frac{\ln x\, dx}{(a+bx)^m} = \frac{1}{b(m-1)}\left[-\frac{\ln x}{(a+bx)^{m-1}} + \int \frac{dx}{x(a+bx)^{m-1}}\right]$ $[m > 1]$

4. $\displaystyle\int \frac{\ln x\, dx}{(a+bx)^{1/2}}$

$$= \frac{2}{b}\left\{(\ln x - 2)(a+bx)^{1/2} - 2a^{1/2}\ln\left[\frac{(a+bx)^{1/2} - a^{1/2}}{x^{1/2}}\right]\right\} \quad [a > 0]$$

$$= \frac{2}{b}\left\{(\ln x - 2)(a+bx)^{1/2} - 2(-a)^{1/2}\arctan\left[\left(\frac{a+bx}{-a}\right)^{1/2}\right]\right\}$$

$$[a < 0]$$

To integrate expressions of the form $\ln(cx)/(a+bx)^m$, use the fact that

$$\ln(cx)/(a+bx)^m = \ln c/(a+bx)^m + \ln x/(a+bx)^m$$

to reexpress the integral in terms of 6.1.3.2.1–4.

6.1.3.3 Integrands involving $x^m \ln(a+bx)$.

1. $\displaystyle\int \ln(a+bx)\, dx = \frac{1}{b}(a+bx)\ln(a+bx) - x$

2. $\displaystyle\int x \ln(a+bx)\, dx = \frac{1}{2}\left(x^2 - \frac{a^2}{b^2}\right)\ln(a+bx) - \frac{1}{2}\left(\frac{x^2}{2} - \frac{ax}{b}\right)$

3. $\displaystyle\int x^2 \ln(a+bx)\, dx = \frac{1}{3}\left(x^3 + \frac{a^3}{b^3}\right)\ln(a+bx) - \frac{1}{3}\left(\frac{x^3}{3} - \frac{ax^2}{2b} + \frac{a^2 x}{b^2}\right)$

4. $\displaystyle\int x^3 \ln(a+bx)\, dx = \frac{1}{4}\left(x^4 - \frac{a^4}{b^4}\right)\ln(a+bx) - \frac{1}{4}\left(\frac{x^4}{4} - \frac{ax^3}{3b} + \frac{a^2 x^2}{2b^2} - \frac{a^3 x}{b^3}\right)$

5. $\displaystyle\int x^m \ln(a+bx)\, dx = \frac{1}{m+1}\left[x^{m+1} - \frac{(-a)^{m+1}}{b^{m+1}}\right]\ln(a+bx)$

$$+ \frac{1}{m+1}\sum_{k=1}^{m+1} \frac{(-1)^k x^{m-k+2} a^{k-1}}{(m-k+2)b^{k-1}}$$

6.1.3.4 Integrands involving $\ln(a+bx)/x^m$.

1. $\displaystyle\int \frac{\ln(a+bx)}{x}\, dx = \ln|a|\ln|x| + \sum_{k=1}^{\infty} \frac{(-1)^{k+1}}{k^2}\left(\frac{bx}{a}\right)^k$ $[|bx| < |a|]$

$$= \frac{1}{2}\ln^2|bx| + \sum_{k=1}^{\infty} \frac{(-1)^k}{k^2}\left(\frac{a}{bx}\right)^2 \quad [|bx| > |a|]$$

2. $\int \dfrac{\ln(a+bx)}{x^2}\,dx = \dfrac{b}{a}\ln x - \left(\dfrac{1}{x}+\dfrac{b}{a}\right)\ln(a+bx)$

3. $\int \dfrac{\ln(a+bx)}{x^3}\,dx = \dfrac{b^2}{2a^2}\ln x + \dfrac{1}{2}\left(\dfrac{b^2}{a^2}-\dfrac{1}{x^2}\right)\ln(a+bx) - \dfrac{b}{2ax}$

6.1.4 Integrands involving $\ln(x^2 \pm a^2)$

6.1.4.1

1. $\int \ln(x^2+a^2)\,dx = x\ln(x^2+a^2) - 2x + 2a\arctan\dfrac{x}{a}$

2. $\int x\ln(x^2+a^2)\,dx = \dfrac{1}{2}[(x^2+a^2)\ln(x^2+a^2)-x^2]$

3. $\int x^2\ln(x^2+a^2)\,dx = \dfrac{1}{3}\left[x^3\ln(x^2+a^2) - \dfrac{2}{3}x^3 + 2a^2x - 2a^3\arctan\dfrac{x}{a}\right]$

4. $\int x^3\ln(x^2+a^2)\,dx = \dfrac{1}{4}\left[(x^4-a^4)\ln(x^2+a^2) - \dfrac{x^4}{2} + a^2x^2\right]$

5. $\int x^{2n}\ln(x^2+a^2)\,dx = \dfrac{1}{2n+1}\left\{x^{2n+1}\ln(x^2+a^2) + (-1)^n 2a^{2n+1}\arctan\dfrac{x}{a}\right.$

$$\left. - 2\sum_{k=0}^{n}\dfrac{(-1)^{n-k}}{2k+1}a^{2n-2k}x^{2k+1}\right\}$$

6. $\int x^{2n+1}\ln(x^2+a^2)\,dx = \dfrac{1}{2n+2}\left\{(x^{2n+2}+(-1)^n a^{2n+2})\ln(x^2+a^2)\right.$

$$\left. + \sum_{k=1}^{n+1}\dfrac{(-1)^{n-k}}{k}a^{2n-2k+2}x^{2k}\right\}$$

7. $\int \ln|x^2-a^2|\,dx = x\ln|x^2-a^2| - 2x + a\ln\left|\dfrac{x+a}{x-a}\right|$

8. $\int x\ln|x^2-a^2|\,dx = \dfrac{1}{2}\{(x^2-a^2)\ln|x^2-a^2| - x^2\}$

9. $\int x^2\ln|x^2-a^2|\,dx = \dfrac{1}{3}\left\{x^3\ln|x^2-a^2| - \dfrac{2}{3}x^3 - 2a^2x + a^3\ln\left|\dfrac{x+a}{x-a}\right|\right\}$

10. $\int x^3\ln|x^2-a^2|\,dx = \dfrac{1}{4}\left\{(x^4-a^4)\ln|x^2-a^2| - \dfrac{x^4}{2} - a^2x^2\right\}$

11. $\displaystyle\int x^{2n}\ln|x^2-a^2|\,dx = \frac{1}{2n+1}\Bigg\{x^{2n+1}\ln|x^2-a^2|$

$$+ a^{2n+1}\ln\left|\frac{x+a}{x-a}\right| - 2\sum_{k=0}^{n}\frac{1}{2k+1}a^{2n-2k}x^{2k+1}\Bigg\}$$

12. $\displaystyle\int x^{2n+1}\ln|x^2-a^2|\,dx = \frac{1}{2n+2}\Bigg\{(x^{2n+2}-a^{2n+2})\ln|x^2-a^2|$

$$- \sum_{k=1}^{n+1}\frac{1}{k}a^{2n-2k+2}x^{2k}\Bigg\}$$

6.1.5 Integrands involving $x^m\ln[x+(x^2\pm a^2)^{1/2}]$

6.1.5.1

1. $\displaystyle\int \ln[x+(x^2+a^2)^{1/2}]\,dx = x\ln[x+(x^2+a^2)^{1/2}] - (x^2+a^2)^{1/2}$

2. $\displaystyle\int x\ln[x+(x^2+a^2)^{1/2}]\,dx = \frac{1}{4}(2x^2+a^2)\ln[x+(x^2+a^2)^{1/2}] - \frac{1}{4}x(x^2+a^2)^{1/2}$

3. $\displaystyle\int x^2\ln[x+(x^2+a^2)^{1/2}]\,dx = \frac{x^3}{3}\ln[x+(x^2+a^2)^{1/2}]$

$$- \frac{1}{9}(x^2+a^2)^{3/2} + \frac{a^2}{3}(x^2+a^2)^{1/2}$$

4. $\displaystyle\int x^3\ln[x+(x^2+a^2)^{1/2}]\,dx = \frac{1}{32}(8x^4-3a^4)\ln[x+(x^2+a^2)^{1/2}]$

$$+ \frac{1}{32}(3a^2x-2x^3)(x^2+a^2)^{1/2}$$

5. $\displaystyle\int \ln[x+(x^2-a^2)^{1/2}]\,dx = x\ln[x+(x^2-a^2)^{1/2}] - (x^2-a^2)^{1/2}$

6. $\displaystyle\int x\ln[x+(x^2-a^2)^{1/2}]\,dx = \frac{1}{4}(2x^2-a^2)\ln[x+(x^2-a^2)^{1/2}] - \frac{1}{4}x(x^2-a^2)^{1/2}$

7. $\displaystyle\int x^2\ln[x+(x^2-a^2)^{1/2}]\,dx = \frac{x^3}{3}\ln[x+(x^2-a^2)^{1/2}]$

$$- \frac{1}{9}(x^2-a^2)^{3/2} - \frac{a^2}{3}(x^2-a^2)^{1/2}$$

8. $\displaystyle\int x^3\ln[x+(x^2-a^2)^{1/2}]\,dx = \frac{1}{32}(8x^4-3a^4)\ln[x+(x^2-a^2)^{1/2}]$

$$- \frac{1}{32}(2x^3+3a^2x)(x^2-a^2)^{1/2}$$

CHAPTER 7

Indefinite Integrals of Hyperbolic Functions

7.1 Basic Results

7.1.1 Integrands involving $\sinh(a + bx)$ and $\cosh(a + bx)$

7.1.1.1

1. $\displaystyle \int \sinh(a + bx)\,dx = \frac{1}{b}\cosh(a + bx)$

2. $\displaystyle \int \cosh(a + bx)\,dx = \frac{1}{b}\sinh(a + bx)$

3. $\displaystyle \int \tanh(a + bx)\,dx = \int \frac{\sinh(a + bx)}{\cosh(a + bx)}\,dx = \frac{1}{b}\ln[\cosh(a + bx)]$

   $\displaystyle \qquad\qquad\qquad\qquad = \frac{1}{b}\ln[\exp(2a + 2bx) + 1] - x$

4. $\displaystyle \int \operatorname{sech}(a + bx)\,dx = \int \frac{dx}{\cosh(a + bx)} = \frac{2}{b}\arctan[\exp(a + bx)]$

175

5. $\displaystyle\int \operatorname{csch}(a+bx)\,dx = \int \frac{dx}{\sinh(a+bx)} = \frac{1}{b}\ln\left|\tanh\frac{1}{2}(a+bx)\right|$

$$= \frac{1}{b}\ln\left|\frac{\exp(a+bx)-1}{\exp(a+bx)+1}\right|$$

6. $\displaystyle\int \coth(a+bx)\,dx = \int \frac{\cosh(a+bx)}{\sinh(a+bx)}\,dx = \frac{1}{b}\ln|\sinh(a+bx)|$

$$= \frac{1}{b}\,|\exp(2a+2bx)-1|-x$$

7.2 Integrands Involving Powers of sinh(bx) or cosh(bx)

7.2.1 Integrands involving powers of sinh(bx)

7.2.1.1

1. $\displaystyle\int \sinh(bx)\,dx = \frac{1}{b}\cosh(bx)$

2. $\displaystyle\int \sinh^2(bx)\,dx = \frac{1}{4b}\sinh(2bx) - \frac{1}{2}x$

3. $\displaystyle\int \sinh^3(bx)\,dx = \frac{1}{12b}\cosh(3bx) - \frac{3}{4b}\cosh(bx)$

4. $\displaystyle\int \sinh^4(bx)\,dx = \frac{1}{32b}\sinh(4bx) - \frac{1}{4b}\sinh(2bx) + \frac{3}{8}x$

5. $\displaystyle\int \sinh^5(bx)\,dx = \frac{1}{80b}\cosh(5bx) - \frac{5}{48b}\cosh(3bx) + \frac{5}{8b}\cosh(bx)$

6. $\displaystyle\int \sinh^6(bx)\,dx = \frac{1}{192b}\sinh(6bx) - \frac{3}{64b}\sinh(4bx) + \frac{15}{64b}\sinh(2bx) - \frac{5}{16}x$

To evaluate integrals of the form $\int \sinh^m(a+bx)\,dx$, make the change of variable $a+bx = u$, and then use the result

$$\int \sinh^m(a+bx)\,dx = \frac{1}{b}\int \sinh^m u\,du,$$

together with 7.2.1.1.1–6.

7.2.2 Integrands involving powers of cosh(bx)

7.2.2.1

1. $\displaystyle\int \cosh(bx)\,dx = \frac{1}{b}\sinh(bx)$

2. $\displaystyle \int \cosh^2(bx)\, dx = \frac{1}{4b} \sinh(2bx) + \frac{x}{2}$

3. $\displaystyle \int \cosh^3(bx)\, dx = \frac{1}{12b} \sinh(3bx) + \frac{3}{4b} \sinh(bx)$

4. $\displaystyle \int \cosh^4(bx)\, dx = \frac{1}{32b} \sinh(4bx) + \frac{1}{4b} \sinh(2bx) + \frac{3}{8}x$

5. $\displaystyle \int \cosh^5(bx)\, dx = \frac{1}{80b} \sinh(5bx) + \frac{5}{48b} \sinh(3bx) + \frac{5}{8b} \sinh(bx)$

6. $\displaystyle \int \cosh^6(bx)\, dx = \frac{1}{192b} \sinh(6bx) + \frac{3}{64b} \sinh(4bx) + \frac{15}{64b} \sinh(2bx) + \frac{5}{16}x$

To evaluate integrals of the form $\int \cosh^m(a + bx)\, dx$, make the change of variable $a + bx = u$, and then use the result

$$\int \cosh^m(a + bx)\, dx = \frac{1}{b} \int \cosh^m u\, du,$$

together with 7.2.2.1.1–6.

7.3 Integrands Involving $(a + bx)^m \sinh(cx)$ or $(a + bx)^m \cosh(cx)$

7.3.1 General results

7.3.1.1

1. $\displaystyle \int (a + bx) \sinh(cx)\, dx = \frac{1}{c}(a + bx)\cosh(cx) - \frac{b}{c^2} \sinh(cx)$

2. $\displaystyle \int (a + bx)^2 \sinh(cx)\, dx = \frac{1}{c}\left((a + bx)^2 + \frac{2b^2}{c^2}\right) \cosh(cx)$

$$- \frac{2b(a + bx)}{c^2} \sinh(cx)$$

3. $\displaystyle \int (a + bx)^3 \sinh(cx)\, dx = \left(\frac{a + bx}{c}\right)\left((a + bx)^2 + \frac{6b^2}{c^2}\right) \cosh(cx)$

$$- \frac{3b}{c^2}\left((a + bx)^2 + \frac{2b^2}{c^2}\right) \sinh(cx)$$

4. $\displaystyle \int (a + bx)^4 \sinh(cx)\, dx = \frac{1}{c}\left((a + bx)^4 + \frac{12b^2}{c^2}(a + bx)^2 + \frac{24b^4}{c^4}\right) \cosh(cx)$

$$- \frac{4b(a + bx)}{c^2}\left((a + bx)^2 + \frac{6b^2}{c^2}\right) \sinh(cx)$$

5. $\int (a + bx) \cosh(cx)\, dx = \frac{1}{c}(a + bx) \sinh(cx) - \frac{b}{c^2} \cosh(cx)$

6. $\int (a + bx)^2 \cosh(cx)\, dx = \frac{1}{c}\left((a + bx)^2 + \frac{2b^2}{c^2}\right) \sinh(cx)$

$$- \frac{2b(a + bx)}{c^2} \cosh(cx)$$

7. $\int (a + bx)^3 \cosh(cx)\, dx = \left(\frac{a + bx}{c}\right)\left((a + bx)^2 + \frac{6b^2}{c^2}\right) \sinh(cx)$

$$- \frac{3b}{c^2}\left((a + bx)^2 + \frac{2b^2}{c^2}\right) \cosh(cx)$$

8. $\int (a + bx)^4 \cosh(cx)\, dx = \frac{1}{c}\left((a + bx)^4 + \frac{12b^2}{c^2}(a + bx)^2 + \frac{24b^4}{c^4}\right) \sinh(cx)$

$$- \frac{4b(a + bx)}{c^2}\left((a + bx)^2 + \frac{6b^2}{c^2}\right) \cosh(cx)$$

7.3.1.2 Special cases $a = 0$, $b = c = 1$.

1. $\int x \sinh x \, dx = x \cosh x - \sinh x$

2. $\int x^2 \sinh x \, dx = (x^2 + 2) \cosh x - 2x \sinh x$

3. $\int x^3 \sinh x \, dx = (x^3 + 6x) \cosh x - (3x^2 + 6) \sinh x$

4. $\int x^4 \sinh x \, dx = (x^4 + 12x^2 + 24) \cosh x - 4x(x^2 + 6) \sinh x$

5. $\int x^n \sinh x \, dx = x^n \cosh x - n \int x^{n-1} \cosh x \, dx$

6. $\int x \cosh x \, dx = x \sinh x - \cosh x$

7. $\int x^2 \cosh x \, dx = (x^2 + 2) \sinh x - 2x \cosh x$

8. $\int x^3 \cosh x \, dx = (x^3 + 6x) \sinh x - (3x^2 + 6) \cosh x$

9. $\int x^4 \cosh x \, dx = (x^4 + 12x^2 + 24) \sinh x - 4x(x^2 + 6) \cosh x$

10. $\int x^n \cosh x \, dx = x^n \sinh x - n \int x^{n-1} \sinh x \, dx$

7.4 Integrands Involving $x^m \sinh^n x$ or $x^m \cosh^n x$

7.4.1 Integrands involving $x^m \sinh^n x$

7.4.1.1

1. $\displaystyle \int x \sinh^2 x \, dx = \frac{1}{4} x \sinh 2x - \frac{1}{8} \cosh 2x - \frac{1}{4} x^2$

2. $\displaystyle \int x^2 \sinh^2 x \, dx = \frac{1}{4} \left(x^2 + \frac{1}{2} \right) \sinh 2x - \frac{1}{4} x \cosh 2x - \frac{1}{6} x^3$

3. $\displaystyle \int x \sinh^3 x \, dx = \frac{3}{4} \sinh x - \frac{1}{36} \sinh 3x - \frac{3}{4} x \cosh x + \frac{1}{12} x \cosh 3x$

4. $\displaystyle \int x^2 \sinh^3 x \, dx = -\left(\frac{3x^2}{4} + \frac{3}{2} \right) \cosh x + \left(\frac{x^2}{12} + \frac{1}{54} \right) \cosh 3x$

$$+ \frac{3}{2} x \sinh x - \frac{1}{18} x \sinh 3x$$

7.4.2 Integrals involving $x^m \cosh^n x$

7.4.2.1

1. $\displaystyle \int x \cosh^2 x \, dx = \frac{1}{4} x \sinh 2x - \frac{1}{8} \cosh 2x + \frac{1}{4} x^2$

2. $\displaystyle \int x^2 \cosh^2 x \, dx = \frac{1}{4} \left(x^2 + \frac{1}{2} \right) \sinh 2x - \frac{1}{4} x \cosh 2x + \frac{1}{6} x^3$

3. $\displaystyle \int x \cosh^3 x \, dx = -\frac{3}{4} \cosh x - \frac{1}{36} \cosh 3x + \frac{3}{4} x \sinh x + \frac{1}{12} x \sinh 3x$

4. $\displaystyle \int x^2 \cosh^3 x \, dx = \left(\frac{3x^2}{4} + \frac{3}{2} \right) \sinh x + \left(\frac{x^2}{12} + \frac{1}{54} \right) \sinh 3x$

$$- \frac{3}{2} x \cosh x - \frac{1}{18} x \cosh 3x$$

7.5 Integrands Involving $x^m \sinh^{-n} x$ or $x^m \cosh^{-n} x$

7.5.1 Integrands involving $x^m \sinh^{-n} x$

7.5.1.1

1. $\displaystyle \int \frac{dx}{\sinh x} = \ln \left| \tanh \frac{x}{2} \right|$

2. $\displaystyle\int \frac{x\,dx}{\sinh x} = \sum_{k=0}^{\infty} \frac{(2 - 2^{2k})B_{2k}}{(2k+1)(2k)!} x^{2k+1}$ $[|x| < \pi]$ (see 1.3.1.1)

3. $\displaystyle\int \frac{x^n\,dx}{\sinh x} = \sum_{k=0}^{\infty} \frac{(2 - 2^{2k})B_{2k}}{(2k+n)(2k)!} x^{2k+n}$ $[|x| < \pi]$ (see 1.3.1.1)

4. $\displaystyle\int \frac{dx}{\sinh^2 x} = -\coth x$

5. $\displaystyle\int \frac{x\,dx}{\sinh^2 x} = -x \coth x + \ln|\sinh x|$

6. $\displaystyle\int \frac{dx}{\sinh^3 x} = -\frac{\cosh x}{2 \sinh^2 x} - \frac{1}{2} \ln\left|\tanh\frac{x}{2}\right|$

7. $\displaystyle\int \frac{x\,dx}{\sinh^3 x} = -\frac{x \cosh x}{2 \sinh^2 x} - \frac{1}{2 \sinh x} - \frac{1}{2}\int \frac{x\,dx}{\sinh x}$

8. $\displaystyle\int \frac{dx}{\sinh^4 x} = \coth x - \frac{1}{3} \coth^3 x$

7.5.2 Integrands involving $x^m \cosh^{-n} x$

7.5.2.1

1. $\displaystyle\int \frac{dx}{\cosh x} = \arctan(\sinh x) = 2 \arctan e^x$

2. $\displaystyle\int \frac{x\,dx}{\cosh x} = \sum_{k=0}^{\infty} \frac{E_{2k}}{(2k+2)(2k)!} x^{2k+2}$ $\left[|x| < \frac{\pi}{2}\right]$ (see 1.3.1.1)

3. $\displaystyle\int \frac{x^n\,dx}{\cosh x} = \sum_{k=0}^{\infty} \frac{E_{2k}}{(2k+n+1)(2k)!} x^{2k+n+1}$ $\left[|x| < \frac{\pi}{2}\right]$ (see 1.3.1.1)

4. $\displaystyle\int \frac{dx}{\cosh^2 x} = \tanh x$

5. $\displaystyle\int \frac{x\,dx}{\cosh^2 x} = x \tanh x - \ln(\cosh x)$

6. $\displaystyle\int \frac{dx}{\cosh^3 x} = \frac{\sinh x}{2 \cosh^2 x} + \frac{1}{2} \arctan(\sinh x)$

7. $\displaystyle\int \frac{x\,dx}{\cosh^3 x} = \frac{x \sinh x}{2 \cosh^2 x} + \frac{1}{2 \cosh x} + \frac{1}{2}\int \frac{x\,dx}{\cosh x}$

8. $\displaystyle\int \frac{dx}{\cosh^4 x} = \tanh x - \frac{1}{3} \tanh^3 x$

7.6 Integrands Involving $(1 \pm \cosh x)^{-m}$

7.6.1 Integrands involving $(1 \pm \cosh x)^{-1}$

7.6.1.1

1. $\displaystyle \int \frac{dx}{1 + \cosh x} = \tanh \frac{x}{2}$

2. $\displaystyle \int \frac{dx}{1 - \cosh x} = \coth \frac{x}{2}$

3. $\displaystyle \int \frac{x \, dx}{1 + \cosh x} = x \tanh \frac{x}{2} - 2 \ln \left(\cosh \frac{x}{2} \right)$

4. $\displaystyle \int \frac{x \, dx}{1 - \cosh x} = x \coth \frac{x}{2} - 2 \ln \left| \sinh \frac{x}{2} \right|$

5. $\displaystyle \int \frac{\cosh x \, dx}{1 + \cosh x} = x - \tanh \frac{x}{2}$

6. $\displaystyle \int \frac{\cosh x \, dx}{1 - \cosh x} = \coth \frac{x}{2} - x$

7.6.2 Integrands involving $(1 \pm \cosh x)^{-2}$

7.6.2.1

1. $\displaystyle \int \frac{dx}{(1 + \cosh x)^2} = \frac{1}{2} \tanh \frac{x}{2} - \frac{1}{6} \tanh^3 \frac{x}{2}$

2. $\displaystyle \int \frac{dx}{(1 - \cosh x)^2} = \frac{1}{2} \coth \frac{x}{2} - \frac{1}{6} \coth^3 \frac{x}{2}$

3. $\displaystyle \int \frac{x \sinh x \, dx}{(1 + \cosh x)^2} = -\frac{x}{\cosh x + 1} + \tanh \frac{x}{2}$

4. $\displaystyle \int \frac{x \sinh x \, dx}{(1 - \cosh x)^2} = -\frac{x}{\cosh x - 1} - \coth \frac{x}{2}$

7.7 Integrands Involving $\sinh(ax) \cosh^{-n} x$ or $\cosh(ax) \sinh^{-n} x$

7.7.1 Integrands involving $\sinh(ax) \cosh^{-n} x$

7.7.1.1

1. $\displaystyle \int \frac{\sinh 2x}{\cosh^n x} \, dx = \left(\frac{2}{2 - n} \right) \cosh^{2-n} x \qquad [n \neq 2]$

2. $\int \dfrac{\sinh 2x}{\cosh^2 x}\, dx = 2\ln(\cosh x)$ [case $n = 2$]

3. $\int \dfrac{\sinh 3x}{\cosh^n x}\, dx = \left(\dfrac{4}{3-n}\right)\cosh^{3-n} x - \left(\dfrac{1}{1-n}\right)\cosh^{1-n} x$ [$n \neq 1, 3$]

4. $\int \dfrac{\sinh 3x}{\cosh x}\, dx = 2\sinh^2 x - \ln(\cosh x)$ [case $n = 1$]

5. $\int \dfrac{\sinh 3x}{\cosh^3 x}\, dx = -\dfrac{1}{2}\tanh^2 x + 4\ln(\cosh x)$ [case $n = 3$]

7.7.2 Integrands involving $\cosh(ax)\sinh^{-n} x$

7.7.2.1

1. $\int \dfrac{\cosh 2x}{\sinh x}\, dx = 2\cosh x + \ln\left|\tanh\dfrac{x}{2}\right|$

2. $\int \dfrac{\cosh 2x}{\sinh^2 x}\, dx = -\coth x + 2x$

3. $\int \dfrac{\cosh 2x}{\sinh^3 x}\, dx = -\dfrac{\cosh x}{2\sinh^2 x} + \dfrac{3}{2}\ln\left|\tanh\dfrac{x}{2}\right|$

4. $\int \dfrac{\cosh 3x}{\sinh^n x}\, dx = \left(\dfrac{4}{3-n}\right)\sinh^{3-n} x + \left(\dfrac{1}{1-n}\right)\sinh^{1-n} x$ [$n \neq 1, 3$]

5. $\int \dfrac{\cosh 3x}{\sinh x}\, dx = 2\sinh^2 x + \ln|\sinh x|$ [case $n = 1$]

6. $\int \dfrac{\cosh 3x}{\sinh^3 x}\, dx = -\dfrac{1}{2}\coth^2 x + 4\ln|\sinh x|$ [case $n = 3$]

7.8 Integrands Involving $\sinh(ax + b)$ and $\cosh(cx + d)$

7.8.1 General case

7.8.1.1

1. $\displaystyle \int \sinh(ax + b)\sinh(cx + d)\, dx = \dfrac{1}{2(a+c)}\sinh[(a+c)x + b + d]$

$$-\dfrac{1}{2(a-c)}\sinh[(a-c)x + b - d]$$

$$[a^2 \neq c^2]$$

2. $\displaystyle\int \sinh(ax + b) \cosh(cx + d)\, dx = \frac{1}{2(a + c)} \cosh[(a + c)x + b + d]$

$$+ \frac{1}{2(a - c)} \cosh[(a - c)x + b - d]$$

$$[a^2 \neq c^2]$$

3. $\displaystyle\int \cosh(ax + b) \cosh(cx + d)\, dx = \frac{1}{2(a + c)} \sinh[(a + c)x + b + d]$

$$+ \frac{1}{2(a - c)} \sinh[(a - c)x + b - d]$$

$$[a^2 \neq c^2]$$

7.8.2 Special case $a = c$

7.8.2.1

1. $\displaystyle\int \sinh(ax + b) \sinh(ax + d)\, dx = -\frac{1}{2}x \cosh(b - d) + \frac{1}{4a} \sinh(2ax + b + d)$

2. $\displaystyle\int \sinh(ax + b) \cosh(ax + d)\, dx = \frac{1}{2}x \sinh(b - d) + \frac{1}{4a} \cosh(2ax + b + d)$

3. $\displaystyle\int \cosh(ax + b) \cosh(ax + d)\, dx = \frac{1}{2}x \cosh(b - d) + \frac{1}{4a} \sinh(2ax + b + d)$

7.8.3 Integrands involving $\sinh^p x \cosh^q x$

7.8.3.1

1. $\displaystyle\int \frac{\sinh^{2m} x}{\cosh x}\, dx = \sum_{k=1}^{m} \frac{(-1)^{m+k}}{2k - 1} \sinh^{2k-1} x + (-1)^m \arctan(\sinh x) \qquad [m \geq 1]$

2. $\displaystyle\int \frac{\sinh^{2m+1} x}{\cosh x}\, dx = \sum_{k=1}^{m} \frac{(-1)^{m+k}}{2k} \sinh^{2k} x + (-1)^m \ln(\cosh x) \qquad [m \geq 1]$

3. $\displaystyle\int \frac{dx}{\sinh^{2m} x \cosh x} = \sum_{k=1}^{m} \frac{(-1)^k \operatorname{cosech}^{2m-2k+1} x}{2m - 2k + 1} + (-1)^m \arctan(\sinh x)$

$$[m \geq 1]$$

4. $\displaystyle\int \frac{dx}{\sinh^{2m+1} x \cosh x} = \sum_{k=1}^{m} \frac{(-1)^k \operatorname{cosech}^{2m-2k+2} x}{2m - 2k + 2} + (-1)^m \ln|\tanh x|$

5. $\displaystyle\int \frac{\cosh^{2m} x}{\sinh x}\, dx = \sum_{k=1}^{m} \frac{\cosh^{2k-1} x}{2k - 1} + \ln\left|\tanh \frac{x}{2}\right|$

6. $\displaystyle\int \frac{\cosh^{2m+1} x}{\sinh x}\, dx = \sum_{k=1}^{m} \frac{\cosh^{2k} x}{2k} + \ln|\sinh x|$

7. $\displaystyle\int \frac{dx}{\sinh x \cosh^{2m} x} = \sum_{k=1}^{m} \frac{\operatorname{sech}^{2m-2k+1} x}{2m - 2k + 1} + \ln\left|\tanh\frac{x}{2}\right|$

8. $\displaystyle\int \frac{dx}{\sinh x \cosh^{2m+1} x} = \sum_{k=1}^{m} \frac{\operatorname{sech}^{2m-2k+2} x}{2m - 2k + 2} + \ln\left|\tanh\frac{x}{2}\right|$

7.9 Integrands Involving tanh *kx* and coth *kx*

7.9.1 Integrands involving tanh *kx*

7.9.1.1

1. $\displaystyle\int \tanh kx \, dx = \frac{1}{k} \ln(\cosh kx)$

2. $\displaystyle\int \tanh^2 kx \, dx = x - \frac{1}{k} \tanh kx$

3. $\displaystyle\int \tanh^3 kx \, dx = \frac{1}{k} \ln(\cosh kx) - \frac{1}{2k} \tanh^2 kx$

4. $\displaystyle\int \tanh^{2n} kx \, dx = x - \frac{1}{k} \sum_{k=1}^{n} \frac{\tanh^{2n-2k+1} kx}{2n - 2k + 1}$

5. $\displaystyle\int \tanh^{2n+1} kx \, dx = \frac{1}{k} \ln(\cosh kx) - \frac{1}{k} \sum_{k=1}^{n} \frac{\tanh^{2n-2k+2} kx}{2n - 2k + 2}$

7.9.2 Integrands involving coth *kx*

7.9.2.1

1. $\displaystyle\int \coth kx \, dx = \frac{1}{k} \ln|\sinh kx|$

2. $\displaystyle\int \coth^2 kx \, dx = x - \frac{1}{k} \coth kx$

3. $\displaystyle\int \coth^3 kx \, dx = \frac{1}{k} \ln|\sinh kx| - \frac{1}{2k} \coth^2 kx$

4. $\displaystyle\int \coth^{2n} kx \, dx = x - \frac{1}{k} \sum_{k=1}^{n} \frac{\coth^{2n-2k+1} kx}{2n - 2k + 1}$

5. $\displaystyle\int \coth^{2n+1} kx \, dx = \frac{1}{k} \ln|\sinh kx| - \frac{1}{k} \sum_{k=1}^{n} \frac{\coth^{2n-2k+2} kx}{2n - 2k + 2}$

7.10 Integrands Involving $(a + bx)^m \sinh kx$ or $(a + bx)^m \cosh kx$

7.10.1 Integrands involving $(a + bx)^m \sinh kx$

7.10.1.1

1. $\displaystyle \int (a + bx) \sinh kx \, dx = \frac{1}{k}(a + bx) \cosh kx - \frac{b}{k^2} \sinh kx$

2. $\displaystyle \int (a + bx)^2 \sinh kx \, dx = \frac{1}{k}\left((a + bx)^2 + \frac{2b^2}{k^2}\right) \cosh kx - \frac{2b(a + bx)}{k^2} \sinh kx$

3. $\displaystyle \int (a + bx)^3 \sinh kx \, dx = \frac{(a + bx)}{k}\left((a + bx)^2 + \frac{6b^2}{k^2}\right) \cosh kx$

   $\displaystyle \qquad\qquad - \frac{3b}{k^2}\left((a + bx)^2 + \frac{2b^2}{k^2}\right) \sinh kx$

7.10.2 Integrands involving $(a + bx)^m \cosh kx$

7.10.2.1

1. $\displaystyle \int (a + bx) \cosh kx \, dx = \frac{1}{k}(a + bx) \sinh kx - \frac{b}{k^2} \cosh kx$

2. $\displaystyle \int (a + bx)^2 \cosh kx \, dx = \frac{1}{k}\left((a + bx)^2 + \frac{2b^2}{k^2}\right) \sinh kx - \frac{2b(a + bx)}{k^2} \cosh kx$

3. $\displaystyle \int (a + bx)^3 \cosh kx \, dx = \frac{(a + bx)}{k}\left((a + bx)^2 + \frac{6b^2}{k^2}\right) \sinh kx$

   $\displaystyle \qquad\qquad - \frac{3b}{k^2}\left((a + bx)^2 + \frac{2b^2}{k^2}\right) \cosh kx$

Indefinite Integrals Involving Inverse Hyperbolic Functions

8.1 Basic Results

8.1.1 Integrands involving products of x^n and arcsinh(x/a) or arccosh(x/a)

8.1.1.1 Integrands involving x^n arcsinh(x/a).

1. $\displaystyle\int \operatorname{arcsinh}\frac{x}{a}\,dx = x\operatorname{arcsinh}\frac{x}{a} - (x^2 + a^2)^{1/2}$ $[a > 0]$

2. $\displaystyle\int x\operatorname{arcsinh}\frac{x}{a}\,dx = \frac{1}{4}(2x^2 + a^2)\operatorname{arcsinh}\frac{x}{a} - \frac{1}{4}x(x^2 + a^2)^{1/2}$ $[a > 0]$

3. $\displaystyle\int x^2\operatorname{arcsinh}\frac{x}{a}\,dx = \frac{1}{3}x^3\operatorname{arcsinh}\frac{x}{a} + \frac{1}{9}(2a^2 - x^2)(x^2 + a^2)^{1/2}$ $[a > 0]$

4. $\displaystyle\int x^3\operatorname{arcsinh}\frac{x}{a}\,dx = \frac{1}{32}(8x^4 - 3a^4)\operatorname{arcsinh}\frac{x}{a} + \frac{1}{32}(3a^2x - 2x^3)(x^2 + a^2)^{1/2}$

$[a > 0]$

187

5. $\displaystyle \int x^n \operatorname{arcsinh} \frac{x}{a}\, dx = \frac{x^{n+1}}{n+1} \operatorname{arcsinh} \frac{x}{a} - \frac{1}{n+1} \int \frac{x^{n+1}\, dx}{(x^2 + a^2)^{1/2}}$ (see 4.3.3)

8.1.1.2 Integrands involving $x^n \operatorname{arccosh}(x/a)$.

1. $\displaystyle \int \operatorname{arccosh} \frac{x}{a}\, dx = x \operatorname{arccosh} \frac{x}{a} - (x^2 + a^2)^{1/2}$ $[\operatorname{arccosh}(x/a) > 0]$

$\displaystyle \qquad\qquad\qquad = x \operatorname{arccosh} \frac{x}{a} + (x^2 - a^2)^{1/2}$ $[\operatorname{arccosh}(x/a) < 0]$

2. $\displaystyle \int x \operatorname{arccosh} \frac{x}{a}\, dx$

$\displaystyle \qquad = \frac{1}{4}(2x^2 - a^2) \operatorname{arccosh} \frac{x}{a} - \frac{1}{4}x(x^2 - a^2)^{1/2}$ $[\operatorname{arccosh}(x/a) > 0]$

$\displaystyle \qquad = \frac{1}{4}(2x^2 - a^2) \operatorname{arccosh} \frac{x}{a} + \frac{1}{4}x(x^2 - a^2)^{1/2}$ $[\operatorname{arccosh}(x/a) < 0]$

3. $\displaystyle \int x^2 \operatorname{arccosh} \frac{x}{a}\, dx$

$\displaystyle \qquad = \frac{1}{3}x^3 \operatorname{arccosh} \frac{x}{a} - \frac{1}{9}(2a^2 + x^2)(x^2 - a^2)^{1/2}$ $[\operatorname{arccosh}(x/a) > 0]$

$\displaystyle \qquad = \frac{1}{3}x^3 \operatorname{arccosh} \frac{x}{a} + \frac{1}{9}(2a^2 + x^2)(x^2 - a^2)^{1/2}$ $[\operatorname{arccosh}(x/a) < 0]$

4. $\displaystyle \int x^3 \operatorname{arccosh} \frac{x}{a}\, dx$

$\displaystyle \qquad = \frac{1}{32}(8x^4 - 3a^4) \operatorname{arccosh} \frac{x}{a} - \frac{1}{32}(3a^2x + 2x^3)(x^2 - a^2)^{1/2}$

$$[\operatorname{arccosh}(x/a) > 0]$$

$\displaystyle \qquad = \frac{1}{32}(8x^4 - 3a^4) \operatorname{arccosh} \frac{x}{a} + \frac{1}{32}(3a^2x + 2x^3)(x^2 - a^2)^{1/2}$

$$[\operatorname{arccosh}(x/a) < 0]$$

5. $\displaystyle \int x^n \operatorname{arccosh} \frac{x}{a}\, dx = \frac{x^{n+1}}{n+1} \operatorname{arccosh} \frac{x}{a} - \frac{1}{n+1} \int \frac{x^{n+1}\, dx}{(x^2 - a^2)^{1/2}}$

$$[\operatorname{arccosh}(x/a) > 0, n \neq -1] \quad \text{(see 4.3.3)}$$

$\displaystyle \qquad = \frac{x^{n+1}}{n+1} \operatorname{arccosh} \frac{x}{a} + \frac{1}{n+1} \int \frac{x^{n+1}\, dx}{(x^2 - a^2)^{1/2}}$

$$[\operatorname{arccosh}(x/a) < 0, n \neq -1] \quad \text{(see 4.3.3)}$$

8.2 Integrands Involving $x^{-n}\operatorname{arcsinh}(x/a)$ or $x^{-n}\operatorname{arccosh}(x/a)$

8.2.1 Integrands involving $x^{-n}\operatorname{arcsinh}(x/a)$

8.2.1.1

1. $\displaystyle\int \frac{1}{x}\operatorname{arcsinh}\frac{x}{a}\,dx$

$$= \frac{x}{a} - \frac{1}{2\cdot 3\cdot 3}\frac{x^3}{a^3} + \frac{1\cdot 3}{2\cdot 4\cdot 5\cdot 5}\frac{x^5}{a^5} - \frac{1\cdot 3\cdot 5}{2\cdot 4\cdot 6\cdot 7\cdot 7}\frac{x^7}{a^7} + \cdots$$

$$[x^2 < a^2]$$

2. $\displaystyle\int \frac{1}{x^2}\operatorname{arcsinh}\frac{x}{a}\,dx = -\frac{1}{x}\operatorname{arcsinh}\frac{x}{a} - \frac{1}{a}\ln\left|\frac{a + (x^2 + a^2)^{1/2}}{x}\right|$

3. $\displaystyle\int \frac{1}{x^3}\operatorname{arcsinh}\frac{x}{a}\,dx = -\frac{1}{2x^2}\operatorname{arcsinh}\frac{x}{a} - \frac{1}{2ax}\left(1 + \frac{x^2}{a^2}\right)^{1/2}$

4. $\displaystyle\int \frac{1}{x^4}\operatorname{arcsinh}\frac{x}{a}\,dx = -\frac{1}{3x^3}\operatorname{arcsinh}\frac{x}{a} + \frac{1}{6a^3}\operatorname{arcsinh}(a/x)$

$$- \frac{1}{6ax^2}\left(1 + \frac{x^2}{a^2}\right)^{1/2}$$

5. $\displaystyle\int \frac{1}{x^n}\operatorname{arcsinh}\frac{x}{a}\,dx = -\frac{1}{(n-1)x^{n-1}}\operatorname{arcsinh}\frac{x}{a} + \frac{1}{n-1}\int \frac{dx}{x^{n-1}(x^2 + a^2)^{1/2}}$

(see 4.3.3)

8.2.2 Integrands involving $x^{-n}\operatorname{arccosh}(x/a)$

8.2.2.1

1. $\displaystyle\int \frac{1}{x}\operatorname{arccosh}\frac{x}{a}\,dx$

$$= \frac{1}{2}\left(\ln\frac{2x}{a}\right)^2 + \frac{1}{2^3}\frac{a^2}{x^2} + \frac{1\cdot 3}{2\cdot 4^3}\frac{a^4}{x^4} + \frac{1\cdot 3\cdot 5}{2\cdot 4\cdot 6^3}\frac{a^6}{x^6} + \cdots$$

$$[\operatorname{arccosh}(x/a) > 0]$$

$$= -\left[\frac{1}{2}\left(\ln\frac{2x}{a}\right)^2 + \frac{1}{2^3}\frac{a^2}{x^2} + \frac{1\cdot 3}{2\cdot 4^3}\frac{a^4}{x^4} + \frac{1\cdot 3\cdot 5}{2\cdot 4\cdot 6^3}\frac{a^6}{x^6} + \cdots\right]$$

$$[\operatorname{arccosh}(x/a) < 0]$$

2. $\displaystyle\int \frac{1}{x^2}\operatorname{arccosh}\frac{x}{a}\,dx = -\frac{1}{x}\operatorname{arccosh}\frac{x}{a} - \frac{1}{a}\arcsin(a/x)$

3. $\displaystyle\int \frac{1}{x^3}\operatorname{arccosh}\frac{x}{a}\,dx = -\frac{1}{2x^2}\operatorname{arccosh}\frac{x}{a} + \frac{1}{2ax}\left(\frac{x^2}{a^2} - 1\right)^{1/2}$ $\qquad [x^2 > a^2]$

4. $\displaystyle\int \frac{1}{x^4} \operatorname{arccosh} \frac{x}{a}\, dx = -\frac{1}{3x^3} \operatorname{arccosh} \frac{x}{a} - \frac{1}{6a^3} \arcsin(a/x)$

$$+ \frac{1}{6ax^2}\left(\frac{x^2}{a^2}-1\right)^{1/2} \qquad [x^2 > a^2]$$

5. $\displaystyle\int \frac{1}{x^n} \operatorname{arccosh} \frac{x}{a}\, dx = -\frac{1}{(n-1)x^{n-1}} \operatorname{arccosh} \frac{x}{a} + \frac{1}{n-1}\int \frac{dx}{x^{n-1}(x^2-a^2)^{1/2}}$

$$[\operatorname{arccosh}(x/a) > 0, n \neq 1] \quad \text{(see 4.3.3)}$$

$$-\frac{1}{(n-1)x^{n-1}} \operatorname{arccosh} \frac{x}{a} - \frac{1}{n-1}\int \frac{dx}{x^{n-1}(x^2-a^2)^{1/2}}$$

$$[\operatorname{arccosh}(x/a) < 0, n \neq 1] \quad \text{(see 4.3.3)}$$

8.3 Integrands Involving $x^n \operatorname{arctanh}(x/a)$ or $x^n \operatorname{arccoth}(x/a)$

8.3.1 Integrands involving $x^n \operatorname{arctanh}(x/a)$

1. $\displaystyle\int \operatorname{arctanh} \frac{x}{a}\, dx = x \operatorname{arctanh} \frac{x}{a} + \frac{1}{2}a \ln(a^2 - x^2) \qquad [x^2 < a^2]$

2. $\displaystyle\int x \operatorname{arctanh} \frac{x}{a}\, dx = \frac{1}{2}(x^2 - a^2) \operatorname{arctanh} \frac{x}{a} + \frac{1}{2}ax \qquad [x^2 < a^2]$

3. $\displaystyle\int x^2 \operatorname{arctanh} \frac{x}{a}\, dx = \frac{1}{3}x^3 \operatorname{arctanh} \frac{x}{a} + \frac{1}{6}ax^2 + \frac{1}{6}a^3 \ln(a^2 - x^2) \qquad [x^2 < a^2]$

4. $\displaystyle\int x^3 \operatorname{arctanh} \frac{x}{a}\, dx = \frac{1}{4}(x^4 - a^4) \operatorname{arctanh} \frac{x}{a} + \frac{1}{12}ax^3 + \frac{1}{4}a^3 x \qquad [x^2 < a^2]$

5. $\displaystyle\int x^4 \operatorname{arctanh} \frac{x}{a}\, dx = \frac{1}{5}x^5 \operatorname{arctanh} \frac{x}{a} + \frac{1}{20}ax^2(2a^2 + x^2) + \frac{1}{10}a^5 \ln(x^2 - a^2)$

$$[x^2 < a^2]$$

6. $\displaystyle\int x^n \operatorname{arctanh} \frac{x}{a}\, dx = -\frac{1}{(n-1)x^{n-1}} \operatorname{arctanh} \frac{x}{a} + \frac{a}{n-1}\int \frac{dx}{x^{n-1}(a^2 - x^2)}$

$$[x^2 < a^2, n \neq 1] \quad \text{(see 4.2.4.9)}$$

8.3.2 Integrands involving $x^n \operatorname{arccoth}(x/a)$

8.3.2.1

1. $\displaystyle\int \operatorname{arccoth} \frac{x}{a}\, dx = x \operatorname{arccoth} \frac{x}{a} + \frac{1}{2}a \ln(x^2 - a^2) \qquad [a^2 < x^2]$

2. $\displaystyle\int x \operatorname{arccoth} \frac{x}{a}\, dx = \frac{1}{2}(x^2 - a^2) \operatorname{arccoth} \frac{x}{a} + \frac{1}{2}ax \qquad [a^2 < x^2]$

3. $\int x^2 \text{ arccoth} \dfrac{x}{a} \, dx = \dfrac{1}{3}x^3 \text{ arccoth} \dfrac{x}{a} + \dfrac{1}{6}ax^2 + \dfrac{1}{6}a^3 \ln(x^2 - a^2)$ $[a^2 < x^2]$

4. $\int x^3 \text{ arccoth} \dfrac{x}{a} \, dx = \dfrac{1}{4}(x^4 - a^4) \text{ arccoth} \dfrac{x}{a} + \dfrac{1}{12}ax^3 + \dfrac{1}{4}a^3 x$ $[a^2 < x^2]$

5. $\int x^4 \text{ arccoth} \dfrac{x}{a} \, dx = \dfrac{1}{5}x^5 \text{ arccoth} \dfrac{x}{a} + \dfrac{1}{20}ax^2(2a^2 + x^2) + \dfrac{1}{10}a^5 \ln(x^2 - a^2)$

$[a^2 < x^2]$

6. $\int x^n \text{ arccoth} \dfrac{x}{a} \, dx = \dfrac{x^{n+1}}{n+1} \text{ arccoth} \dfrac{x}{a} - \dfrac{a}{n+1} \int \dfrac{x^{n+1} dx}{a^2 - x^2}$

$[a^2 < x^2, n \neq -1]$ (see 4.2.4)

8.4 Integrands Involving x^{-n} arctanh(x/a) or x^{-n} arccoth(x/a)

8.4.1 Integrands involving x^{-n} arctanh(x/a)

8.4.1.1

1. $\int \dfrac{1}{x} \text{ arctanh} \dfrac{x}{a} \, dx = \dfrac{x}{a} + \dfrac{x^3}{3^2 a^3} + \dfrac{x^5}{5^2 a^5} + \dfrac{x^7}{7^2 a^7} + \cdots$ $[x^2 < a^2]$

2. $\int \dfrac{1}{x^2} \text{ arctanh} \dfrac{x}{a} \, dx = -\dfrac{1}{x} \text{ arctanh} \dfrac{x}{a} - \dfrac{1}{2a} \ln\left(\dfrac{a^2 - x^2}{x^2}\right)$ $[x^2 < a^2]$

3. $\int \dfrac{1}{x^3} \text{ arctanh} \dfrac{x}{a} \, dx = \dfrac{1}{2}\left(\dfrac{1}{a^2} - \dfrac{1}{x^2}\right) \text{ arctanh} \dfrac{x}{a} - \dfrac{1}{2ax}$ $[x^2 < a^2]$

4. $\int \dfrac{1}{x^n} \text{ arctanh} \dfrac{x}{a} \, dx = -\dfrac{1}{(n-1)x^{n-1}} \text{ arctanh} \dfrac{x}{a} + \dfrac{a}{n-1} \int \dfrac{dx}{x^{n-1}(a^2 - x^2)}$

$[x^2 < a^2, n \neq 1]$ (see 4.2.4.9)

8.4.2 Integrands involving x^{-n} arccoth(x/a)

1. $\int \dfrac{1}{x} \text{ arccoth} \dfrac{x}{a} \, dx = -\dfrac{a}{x} - \dfrac{a^3}{3^2 x^3} - \dfrac{a^5}{5^2 x^5} - \dfrac{a^7}{7^2 x^7} - \cdots$ $[a^2 < x^2]$

2. $\int \dfrac{1}{x^2} \text{ arccoth} \dfrac{x}{a} \, dx = -\dfrac{1}{x} \text{ arccoth} \dfrac{x}{a} - \dfrac{1}{2a} \ln\left(\dfrac{x^2 - a^2}{x^2}\right)$ $[a^2 < x^2]$

3. $\int \dfrac{1}{x^3} \text{ arccoth} \dfrac{x}{a} \, dx = \dfrac{1}{2}\left(\dfrac{1}{a^2} - \dfrac{1}{x^2}\right) \text{ arccoth} \dfrac{x}{a} - \dfrac{1}{2ax}$ $[a^2 < x^2]$

4. $\int \dfrac{1}{x^n} \text{ arccoth} \dfrac{x}{a} \, dx = -\dfrac{1}{(n-1)x^{n-1}} \text{ arccoth} \dfrac{x}{a} + \dfrac{a}{n-1} \int \dfrac{dx}{x^{n-1}(a^2 - x^2)}$

$[a^2 < x^2, n \neq 1]$ (see 4.2.4.9)

9

Indefinite Integrals of Trigonometric Functions

9.1 Basic Results

9.1.1 Simplification by means of substitutions

9.1.1.1   Integrals of the form $\int R(\sin x, \cos x, \tan x, \cot x)\,dx$, in which R is a rational function in terms of the functions $\sin x$, $\cos x$, $\tan x$, and $\cot x$, but in which x does not appear explicitly, can always be reduced to an integral of a rational function of t by means of the substitution $t = \tan(x/2)$. In terms of this substitution it follows that

$$t = \tan \frac{x}{2}, \quad \sin x = \frac{2t}{1+t^2}, \quad \cos t = \frac{1-t^2}{1+t^2}, \quad \tan x = \frac{2t}{1-t^2},$$

$$\cot x = \frac{1-t^2}{2t}, \quad \text{and} \quad dx = \frac{2\,dt}{1+t^2}.$$

Thus, for example,

$$\int \frac{\cos x \, dx}{2 + \sin x} = \int \frac{(1 - t^2) \, dt}{(1 + t^2)(1 + t + t^2)}$$

$$= - \int \frac{2t \, dt}{1 + t^2} + \int \frac{(1 + 2t) \, dt}{1 + t + t^2}$$

$$= - \ln(1 + t^2) + \ln(1 + t + t^2) + C$$

$$= \ln\left(\frac{1 + t + t^2}{1 + t^2}\right) + C = \ln\left(1 + \frac{t}{1 + t^2}\right) + C$$

$$= \ln\left(1 + \frac{1}{2} \sin x\right) + C.$$

Since this can be written

$$\ln\left[\frac{1}{2}(2 + \sin x)\right] + C = \ln(2 + \sin x) + \ln\frac{1}{2} + C,$$

the term $\ln\frac{1}{2}$ can be combined with the arbitrary constant C, showing that the required indefinite integral can either be written as

$$\ln\left(1 + \frac{1}{2} \sin x\right) + C \quad \text{or as} \quad \ln(2 + \sin x) + C.$$

Other substitutions that are useful in special cases are listed below.

1. If

$$R(\sin x, \cos x) = -R(-\sin x, \cos x),$$

setting $t = \cos x$ and using the results

$$\sin x = (1 - t^2)^{1/2} \quad \text{and} \quad dx = \frac{dt}{(1 - t^2)^{1/2}}$$

gives

$$\int R(\sin x, \cos x) \, dx = \int R((1 - t^2)^{1/2}, t) \frac{dt}{(1 - t^2)^{1/2}}.$$

2. If

$$R(\sin x, \cos x) = -R(\sin x, -\cos x),$$

setting $t = \sin x$ and using the results

$$\cos x = (1 - t^2)^{1/2} \quad \text{and} \quad dx = \frac{dt}{(1 - t^2)^{1/2}}$$

gives

$$\int R(\sin x, \cos x) \, dx = \int R(t, (1 - t^2)^{1/2}) \frac{dt}{(1 - t^2)^{1/2}}.$$

3. If

$$R(\sin x, \cos x) = R(-\sin x, -\cos x),$$

setting $t = \tan x$ and using the results

$$\sin x = \frac{t}{(1+t^2)^{1/2}}, \quad \cos x = \frac{1}{(1+t^2)^{1/2}}, \quad dx = \frac{dt}{1+t^2}$$

gives

$$\int R(\sin x, \cos x)\, dx = \int R\left(\frac{t}{(1+t^2)^{1/2}}, \frac{1}{(1+t^2)^{1/2}}\right) \frac{dt}{1+t^2}.$$

9.2 Integrands Involving Powers of *x* and Powers of sin *x* or cos *x*

9.2.1 Integrands involving $x^n \sin^m x$

9.2.1.1

1. $\displaystyle\int \sin x\, dx = -\cos x$

2. $\displaystyle\int \sin^2 x\, dx = -\frac{1}{4}\sin 2x + \frac{1}{2}x = -\frac{1}{2}\sin x \cos x + \frac{1}{2}x$

3. $\displaystyle\int \sin^3 x\, dx = \frac{1}{12}\cos 3x - \frac{3}{4}\cos x = \frac{1}{3}\cos^3 x - \cos x$

4. $\displaystyle\int \sin^4 x\, dx = \frac{1}{32}\sin 4x - \frac{1}{4}\sin 2x + \frac{3}{8}x$

$$= -\frac{1}{4}\sin^3 x \cos x - \frac{3}{8}\sin x \cos x + \frac{3}{8}x$$

5. $\displaystyle\int \sin^5 x\, dx = -\frac{1}{80}\cos 5x + \frac{5}{48}\cos 3x - \frac{5}{8}\cos x$

$$= -\frac{1}{5}\sin^4 x \cos x + \frac{4}{15}\cos^3 x - \frac{4}{5}\cos x$$

6. $\displaystyle\int \sin^{2n} x\, dx = \frac{1}{2^{2n}}\binom{2n}{n}x + \frac{(-1)^n}{2^{2n-1}}\sum_{k=0}^{n-1}(-1)^k\binom{2n}{k}\frac{\sin(2n-2k)x}{2n-2k}$

7. $\displaystyle\int \sin^{2n+1} x\, dx = \frac{1}{2^{2n}}(-1)^{n+1}\sum_{k=0}^{n}(-1)^k\binom{2n+1}{k}\frac{\cos(2n+1-2k)x}{2n+1-2k}$

8. $\displaystyle\int x \sin x\, dx = \sin x - x \cos x$

9. $\displaystyle\int x^2 \sin x\, dx = 2x \sin x - (x^2 - 2)\cos x$

10. $\int x^3 \sin x \, dx = (3x^2 - 6) \sin x - (x^3 - 6x) \cos x$

11. $\int x^4 \sin x \, dx = (4x^3 - 24x) \sin x - (x^4 - 12x^2 + 24) \cos x$

12. $\int x^{2n} \sin x \, dx = (2n)! \left\{ \sum_{k=0}^{n} (-1)^{k+1} \frac{x^{2n-2k}}{(2n-2k)!} \cos x \right.$

$$\left. + \sum_{k=0}^{n-1} (-1)^k \frac{x^{2n-2k-1}}{(2n-2k-1)!} \sin x \right\}$$

13. $\int x^{2n+1} \sin x \, dx = (2n+1)! \left\{ \sum_{k=0}^{n} (-1)^{k+1} \frac{x^{2n-2k+1}}{(2n-2k+1)!} \cos x \right.$

$$\left. + \sum_{k=0}^{n} (-1)^k \frac{x^{2n-2k}}{(2n-2k)!} \sin x \right\}$$

14. $\int \sin^2 x \, dx = \frac{1}{2}x - \frac{1}{4} \sin 2x$

15. $\int x \sin^2 x \, dx = \frac{1}{4}x^2 - \frac{1}{4}x \sin 2x - \frac{1}{8} \cos 2x$

16. $\int x^2 \sin^2 x \, dx = \frac{1}{6}x^3 - \frac{1}{4}x \cos 2x - \frac{1}{4}\left(x^2 - \frac{1}{2}\right) \sin 2x$

17. $\int x^m \sin^n x \, dx$

$$= \frac{x^{m-1} \sin^{n-1} x}{n^2} [m \sin x - nx \cos x] + \left(\frac{n-1}{n}\right) \int x^m \sin^{n-2} x \, dx$$

$$- \frac{m(m-1)}{n^2} \int x^{m-2} \sin^n x \, dx$$

9.2.2 Integrands involving $x^{-n} \sin^m x$

9.2.2.1

1. $\int \frac{\sin x}{x} dx = x - \frac{x^3}{3 \cdot 3!} + \frac{x^5}{5 \cdot 5!} - \frac{x^7}{7 \cdot 7!} + \cdots = \sum_{n=0}^{\infty} (-1)^n \frac{x^{2n+1}}{(2n+1)(2n+1)!}$

2. $\int \frac{\sin x}{x^2} dx = -\frac{\sin x}{x} + \int \frac{\cos x}{x} dx$

3. $\int \frac{\sin x}{x^3} dx = -\frac{\sin x}{2x^2} - \frac{\cos x}{2x} - \frac{1}{2} \int \frac{\sin x}{x} dx$

4. $\displaystyle \int \frac{\sin x}{x^n} \, dx = -\frac{\sin x}{(n-1)x^{n-1}} - \frac{\cos x}{(n-1)(n-2)x^{n-2}}$

$$-\frac{1}{(n-1)(n-2)} \int \frac{\sin x}{x^{n-2}} \, dx \qquad [n > 2]$$

5. $\displaystyle \int \frac{\sin^m x}{x^n} \, dx = -\frac{\sin^{m-1} x[(n-2)\sin x + mx \cos x]}{(n-1)(n-2)x^{n-1}} - \frac{m^2}{(n-1)(n-2)}$

$$\times \int \frac{\sin^m x}{x^{n-2}} \, dx + \frac{m(m-1)}{(n-1)(n-2)} \int \frac{\sin^{m-2} x}{x^{n-2}} \, dx$$

$$[n \neq 1, 2]$$

9.2.3 Integrands involving $x^n \sin^{-m} x$

9.2.3.1

1. $\displaystyle \int \frac{dx}{\sin x} = \ln\left|\tan \frac{x}{2}\right| = -\frac{1}{2} \ln\left(\frac{1 + \cos x}{1 - \cos x}\right) \qquad [|x| < \pi]$

2. $\displaystyle \int \frac{dx}{\sin^2 x} = -\cot x \qquad [|x| < \pi]$

3. $\displaystyle \int \frac{dx}{\sin^3 x} = -\frac{\cos x}{2\sin^2 x} + \frac{1}{2}\ln\left|\tan\frac{x}{2}\right| \qquad [|x| < \pi]$

4. $\displaystyle \int \frac{dx}{\sin^n x} = -\frac{\cos x}{(n-1)\sin^{n-1} x} + \left(\frac{n-2}{n-1}\right)\int \frac{dx}{\sin^{n-2} x} \qquad [|x| < \pi, n > 1]$

5. $\displaystyle \int \frac{x \, dx}{\sin x} = x + \sum_{k=1}^{\infty} (-1)^{k+1} \frac{2(2^{2k-1} - 1)}{(2k+1)(2k)!} B_{2k} x^{2k+1} \qquad [|x| < \pi]$

6. $\displaystyle \int \frac{x^2 \, dx}{\sin x} = \frac{x^2}{2} + \sum_{k=1}^{\infty} (-1)^{k+1} \frac{2(2^{2k-1} - 1)}{(2k+2)(2k)!} B_{2k} x^{2k+2} \qquad [|x| < \pi]$

7. $\displaystyle \int \frac{x^n}{\sin x} \, dx = \frac{x^n}{n} + \sum_{k=1}^{\infty} (-1)^{k+1} \frac{2(2^{2k-1} - 1)}{(2k+n)(2k)!} B_{2k} x^{2k+n} \qquad [|x| < \pi, n > 0]$

8. $\displaystyle \int \frac{x^n \, dx}{\sin^2 x} = -x^n \cot x + \left(\frac{n}{n-1}\right) x^{n-1} + n \sum_{k=1}^{\infty} (-1)^k \frac{2^{2k} x^{n+2k-1}}{(n+2k-1)(2k)!} B_{2k}$

$$[|x| < \pi, n > 1]$$

9.2.4 Integrands involving $x^n \cos^m x$

9.2.4.1

1. $\displaystyle \int \cos x \, dx = \sin x$

2. $\displaystyle\int \cos^2 x \, dx = \frac{1}{4} \sin 2x + \frac{1}{2}x = \frac{1}{2} \sin x \cos x + \frac{1}{2}x$

3. $\displaystyle\int \cos^3 x \, dx = \frac{1}{12} \sin 3x + \frac{3}{4} \sin x = \sin x - \frac{1}{3} \sin^3 x$

4. $\displaystyle\int \cos^4 x \, dx = \frac{1}{32} \sin 4x + \frac{1}{4} \sin 2x + \frac{3}{8}x = \frac{1}{4} \sin x \cos^3 x + \frac{3}{8} \sin x \cos x + \frac{3}{8}x$

5. $\displaystyle\int \cos^5 x \, dx = \frac{1}{80} \sin 5x + \frac{5}{48} \sin 3x + \frac{5}{8} \sin x$

$\displaystyle = \frac{1}{5} \cos^4 x \sin x - \frac{4}{15} \sin^3 x + \frac{4}{5} \sin x$

6. $\displaystyle\int \cos^{2n} x \, dx = \frac{1}{2^{2n}} \binom{2n}{n} x + \frac{1}{2^{2n-1}} \sum_{k=0}^{n-1} \binom{2n}{k} \frac{\sin(2n - 2k)x}{2n - 2k}$

7. $\displaystyle\int \cos^{2n+1} x \, dx = \frac{1}{2^{2n}} \sum_{k=0}^{n} \binom{2n + 1}{k} \frac{\sin(2n - 2k + 1)x}{2n - 2k + 1}$

8. $\displaystyle\int x \cos x \, dx = \cos x + x \sin x$

9. $\displaystyle\int x^2 \cos x \, dx = 2x \cos x + (x^2 - 2) \sin x$

10. $\displaystyle\int x^3 \cos x \, dx = (3x^2 - 6) \cos x + (x^3 - 6x) \sin x$

11. $\displaystyle\int x^4 \cos x \, dx = (4x^3 - 24x) \cos x + (x^4 - 12x^2 + 24) \sin x$

12. $\displaystyle\int x^{2n} \cos x \, dx = (2n)! \left\{ \sum_{k=0}^{n} (-1)^k \frac{x^{2n-2k}}{(2n - 2k)!} \sin x \right.$

$\displaystyle \left. + \sum_{k=0}^{n-1} (-1)^k \frac{x^{2n-2k-1}}{(2n - 2k - 1)!} \cos x \right\}$

13. $\displaystyle\int x^{2n+1} \cos x \, dx = (2n + 1)! \left\{ \sum_{k=0}^{n} (-1)^k \frac{x^{2n-2k+1}}{(2n - 2k + 1)!} \sin x \right.$

$\displaystyle \left. + \sum_{k=0}^{n} (-1)^k \frac{x^{2n-2k}}{(2n - 2k)!} \cos x \right\}$

14. $\displaystyle\int \cos^2 x \, dx = \frac{1}{2}x + \frac{1}{4} \sin 2x$

15. $\displaystyle\int x \cos^2 x \, dx = \frac{1}{4}x^2 + \frac{1}{4}x \sin 2x + \frac{1}{8} \cos 2x$

16. $\displaystyle \int x^2 \cos^2 x \, dx = \frac{1}{6}x^3 + \frac{1}{4}x \cos 2x + \frac{1}{4}\left(x^2 - \frac{1}{2}\right)\sin 2x$

17. $\displaystyle \int x^m \cos^n x \, dx = \frac{x^{m-1} \cos^{n-1} x}{n^2}[m \cos x + nx \sin x]$

$$+ \left(\frac{n-1}{n}\right)\int x^m \cos^{n-2} x \, dx$$

$$- \frac{m(m-1)}{n^2}\int x^{m-2} \cos^n x \, dx$$

9.2.5 Integrands involving $x^{-n} \cos^m x$

9.2.5.1

1. $\displaystyle \int \frac{\cos x}{x} \, dx = \ln|x| - \frac{x^2}{2.2!} + \frac{x^4}{4.4!} - \frac{x^6}{6.6!} + \cdots$

2. $\displaystyle \int \frac{\cos x}{x^2} \, dx = -\frac{\cos x}{x} - \int \frac{\sin x}{x} \, dx$

3. $\displaystyle \int \frac{\cos x}{x^3} \, dx = -\frac{\cos x}{2x^2} + \frac{\sin x}{2x} - \frac{1}{2}\int \frac{\cos x}{x} \, dx$

4. $\displaystyle \int \frac{\cos x}{x^n} \, dx = -\frac{\cos x}{(n-1)x^{n-1}} + \frac{\sin x}{(n-1)(n-2)x^{n-2}}$

$$- \frac{1}{(n-1)(n-2)}\int \frac{\cos x}{x^{n-2}} \, dx \qquad [n > 2]$$

5. $\displaystyle \int \frac{\cos^m x}{x^n} \, dx = -\frac{\cos^{m-1} x[(n-2)\cos x - mx \sin x]}{(n-1)(n-2)x^{n-1}}$

$$- \frac{m^2}{(n-1)(n-2)}\int \frac{\cos^m x}{x^{n-2}} \, dx$$

$$+ \frac{m(m-1)}{(n-1)(n-2)}\int \frac{\cos^{m-2} x}{x^{n-2}} \, dx \qquad [n \neq 1, 2]$$

9.2.6 Integrands involving $x^n \cos^{-m} x$

9.2.6.1

1. $\displaystyle \int \frac{dx}{\cos x} = \ln|\sec x + \tan x| = \ln\left|\tan\left(\frac{\pi}{4} + \frac{x}{2}\right)\right|$

$$= \frac{1}{2}\ln\left(\frac{1 + \sin x}{1 - \sin x}\right) \qquad \left[|x| < \frac{\pi}{2}\right]$$

2. $\displaystyle\int \frac{dx}{\cos^2 x} = \tan x \qquad \left[|x| < \frac{\pi}{2}\right]$

3. $\displaystyle\int \frac{dx}{\cos^3 x} = \frac{1}{2}\frac{\sin x}{\cos^2 x} + \frac{1}{2}\ln\left|\tan\left(\frac{\pi}{4}+\frac{x}{2}\right)\right| \qquad \left[|x| < \frac{\pi}{2}\right]$

4. $\displaystyle\int \frac{dx}{\cos^n x} = \frac{\sin x}{(n-1)\cos^{n-1} x} + \left(\frac{n-2}{n-1}\right)\int \frac{dx}{\cos^{n-2} x} \qquad \left[|x| < \frac{\pi}{2}, n > 1\right]$

5. $\displaystyle\int \frac{x\,dx}{\cos x} = \sum_{k=0}^{\infty} \frac{|E_{2k}|x^{2k+2}}{(2k+2)(2k)!} \qquad \left[|x| < \frac{\pi}{2}\right]$

6. $\displaystyle\int \frac{x^2 dx}{\cos x} = \sum_{k=0}^{\infty} \frac{|E_{2k}|x^{2k+3}}{(2k+3)(2k)!} \qquad \left[|x| < \frac{\pi}{2}, n > 0\right]$

7. $\displaystyle\int \frac{x^n dx}{\cos x} = \sum_{k=0}^{\infty} \frac{|E_{2k}|x^{2k+n+1}}{(2k+n+1)(2k)!} \qquad \left[|x| < \frac{\pi}{2}, n > 0\right]$

8. $\displaystyle\int \frac{x^n}{\cos^2 x}\,dx = x^n \tan x + n\sum_{k=1}^{\infty}(-1)^k\frac{2^{2k}(2^{2k}-1)x^{2k+n-1}}{(2k+n-1)(2k)!}B_{2k}$

$$\left[|x| < \frac{\pi}{2}, n > 1\right]$$

9.2.7 Integrands involving $x^n \sin x/(a + b\cos x)^m$ or $x^n \cos x/(a + b\sin x)^m$

9.2.7.1

1. $\displaystyle\int \frac{dx}{1 + \sin x} = -\tan\left(\frac{\pi}{4}-\frac{x}{2}\right)$

2. $\displaystyle\int \frac{dx}{1 - \sin x} = \tan\left(\frac{\pi}{4}+\frac{x}{2}\right)$

3. $\displaystyle\int \frac{x\,dx}{1 + \sin x} = -x\tan\left(\frac{\pi}{4}-\frac{x}{2}\right) + 2\ln\left[\cos\left(\frac{\pi}{4}-\frac{x}{2}\right)\right]$

4. $\displaystyle\int \frac{x\,dx}{1 - \sin x} = x\cot\left(\frac{\pi}{4}-\frac{x}{2}\right) + 2\ln\left[\sin\left(\frac{\pi}{4}-\frac{x}{2}\right)\right]$

5. $\displaystyle\int \frac{x\,dx}{1 + \cos x} = x\tan\frac{x}{2} + 2\ln\left[\cos\frac{x}{2}\right]$

6. $\displaystyle\int \frac{x\,dx}{1 - \cos x} = -x\cot\frac{x}{2} + 2\ln\left[\sin\frac{x}{2}\right]$

7. $\displaystyle\int \frac{x\cos x\,dx}{(1 + \sin x)^2} = -\frac{x}{1+\sin x} + \tan\left(\frac{x}{2}-\frac{\pi}{4}\right)$

8. $\displaystyle\int \frac{x \cos x \, dx}{(1 - \sin x)^2} = \frac{x}{1 - \sin x} + \tan\left(\frac{x}{2} + \frac{\pi}{4}\right)$

9. $\displaystyle\int \frac{x \sin x \, dx}{(1 + \cos x)^2} = \frac{x}{1 + \cos x} - \tan\frac{x}{2}$

10. $\displaystyle\int \frac{x \sin x \, dx}{(1 - \cos x)^2} = -\frac{x}{1 - \cos x} - \cot\frac{x}{2}$

11. $\displaystyle\int \frac{dx}{a + b \sin x} = \frac{2}{(a^2 - b^2)^{1/2}} \arctan\left[\frac{a \tan\frac{x}{2} + b}{(a^2 - b^2)^{1/2}}\right] \quad [a^2 > b^2]$

$\displaystyle = \frac{1}{(b^2 - a^2)^{1/2}} \ln\left[\frac{a \tan\frac{x}{2} + b - (b^2 - a^2)^{1/2}}{a \tan\frac{x}{2} + b + (b^2 - a^2)^{1/2}}\right]$

$[a^2 < b^2]$

12. $\displaystyle\int \frac{dx}{a + b \cos x} = \frac{2}{(a^2 - b^2)^{1/2}} \arctan\left[\frac{(a^2 - b^2)^{1/2} \tan\frac{x}{2}}{a + b}\right] \quad [a^2 > b^2]$

$\displaystyle = \frac{1}{(b^2 - a^2)^{1/2}} \ln\left[\frac{(b^2 - a^2)^{1/2} \tan\frac{x}{2} + a + b}{(b^2 - a^2)^{1/2} \tan\frac{x}{2} - a - b}\right] \quad [a^2 < b^2]$

13. $\displaystyle\int \frac{x^n \sin x \, dx}{(a + b \cos x)^m} = \frac{x^n}{(m - 1)b(a + b \cos x)^{m-1}}$

$\displaystyle - \frac{n}{(m - 1)b} \int \frac{x^{n-1} \, dx}{(a + b \cos x)^{m-1}} \quad [m \neq 1]$

14. $\displaystyle\int \frac{x^n \cos x \, dx}{(a + b \sin x)^m} = -\frac{x^n}{(m - 1)b(a + b \sin x)^{m-1}}$

$\displaystyle + \frac{n}{(m - 1)b} \int \frac{x^{n-1} \, dx}{(a + b \sin x)^{m-1}} \quad [m \neq 1]$

9.3 Integrands Involving tan x and/or cot x

9.3.1 Integrands involving $\tan^n x$ or $\tan^n x / (\tan x \pm 1)$

9.3.1.1

1. $\displaystyle\int \tan x \, dx = -\ln \cos x$

2. $\displaystyle\int \tan^2 x \, dx = \tan x - x$

3. $\displaystyle\int \tan^3 x \, dx = \frac{1}{2} \tan^2 x + \ln \cos x$

4. $\displaystyle\int \tan^4 x \, dx = \frac{1}{3} \tan^3 x - \tan x + x$

5. $\displaystyle\int \tan^{2n} x \, dx = \sum_{k=1}^{n} (-1)^{k-1} \frac{\tan^{2n-2k+1} x}{(2n-2k+1)} + (-1)^n x$

6. $\displaystyle\int \tan^{2n+1} x \, dx = \sum_{k=1}^{n} (-1)^{k-1} \frac{\tan^{2n-2k+2} x}{(2n-2k+2)} - (-1)^n \ln \cos x$

7. $\displaystyle\int \frac{dx}{\tan x + 1} = \int \frac{\cot x \, dx}{1 + \cot x} = \frac{1}{2}x + \frac{1}{2} \ln|\sin x + \cos x|$

8. $\displaystyle\int \frac{dx}{\tan x - 1} = \int \frac{\cot x \, dx}{1 - \cot x} = -\frac{1}{2}x + \frac{1}{2} \ln|\sin x - \cos x|$

9. $\displaystyle\int \frac{\tan x \, dx}{\tan x + 1} = \int \frac{dx}{1 + \cot x} = \frac{1}{2}x - \frac{1}{2} \ln|\sin x + \cos x|$

10. $\displaystyle\int \frac{\tan x \, dx}{\tan x - 1} = \int \frac{dx}{1 - \cot x} = \frac{1}{2}x + \frac{1}{2} \ln|\sin x - \cos x|$

9.3.2 Integrands involving $\cot^n x$ or $\tan x$ and $\cot x$

9.3.2.1

1. $\displaystyle\int \cot x \, dx = \ln|\sin x|$

2. $\displaystyle\int \cot^2 x \, dx = -\cot x - x$

3. $\displaystyle\int \cot^3 x \, dx = -\frac{1}{2} \cot^2 x - \ln|\sin x|$

4. $\displaystyle\int \cot^4 x \, dx = -\frac{1}{3} \cot^3 x + \cot x + x$

5. $\displaystyle\int \cot^n x \, dx = -\frac{\cot^{n-1} x}{n-1} - \int \cot^{n-2} x \, dx \qquad [n \neq 1]$

6. $\displaystyle\int (1 + \tan^2 x) \cot x \, dx = \ln|\tan x|$

7. $\displaystyle\int (1 + \tan^2 x) \cot^2 x \, dx = -\cot x$

8. $\displaystyle\int (1 + \tan^2 x) \cot^3 x \, dx = -\frac{1}{2} \cot^2 x$

9. $\displaystyle\int (1 + \tan^2 x) \cot^n x \, dx = -\frac{\cot^{n-1} x}{n-1}$

9.4 Integrands Involving sin *x* and cos *x*

9.4.1 Integrands involving $\sin^m x \cos^n x$

9.4.1.1

1. $\displaystyle\int \sin x \cos x \, dx = \frac{1}{2} \sin^2 x$

2. $\displaystyle\int \sin x \cos^2 x \, dx = -\frac{1}{4}\left[\frac{1}{3} \cos 3x + \cos x \right] = -\frac{1}{3} \cos^3 x$

3. $\displaystyle\int \sin x \cos^4 x \, dx = -\frac{1}{5} \cos^5 x$

4. $\displaystyle\int \sin^2 x \cos x \, dx = -\frac{1}{4}\left[\frac{1}{3} \sin 3x - \sin x \right] = \frac{1}{3} \sin^3 x$

5. $\displaystyle\int \sin^2 x \cos^2 x \, dx = -\frac{1}{8}\left[\frac{1}{4} \sin 4x - x \right]$

6. $\displaystyle\int \sin^2 x \cos^3 x \, dx = -\frac{1}{16}\left[\frac{1}{5} \sin 5x + \frac{1}{3} \sin 3x - 2 \sin x \right]$

7. $\displaystyle\int \sin^2 x \cos^4 x \, dx = \frac{1}{16}x + \frac{1}{64} \sin 2x - \frac{1}{64} \sin 4x - \frac{1}{192} \sin 6x$

8. $\displaystyle\int \sin^3 x \cos x \, dx = \frac{1}{8}\left(\frac{1}{4} \cos 4x - \cos 2x \right) = \frac{1}{4} \sin^4 x$

9. $\displaystyle\int \sin^3 x \cos^2 x \, dx = \frac{1}{16}\left(\frac{1}{5} \cos 5x - \frac{1}{3} \cos 3x - 2 \cos x \right)$

10. $\displaystyle\int \sin^3 x \cos^3 x \, dx = \frac{1}{32}\left(\frac{1}{6} \cos 6x - \frac{3}{2} \cos 2x \right)$

9.4.2 Integrands involving $\sin^{-n} x$

9.4.2.1

1. $\displaystyle\int \frac{dx}{\sin x} = \ln\left| \tan\frac{x}{2} \right|$

2. $\displaystyle\int \frac{dx}{\sin^2 x} = -\cot x$

3. $\displaystyle\int \frac{dx}{\sin^3 x} = -\frac{1}{2}\frac{\cos x}{\sin^2 x} + \frac{1}{2} \ln\left| \tan\frac{x}{2} \right|$

4. $\int \dfrac{dx}{\sin^4 x} = -\dfrac{\cos x}{3 \sin^3 x} - \dfrac{2}{3}\cot x$

5. $\int \dfrac{dx}{\sin^5 x} = -\dfrac{\cos x}{4 \sin^4 x} - \dfrac{3}{8}\dfrac{\cos x}{\sin^2 x} + \dfrac{3}{8}\ln\left|\tan\dfrac{x}{2}\right|$

9.4.3 Integrands involving $\cos^{-n} x$

9.4.3.1

1. $\int \dfrac{dx}{\cos x} = \ln\left|\tan\left(\dfrac{\pi}{4} + \dfrac{x}{2}\right)\right| = \dfrac{1}{2}\ln\left(\dfrac{1 + \sin x}{1 - \sin x}\right)$

2. $\int \dfrac{dx}{\cos^2 x} = \tan x$

3. $\int \dfrac{dx}{\cos^3 x} = \dfrac{1}{2}\dfrac{\sin x}{\cos^2 x} + \dfrac{1}{2}\ln\left|\tan\left(\dfrac{\pi}{4} + \dfrac{x}{2}\right)\right|$

4. $\int \dfrac{dx}{\cos^4 x} = \dfrac{\sin x}{3 \cos^3 x} + \dfrac{2}{3}\tan x$

5. $\int \dfrac{dx}{\cos^5 x} = \dfrac{\sin x}{4 \cos^4 x} + \dfrac{3}{8}\dfrac{\sin x}{\cos^2 x} + \dfrac{3}{8}\ln\left|\tan\left(\dfrac{\pi}{4} + \dfrac{x}{2}\right)\right|$

9.4.4 Integrands involving $\sin^m x / \cos^n x$ or $\cos^m x / \sin^n x$

9.4.4.1

1. $\int \dfrac{\sin x}{\cos x}\, dx = -\ln\cos x$

2. $\int \dfrac{\sin x}{\cos^2 x}\, dx = \dfrac{1}{\cos x}$

3. $\int \dfrac{\sin x}{\cos^3 x}\, dx = \dfrac{1}{2 \cos^2 x}$

4. $\int \dfrac{\sin x}{\cos^4 x}\, dx = \dfrac{1}{3 \cos^3 x}$

5. $\int \dfrac{\sin x}{\cos^n x}\, dx = \dfrac{1}{(n - 1) \cos^{n-1} x}$

6. $\int \dfrac{\sin^2 x}{\cos x}\, dx = -\sin x + \ln\left|\tan\left(\dfrac{\pi}{4} + \dfrac{x}{2}\right)\right|$

7. $\displaystyle\int \frac{\sin^2 x}{\cos^2 x}\, dx = \tan x - x$

8. $\displaystyle\int \frac{\sin^2 x}{\cos^3 x}\, dx = \frac{\sin x}{2\cos^2 x} - \frac{1}{2}\ln\left|\tan\left(\frac{\pi}{4} + \frac{x}{2}\right)\right|$

9. $\displaystyle\int \frac{\sin^2 x}{\cos^4 x}\, dx = \frac{1}{3}\tan^3 x$

10. $\displaystyle\int \frac{\sin^3 x}{\cos x}\, dx = -\frac{1}{2}\sin^2 x - \ln\cos x$

11. $\displaystyle\int \frac{\sin^3 x}{\cos^2 x}\, dx = \cos x + \frac{1}{\cos x}$

12. $\displaystyle\int \frac{\sin^3 x}{\cos^3 x}\, dx = \frac{1}{2\cos^2 x} + \ln\cos x$

13. $\displaystyle\int \frac{\sin^3 x}{\cos^4 x}\, dx = -\frac{1}{\cos x} + \frac{1}{3\cos^3 x}$

14. $\displaystyle\int \frac{\sin^4 x}{\cos x}\, dx = -\frac{1}{3}\sin^3 x - \sin x + \ln\left|\tan\left(\frac{\pi}{4} + \frac{x}{2}\right)\right|$

15. $\displaystyle\int \frac{\sin^4 x}{\cos^2 x}\, dx = \tan x + \frac{1}{2}\sin x \cos x - \frac{3}{2}x$

16. $\displaystyle\int \frac{\sin^4 x}{\cos^3 x}\, dx = \frac{\sin x}{2\cos^2 x} + \sin x - \frac{3}{2}\ln\left|\tan\left(\frac{\pi}{4} + \frac{x}{2}\right)\right|$

17. $\displaystyle\int \frac{\sin^4 x}{\cos^4 x}\, dx = \frac{1}{3}\tan^3 x - \tan x + x$

18. $\displaystyle\int \frac{\cos x}{\sin x}\, dx = \ln|\sin x|$

19. $\displaystyle\int \frac{\cos x}{\sin^2 x}\, dx = -\frac{1}{\sin x}$

20. $\displaystyle\int \frac{\cos x}{\sin^3 x}\, dx = -\frac{1}{2\sin^2 x}$

21. $\displaystyle\int \frac{\cos x}{\sin^4 x}\, dx = -\frac{1}{3\sin^3 x}$

22. $\displaystyle\int \frac{\cos x}{\sin^n x}\, dx = -\frac{1}{(n-1)\sin^{n-1} x}$

23. $\displaystyle\int \frac{\cos^2 x}{\sin x}\, dx = \cos x + \ln\left|\tan\frac{x}{2}\right|$

24. $\displaystyle\int \frac{\cos^2 x}{\sin^2 x}\, dx = -\cot x - x$

25. $\displaystyle\int \frac{\cos^2 x}{\sin^3 x}\, dx = -\frac{\cos x}{2\sin^2 x} - \frac{1}{2}\ln\left|\tan\frac{x}{2}\right|$

26. $\displaystyle\int \frac{\cos^2 x}{\sin^4 x}\, dx = -\frac{1}{3}\cot^3 x$

27. $\displaystyle\int \frac{\cos^3 x}{\sin x}\, dx = \frac{1}{2}\cos^2 x + \ln|\sin x|$

28. $\displaystyle\int \frac{\cos^3 x}{\sin^2 x}\, dx = -\sin x - \frac{1}{\sin x}$

29. $\displaystyle\int \frac{\cos^3 x}{\sin^3 x}\, dx = -\frac{1}{2\sin^2 x} - \ln|\sin x|$

30. $\displaystyle\int \frac{\cos^3 x}{\sin^4 x}\, dx = \frac{1}{\sin x} - \frac{1}{3\sin^3 x}$

31. $\displaystyle\int \frac{\cos^4 x}{\sin x}\, dx = \frac{1}{3}\cos^3 x + \cos x + \ln\left|\tan\frac{x}{2}\right|$

32. $\displaystyle\int \frac{\cos^4 x}{\sin^2 x}\, dx = -\cot x - \frac{1}{2}\sin x \cos x - \frac{3}{2}x$

33. $\displaystyle\int \frac{\cos^4 x}{\sin^3 x}\, dx = -\frac{\cos x}{2\sin^2 x} - \cos x - \frac{3}{2}\ln\left|\tan\frac{x}{2}\right|$

34. $\displaystyle\int \frac{\cos^4 x}{\sin^4 x}\, dx = -\frac{1}{3}\cot^3 x + \cot x + x$

9.4.5 Integrands involving $\sin^{-m} x \cos^{-n} x$

9.4.5.1

1. $\displaystyle\int \frac{dx}{\sin x \cos x} = \ln(\tan x)$

2. $\displaystyle\int \frac{dx}{\sin x \cos^2 x} = \frac{1}{\cos x} + \ln\left|\tan\frac{x}{2}\right|$

3. $\displaystyle\int \frac{dx}{\sin x \cos^3 x} = \frac{1}{2\cos^2 x} + \ln|\tan x|$

4. $$\int \frac{dx}{\sin x \cos^4 x} = \frac{1}{\cos x} + \frac{1}{3 \cos^3 x} + \ln\left|\tan\frac{x}{2}\right|$$

5. $$\int \frac{dx}{\sin^2 x \cos x} = \ln\left|\tan\left(\frac{\pi}{4} + \frac{x}{2}\right)\right| - \operatorname{cosec} x$$

6. $$\int \frac{dx}{\sin^2 x \cos^2 x} = -2 \cot 2x$$

7. $$\int \frac{dx}{\sin^2 x \cos^3 x} = \left(\frac{1}{2 \cos^2 x} - \frac{3}{2}\right)\frac{1}{\sin x} + \frac{3}{2}\ln\left|\tan\left(\frac{\pi}{4} + \frac{x}{2}\right)\right|$$

8. $$\int \frac{dx}{\sin^2 x \cos^4 x} = \frac{1}{3 \sin x \cos^3 x} - \frac{8}{3}\cot 2x$$

9. $$\int \frac{dx}{\sin^{2m} x \cos^{2n} x} = \sum_{k=0}^{m+n-1} \binom{m+n-1}{k} \frac{\tan^{2k-2m+1} x}{(2k - 2m + 1)}$$

10. $$\int \frac{dx}{\sin^{2m+1} x \cos^{2n+1} x} = \sum_{k=0}^{m+n} \binom{m+n}{k} \frac{\tan^{2k-2m} x}{(2k - 2m)} + \binom{m+n}{m}\ln|\tan x|$$

9.5 Integrands Involving Sines and Cosines with Linear Arguments and Powers of *x*

9.5.1 Integrands involving products of $(ax + b)^n$, $\sin(cx + d)$, and/or $\cos(px + q)$

9.5.1.1

1. $$\int \sin(ax + b)\, dx = -\frac{1}{a}\cos(ax + b)$$

2. $$\int \cos(ax + b)\, dx = \frac{1}{a}\sin(ax + b)$$

3. $$\int \sin(ax + b)\sin(cx + d)\, dx = \frac{\sin[(a - c)x + b - d]}{2(a - c)}$$

$$- \frac{\sin[(a + c)x + b + d]}{2(a + c)} \qquad [a^2 \neq c^2]$$

4. $$\int \sin(ax + b)\cos(cx + d)\, dx = -\frac{\cos[(a - c)x + b - d]}{2(a - c)}$$

$$- \frac{\cos[(a + c)x + b + d]}{2(a + c)} \qquad [a^2 \neq c^2]$$

5. $\int \cos(ax+b)\cos(cx+d)\,dx = \dfrac{\sin[(a-c)x+b-d]}{2(a-c)}$

$+ \dfrac{\sin[(a+c)x+b+d]}{2(a+c)} \qquad [a^2 \neq c^2]$

Special case a = c.

6. $\int \sin(ax+b)\sin(ax+d)\,dx = \dfrac{x}{2}\cos(b-d) - \dfrac{\sin(2ax+b+d)}{4a}$

7. $\int \sin(ax+b)\cos(ax+d)\,dx = \dfrac{x}{2}\sin(b-d) - \dfrac{\cos(2ax+b+d)}{4a}$

8. $\int \cos(ax+b)\cos(ax+d)\,dx = \dfrac{x}{2}\cos(b-d) + \dfrac{\sin(2ax+b+d)}{4a}$

9. $\int (a+bx)\sin kx\,dx = -\dfrac{1}{k}(a+bx)\cos kx + \dfrac{b}{k^2}\sin kx$

10. $\int (a+bx)\cos kx\,dx = \dfrac{1}{k}(a+bx)\sin kx + \dfrac{b}{k^2}\cos kx$

11. $\int (a+bx)^2 \sin kx\,dx = \dfrac{1}{k}\left[\dfrac{2b^2}{k^2} - (a+bx)^2\right]\cos kx + \dfrac{2b(a+bx)}{k^2}\sin kx$

12. $\int (a+bx)^2 \cos kx\,dx = \dfrac{1}{k}\left[(a+bx)^2 - \dfrac{2b^2}{k^2}\right]\sin kx + \dfrac{2b(a+bx)}{k^2}\cos kx$

9.5.2 Integrands involving $x^n \sin^m x$ or $x^n \cos^m x$

9.5.2.1

1. $\int x\sin x\,dx = \sin x - x\cos x$

2. $\int x\cos x\,dx = \cos x + x\sin x$

3. $\int x^2 \sin x\,dx = 2x\sin x - (x^2 - 2)\cos x$

4. $\int x^2 \cos x\,dx = 2x\cos x + (x^2 - 2)\sin x$

5. $\int x^3 \sin x\,dx = (3x^2 - 6)\sin x - (x^3 - 6x)\cos x$

6. $\int x^3 \cos x\,dx = (3x^2 - 6)\cos x + (x^3 - 6x)\sin x$

7. $\displaystyle \int x^n \sin x \, dx = -x^n \cos x + n \int x^{n-1} \cos x \, dx$

8. $\displaystyle \int x^n \cos x \, dx = x^n \sin x - n \int x^{n-1} \sin x \, dx$

9. $\displaystyle \int x \sin^2 x \, dx = \frac{1}{4}x^2 - \frac{1}{4}x \sin 2x - \frac{1}{8} \cos 2x$

10. $\displaystyle \int x \cos^2 x \, dx = \frac{1}{4}x^2 + \frac{1}{4}x \sin 2x + \frac{1}{8} \cos 2x$

11. $\displaystyle \int x^2 \sin^2 x \, dx = \frac{1}{6}x^3 - \frac{1}{4}x \cos 2x - \frac{1}{4}\left(x^2 - \frac{1}{2}\right) \sin 2x$

12. $\displaystyle \int x^2 \cos^2 x \, dx = \frac{1}{6}x^3 + \frac{1}{4}x \cos 2x + \frac{1}{4}\left(x^2 - \frac{1}{2}\right) \sin 2x$

13. $\displaystyle \int x \sin^3 x \, dx = \frac{3}{4} \sin x - \frac{1}{36} \sin 3x - \frac{3}{4}x \cos x + \frac{1}{12}x \cos 3x$

14. $\displaystyle \int x \cos^3 x \, dx = \frac{3}{4} \cos x + \frac{1}{36} \cos 3x + \frac{3}{4}x \sin x + \frac{1}{12}x \sin 3x$

15. $\displaystyle \int x^2 \sin^3 x \, dx = -\left(\frac{3}{4}x^2 - \frac{3}{2}\right) \cos x + \left(\frac{x^2}{12} - \frac{1}{54}\right) \cos 3x$

$$+ \frac{3}{2}x \sin x - \frac{1}{18}x \sin 3x$$

16. $\displaystyle \int x^2 \cos^3 x \, dx = \left(\frac{3}{4}x^2 - \frac{3}{2}\right) \sin x + \left(\frac{x^2}{12} - \frac{1}{54}\right) \sin 3x$

$$+ \frac{3}{2}x \cos x + \frac{1}{18}x \cos 3x$$

10

Indefinite Integrals of Inverse Trigonometric Functions

10.1 Integrands Involving Powers of x and Powers of Inverse Trigonometric Functions

10.1.1 Integrands involving $x^n \arcsin^m(x/a)$

10.1.1.1

1. $\displaystyle\int \arcsin \frac{x}{a}\, dx = \arcsin \frac{x}{a} + (a^2 - x^2)^{1/2}$ $[|x/a| \le 1]$

2. $\displaystyle\int \arcsin^2 \frac{x}{a}\, dx = x \arcsin^2 \frac{x}{a} + 2(a^2 - x^2)^{1/2} \arcsin \frac{x}{a} - 2x$ $[|x/a| \le 1]$

3. $\displaystyle\int \arcsin^3 \frac{x}{a}\, dx = x \arcsin^3 \frac{x}{a} + 3(a^2 - x^2)^{1/2} \arcsin^2 \frac{x}{a}$

$$- 6x \arcsin \frac{x}{a} - 6(a^2 - x^2)^{1/2} \quad [|x/a| \le 1]$$

4. $\int \arcsin^n \dfrac{x}{a}\, dx = x \arcsin^n \dfrac{x}{a} + n(a^2 - x^2)^{1/2} \arcsin^{n-1} \dfrac{x}{a}$

$$- n(n - 1) \int \arcsin^{n-2} \dfrac{x}{a}\, dx \qquad [|x/a| \leq 1,\, n \neq 1]$$

5. $\int x \arcsin \dfrac{x}{a}\, dx = \left(\dfrac{x^2}{2} - \dfrac{a^2}{4} \right) \arcsin \dfrac{x}{a} + \dfrac{x}{4}(a^2 - x^2)^{1/2} \qquad [|x/a| \leq 1]$

6. $\int x^2 \arcsin \dfrac{x}{a}\, dx = \dfrac{x^3}{3} \arcsin \dfrac{x}{a} + \dfrac{1}{9}(x^2 + 2a^2)(a^2 - x^2)^{1/2} \qquad [|x/a| \leq 1]$

7. $\int x^3 \arcsin \dfrac{x}{a}\, dx = \left(\dfrac{x^4}{4} - \dfrac{3a^4}{32} \right) \arcsin \dfrac{x}{a} + \dfrac{1}{32}(2x^3 + 3a^2 x)(a^2 - x^2)^{1/2}$

$$[|x/a| \leq 1]$$

8. $\int x^n \arcsin \dfrac{x}{a}\, dx = \dfrac{x^{n+1}}{n+1} \arcsin \dfrac{x}{a} - \dfrac{1}{n+1} \int \dfrac{x^{n+1} dx}{(a^2 - x^2)^{1/2}} \qquad [|x/a| \leq 1]$

10.1.2 Integrands involving $x^{-n} \arcsin(x/a)$

10.1.2.1

1. $\int \dfrac{1}{x} \arcsin \dfrac{x}{a}\, dx = \sum_{k=0}^{\infty} \dfrac{(2k - 1)!!}{(2k)!!(2k + 1)^2} \left(\dfrac{x}{a} \right)^{2k+1}$

$$[|x/a| < 1,\, (2k)!! = 2 \cdot 4 \cdot 6 \cdots (2k)]$$

2. $\int \dfrac{1}{x^2} \arcsin \dfrac{x}{a}\, dx = -\dfrac{1}{x} \arcsin \dfrac{x}{a} - \dfrac{1}{a} \ln \left| \dfrac{a + (a^2 - x^2)^{1/2}}{x} \right| \qquad [|x/a| < 1]$

3. $\int \dfrac{1}{x^3} \arcsin \dfrac{x}{a}\, dx = -\dfrac{1}{2x^2} \arcsin \dfrac{x}{a} - \dfrac{(a^2 - x^2)^{1/2}}{2a^2 x} \qquad [|x/a| < 1]$

4. $\int \dfrac{1}{x^n} \arcsin \dfrac{x}{a}\, dx = \dfrac{-1}{(n - 1)x^{n-1}} \arcsin \dfrac{x}{a} + \dfrac{1}{n - 1} \int \dfrac{dx}{x^{n-1}(a^2 - x^2)^{1/2}}$

$$[|x/a| < 1,\, n \neq 1]$$

10.1.3 Integrands involving $x^n \arccos^m(x/a)$

10.1.3.1

1. $\int \arccos \dfrac{x}{a}\, dx = x \arccos \dfrac{x}{a} - (a^2 - x^2)^{1/2} \qquad [|x/a| \leq 1]$

2. $\int \arccos^2 \dfrac{x}{a}\, dx = x \arccos^2 \dfrac{x}{a} - 2(a^2 - x^2)^{1/2} \arccos \dfrac{x}{a} - 2x \qquad [|x/a| \leq 1]$

3. $\int \arccos^3 \dfrac{x}{a}\, dx = x \arccos^3 \dfrac{x}{a} - 3(a^2 - x^2)^{1/2} \arccos^2 \dfrac{x}{a}$

$$- 6x \arccos \dfrac{x}{a} + 6(a^2 - x^2)^{1/2} \qquad [|x/a| \leq 1]$$

4. $\displaystyle\int \arccos^n \frac{x}{a}\, dx = x \arccos^n \frac{x}{a} - n(a^2 - x^2)^{1/2} \arccos^{n-1} \frac{x}{a}$

$$- n(n-1) \int \arccos^{n-2} \frac{x}{a}\, dx \qquad [|x/a| \le 1, n \ne 1]$$

5. $\displaystyle\int x \arccos \frac{x}{a}\, dx = \left(\frac{x^2}{2} - \frac{a^2}{4}\right) \arccos \frac{x}{a} - \frac{x}{4}(a^2 - x^2)^{1/2} \qquad [|x/a| < 1]$

6. $\displaystyle\int x^2 \arccos \frac{x}{a}\, dx = \frac{x^3}{3} \arccos \frac{x}{a} - \frac{1}{9}(x^2 + 2a^2)(a^2 - x^2)^{1/2} \qquad [|x/a| \le 1]$

7. $\displaystyle\int x^3 \arccos \frac{x}{a}\, dx = \left(\frac{x^4}{4} - \frac{3a^4}{32}\right) \arccos \frac{x}{a} - \frac{1}{32}(2x^3 + 3a^2 x)(a^2 - x^2)^{1/2}$

$$[|x/a| \le 1]$$

8. $\displaystyle\int x^n \arccos \frac{x}{a}\, dx = \frac{x^{n+1}}{n+1} \arccos \frac{x}{a} + \frac{1}{n+1} \int \frac{x^{n+1}\, dx}{(a^2 - x^2)^{1/2}}$

$$[|x/a| \le 1, n \ne -1]$$

10.1.4 Integrands involving $x^{-n} \arccos(x/a)$

10.1.4.1

1. $\displaystyle\int \frac{1}{x} \arccos \frac{x}{a}\, dx = \frac{\pi}{2} \ln|x| - \sum_{k=0}^{\infty} \frac{(2k-1)!!}{(2k)!!(2k+1)^2} \left(\frac{x}{a}\right)^{2k+1}$

$$[|x/a| < 1, (2k)!! = 2 \cdot 4 \cdot 6 \cdots (2k), (2k-1)!! = 1 \cdot 3 \cdot 5 \cdots (2k-1)]$$

2. $\displaystyle\int \frac{1}{x^2} \arccos \frac{x}{a}\, dx = -\frac{1}{x} \arccos \frac{x}{a} + \frac{1}{a} \ln\left|\frac{a + (a^2 - x^2)^{1/2}}{x}\right| \qquad [|x/a| \le 1]$

3. $\displaystyle\int \frac{1}{x^3} \arccos \frac{x}{a}\, dx = -\frac{1}{2x^2} \arccos \frac{x}{a} + \frac{(a^2 - x^2)^{1/2}}{2a^2 x} \qquad [|x/a| \le 1]$

4. $\displaystyle\int \frac{1}{x^n} \arccos \frac{x}{a}\, dx = \frac{-1}{(n-1)x^{n-1}} \arccos \frac{x}{a} - \frac{1}{n-1} \int \frac{dx}{x^{n-1}(a^2 - x^2)^{1/2}}$

$$[|x/a| \le 1, n \ne 1]$$

10.1.5 Integrands involving $x^n \arctan(x/a)$

10.1.5.1

1. $\displaystyle\int \arctan \frac{x}{a}\, dx = x \arctan \frac{x}{a} - \frac{a}{2} \ln(x^2 + a^2)$

2. $\displaystyle\int x \arctan \frac{x}{a}\, dx = \frac{1}{2}(x^2 + a^2) \arctan \frac{x}{a} - \frac{1}{2}ax$

3. $\displaystyle\int x^2 \arctan \frac{x}{a}\, dx = \frac{1}{3}x^3 \arctan \frac{x}{a} - \frac{1}{6}ax^2 + \frac{1}{6}a^3 \ln(x^2 + a^2)$

4. $\displaystyle\int x^3 \arctan \frac{x}{a}\, dx = \frac{1}{4}(x^4 - a^4) \arctan \frac{x}{a} - \frac{1}{12}ax^3 + \frac{1}{4}a^3 x$

5. $\displaystyle\int x^n \arctan \frac{x}{a}\, dx = \frac{x^{n+1}}{n+1} \arctan \frac{x}{a} - \frac{a}{n+1} \int \frac{x^{n+1}\, dx}{x^2 + a^2}$

10.1.6 Integrands involving $x^{-n} \arctan(x/a)$

10.1.6.1

1. $\displaystyle\int \frac{1}{x} \arctan \frac{x}{a}\, dx = \sum_{k=0}^{\infty} \frac{(-1)^k}{(2k+1)^2} \left(\frac{x}{a}\right)^{2k+1}$ $[|x/a| < 1]$

   $\displaystyle = \frac{\pi}{2} \ln|x| + \sum_{k=0}^{\infty} \frac{(-1)^k}{(2k+1)^2} \left(\frac{a}{x}\right)^{2k+1}$ $[x/a > 1]$

   $\displaystyle = -\frac{\pi}{2} \ln|x| + \sum_{k=0}^{\infty} \frac{(-1)^k}{(2k+1)^2} \left(\frac{a}{x}\right)^{2k+1}$ $[x/a < -1]$

2. $\displaystyle\int \frac{1}{x^2} \arctan \frac{x}{a}\, dx = -\frac{1}{x} \arctan \frac{x}{a} + \frac{1}{2a} \ln\left(\frac{x^2}{x^2 + a^2}\right)$

3. $\displaystyle\int \frac{1}{x^3} \arctan \frac{x}{a}\, dx = -\frac{1}{2}\left(\frac{1}{x^2} + \frac{1}{a^2}\right) \arctan \frac{x}{a} - \frac{1}{2ax}$

4. $\displaystyle\int \frac{1}{x^n} \arctan \frac{x}{a}\, dx = -\frac{1}{(n-1)x^{n-1}} \arctan \frac{x}{a} + \frac{a}{n-1} \int \frac{dx}{x^{n-1}(x^2 + a^2)}$

   $[n \neq 1]$

10.1.7 Integrands involving $x^n \operatorname{arccot}(x/a)$

10.1.7.1

1. $\displaystyle\int \operatorname{arccot} \frac{x}{a}\, dx = x \operatorname{arccot} \frac{x}{a} + \frac{1}{2}a \ln(x^2 + a^2)$

2. $\displaystyle\int x \operatorname{arccot} \frac{x}{a}\, dx = \frac{1}{2}(x^2 + a^2) \operatorname{arccot} \frac{x}{a} + \frac{1}{2}ax$

3. $\displaystyle\int x^2 \operatorname{arccot} \frac{x}{a}\, dx = \frac{1}{3}x^3 \operatorname{arccot} \frac{x}{a} + \frac{1}{6}ax^2 - \frac{1}{6}a^3 \ln(x^2 + a^2)$

4. $\displaystyle\int x^3 \operatorname{arccot} \frac{x}{a}\, dx = \frac{1}{4}(x^4 - a^4) \operatorname{arccot} \frac{x}{a} + \frac{1}{12}ax^3 - \frac{1}{4}a^3 x$

5. $\int x^n \operatorname{arccot} \dfrac{x}{a}\, dx = \dfrac{x^{n+1}}{n+1} \operatorname{arccot} \dfrac{x}{a} + \dfrac{a}{n+1} \int \dfrac{x^{n+1}\, dx}{x^2 + a^2}$

10.1.8 Integrands involving $x^{-n} \operatorname{arccot}(x/a)$

10.1.8.1

1. $\int \dfrac{1}{x} \operatorname{arccot} \dfrac{x}{a}\, dx = \dfrac{\pi}{2} \ln|x| - \sum\limits_{k=0}^{\infty} \dfrac{(-1)^k}{(2k+1)^2} \left(\dfrac{x}{a}\right)^{2k+1} \qquad [|x/a| < 1]$

$= -\sum\limits_{k=0}^{\infty} \dfrac{(-1)^k}{(2k+1)^2} \left(\dfrac{a}{x}\right)^{2k+1} \qquad [x/a > 1]$

$= \pi \ln|x| - \sum\limits_{k=0}^{\infty} \dfrac{(-1)^k}{(2k+1)^2} \left(\dfrac{a}{x}\right)^{2k+1} \qquad [x/a < -1]$

2. $\int \dfrac{1}{x^2} \operatorname{arccot} \dfrac{x}{a}\, dx = -\dfrac{1}{x} \operatorname{arccot} \dfrac{x}{a} - \dfrac{1}{2a} \ln\left(\dfrac{x^2}{x^2 + a^2}\right)$

3. $\int \dfrac{1}{x^3} \operatorname{arccot} \dfrac{x}{a}\, dx = -\dfrac{1}{2}\left(\dfrac{1}{x^2} + \dfrac{1}{a^2}\right) \operatorname{arccot} \dfrac{x}{a} + \dfrac{1}{2ax}$

4. $\int \dfrac{1}{x^n} \operatorname{arccot} \dfrac{x}{a}\, dx = \dfrac{-1}{(n-1)x^{n-1}} \operatorname{arccot} \dfrac{x}{a} - \dfrac{a}{n-1} \int \dfrac{dx}{x^{n-1}(x^2 + a^2)}$

$[n \neq 1]$

10.1.9 Integrands involving products of rational functions and $\operatorname{arccot}(x/a)$

10.1.9.1

1. $\int \dfrac{1}{(x^2 + a^2)} \operatorname{arccot} \dfrac{x}{a}\, dx = -\dfrac{1}{2a} \operatorname{arccot}^2 \dfrac{x}{a}$

2. $\int \left(\dfrac{x^2}{x^2 + a^2}\right) \operatorname{arccot} \dfrac{x}{a}\, dx = x \operatorname{arccot} \dfrac{x}{a} + \dfrac{1}{2}a \ln(x^2 + a^2) + \dfrac{1}{2}a \operatorname{arccot}^2 \dfrac{x}{a}$

3. $\int \dfrac{1}{(x^2 + a^2)^2} \operatorname{arccot} \dfrac{x}{a}\, dx = \dfrac{x}{2a^2(x^2 + a^2)} \operatorname{arccot} \dfrac{x}{a} - \dfrac{1}{4a^3} \operatorname{arccot}^2 \dfrac{x}{a}$

$- \dfrac{1}{4a(x^2 + a^2)}$

4. $\int \dfrac{1}{(x^2 + a^2)} \operatorname{arccot}^n \dfrac{x}{a}\, dx = -\dfrac{1}{(n+1)a} \operatorname{arccot}^{n+1}\left(\dfrac{x}{a}\right)$

11

The Gamma, Beta, Pi, and Psi Functions

11.1 The Euler Integral and Limit and Infinite Product Representations for $\Gamma(x)$

11.1.1 Definitions and notation

11.1.1.1 The **gamma function,** denoted by $\Gamma(x)$, provides a generalization of factorial n to the case in which n is not an integer. It is defined by the **Euler integral:**

1. $\Gamma(x) = \displaystyle\int_0^\infty t^{x-1} e^{-t} dt,$

and for the integral values of x has the property that

2. $\Gamma(n+1) = 1 \cdot 2 \cdot 3 \cdots (n-1)n = n!$

and, in particular,

3. $\Gamma(1) = \Gamma(2) = 1.$

A related notation defines the **pi function** $\Pi(x)$ as

4. $\Pi(x) = \Gamma(x+1) = \displaystyle\int_0^\infty t^x e^{-t} dt.$

It follows from this that when n is an integer

5. $\Pi(n) = n!$ and $\Pi(n) = n\Pi(n-1).$

Alternative definitions of the gamma function are the Euler definition in terms of a limit

6. $\Gamma(x) = \displaystyle\lim_{n\to\infty} \frac{n! n^x}{x(x+1)\cdots(x+n)}$ $[n \neq 0, -1, -2, \ldots]$

and the Weierstrass infinite product representation

7. $\dfrac{1}{\Gamma(x)} = xe^{\gamma x} \displaystyle\prod_{n=1}^{\infty} \left[\left(1 + \frac{x}{n}\right)e^{-x/n}\right],$

where γ is **Euler's constant** (also known as the **Euler–Mascheroni constant**) defined as

$$\gamma = \lim_{n\to\infty} \left(1 + \frac{1}{2} + \frac{1}{3} + \cdots + \frac{1}{n} - \ln n\right) = 0.57721566\ldots.$$

The symbol C is also used to denote this constant instead of γ.

11.1.2 Special properties of $\Gamma(x)$

11.1.2.1   The two most important recurrence formulas involving the gamma function are

1. $\Gamma(x+1) = x\Gamma(x)$

2. $\Gamma(x+n) = (x+n-1)(x+n-2)\cdots(x+1)\Gamma(x+1).$

The **reflection formula** for the gamma function is

3. $\Gamma(x)\Gamma(1-x) = -x\Gamma(-x)\Gamma(x) = \pi \operatorname{cosec} \pi x$

4. $\Gamma(x)\Gamma(-x) = -\dfrac{\pi}{x \sin \pi x}$

5. $\Gamma\left(\dfrac{1}{2} + x\right)\Gamma\left(\dfrac{1}{2} - x\right) = \dfrac{\pi}{\cos \pi x}.$

6.   A result involving $\Gamma(2n)$ is

$$\Gamma(2n) = \pi^{-1/2} 2^{2n-1} \Gamma(n)\Gamma\left(n + \frac{1}{2}\right).$$

When working with series it is sometimes useful to express the binomial coefficient $\binom{n}{m}$ in terms of the gamma function by using the result

7. $\dbinom{n}{m} = \dfrac{n!}{m!(n-m)!} = \dfrac{\Gamma(n+1)}{\Gamma(m+1)\Gamma(n-m+1)}.$

11.1.3 Asymptotic representations of $\Gamma(x)$ and $n!$

11.1.3.1   The **Stirling formulas** that yield asymptotic approximations to $\Gamma(x)$ and $n!$ for large values of x and n, respectively, are

1. $\Gamma(x) \sim e^{-x}x^{x-(1/2)}(2\pi)^{1/2}\left(1 + \dfrac{1}{12x} + \dfrac{1}{288x^2} - \dfrac{139}{51840x^3} - \cdots\right)$ $[x \gg 0]$

2. $n! \sim (2\pi)^{1/2}n^{n+(1/2)}e^{-n}$ $[n \gg 0]$.

Two other useful asymptotic results are

3. $\Gamma(ax + b) \sim (2\pi)^{1/2}e^{-ax}(ax)^{ax+b-(1/2)}$ $[x \gg 0, a > 0]$

4. $\ln\Gamma(x) \sim \left(x - \dfrac{1}{2}\right)\ln x - x + \dfrac{1}{2}\ln(2\pi) + \displaystyle\sum_{n=1}^{\infty}\dfrac{B_{2n}}{2n(2n-1)x^{2n-1}}$ $[x \gg 0]$.

11.1.4 Special values of $\Gamma(x)$

11.1.4.1

1. $\Gamma(1/4) = 3.62560990\ldots$

2. $\Gamma(1/2) = \pi^{1/2} = 1.77245385\ldots$

3. $\Gamma(3/4) = 1.22541670\ldots$

4. $\Gamma(3/2) = \frac{1}{2}\pi^{1/2} = 0.88622692\ldots$

5. $\Gamma\left(n + \dfrac{1}{4}\right) = \dfrac{1 \cdot 5 \cdot 9 \cdot 13 \cdots (4n-3)}{4^n}\Gamma(1/4)$

6. $\Gamma\left(n + \dfrac{1}{2}\right) = \dfrac{1 \cdot 3 \cdot 5 \cdot 7 \cdots (2n-1)}{2^n}\Gamma(1/2)$

7. $\Gamma\left(n + \dfrac{3}{4}\right) = \dfrac{3 \cdot 7 \cdot 11 \cdot 15 \cdots (4n-1)}{4^n}\Gamma(3/4)$

8. $\Gamma\left(\dfrac{1}{2}\right)\Gamma\left(-\dfrac{1}{2}\right) = -2\pi$

11.1.5 The gamma function in the complex plane

11.1.5.1   For complex $z = x + iy$, the gamma function is defined as

1. $\Gamma(z) = \displaystyle\int_0^\infty t^{z-1}e^{-t}dt$ $[Re\{z\} > 0]$.

The following properties hold for $\Gamma(z)$:

2. $\Gamma(\bar{z}) = \overline{\Gamma(z)}$

3. $\ln\Gamma(\bar{z}) = \overline{\ln\Gamma(z)}$

4. $\arg\Gamma(z + 1) = \arg\Gamma(z) + \arctan\dfrac{y}{x}$

5. $|\Gamma(z)| \le |\Gamma(x)|$

11.1.6 The psi (digamma) function

11.1.6.1   The **psi function,** written $\psi(z)$ and also called the **digamma function,** is defined as

1. $\psi(z) = \dfrac{d}{dz}[\ln \Gamma(z)] = \Gamma'(z)/\Gamma(z).$

The psi function has the following properties:

2. $\psi(z+1) = \psi(z) + \dfrac{1}{z}$ (recurrence formula)

3. $\psi(1-z) = \psi(z) + \pi \cot \pi z$ (reflection formula)

4. $\psi(\bar{z}) = \overline{\psi(z)}$

5. $\psi(1) = -\gamma$

6. $\psi(n) = -\gamma + \displaystyle\sum_{k=1}^{n-1} 1/k \qquad [n \geq 2]$

7. $\psi\left(\dfrac{1}{2}\right) = -\gamma - 2\ln 2 = -1.96351002\ldots$

8. $\psi\left(n + \dfrac{1}{2}\right) = -\gamma - 2\ln 2 + 2\left(1 + \dfrac{1}{3} + \dfrac{1}{5} + \cdots + \dfrac{1}{2n-1}\right) \qquad [n \geq 1]$

9. $\psi(z+n) = \dfrac{1}{(n-1)+z} + \dfrac{1}{(n-2)+z} + \cdots + \dfrac{1}{2+z} + \dfrac{1}{1+z} + \psi(z+1)$

10. $\psi(x) = \displaystyle\sum_{n=1}^{\infty} \left(\dfrac{1}{n} - \dfrac{1}{x-1+n}\right) - \gamma$

11. $\psi(z) \sim \ln z - \dfrac{1}{2z} - \displaystyle\sum_{n=1}^{\infty} \dfrac{B_{2n}}{2nz^{2n}} \qquad [z \to \infty \text{ in } |\arg z| < \pi]$

Care must be exercised when using the psi function in conjunction with other reference sources because on occasions, instead of the definition used in 11.1.6.1.1, $\psi(z)$ is defined as $d[\ln \Gamma(z+1)]/dz$.

11.1.7 The beta function

11.1.7.1   The **beta function,** denoted by $B(x, y)$, is defined as

1. $B(x, y) = \displaystyle\int_0^1 t^{x-1}(1-t)^{y-1}dt = \int_0^{\infty} \dfrac{t^{x-1}}{(1+t)^{x+y}}\,dt \qquad [x > 0, y > 0].$

$B(x, y)$ has the following properties:

2. $B(x, y) = \dfrac{\Gamma(x)\Gamma(y)}{\Gamma(x+y)} \qquad [x > 0, y > 0]$

3. $B(x, y) = B(y, x)$ (symmetry)

4. $B(m, n) = \dfrac{(m-1)!(n-1)!}{(m+n-1)!}$ (m, n nonnegative integers)

5. $B(x, y)B(x+y, w) = B(y, w)B(y+w, x)$

6. $\dfrac{1}{B(m, n)} = m\dbinom{n+m-1}{n-1} = n\dbinom{n+m-1}{m-1}$ $[m, n = 1, 2, \ldots]$

11.1.8 Graph of $\Gamma(x)$ and tabular values of $\Gamma(x)$ and $\ln \Gamma(x)$

11.1.8.1   Figure 11.1 shows the behavior of $\Gamma(x)$ for $-4 < x < 4$. The gamma function becomes infinite at $x = 0, -1, -2, \ldots$.

Table 11.1 lists numerical values of $\Gamma(x)$ and $\ln \Gamma(x)$ for $0 \le x \le 2$. If required, the table may be extended by using these results:

1. $\Gamma(x+1) = x\,\Gamma(x)$,

2. $\ln \Gamma(x+1) = \ln x + \ln \Gamma(x)$.

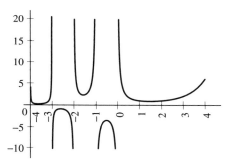

FIGURE 11.1 ▌ Graph of $\Gamma(x)$ for $-4 < x < 4$.

TABLE 11.1   Tables of $\Gamma(x)$ and $\ln \Gamma(x)$

x	$\Gamma(x)$	$\ln \Gamma(x)$	x	$\Gamma(x)$	$\ln \Gamma(x)$
0	∞	∞	0.46	1.925227	0.655044
0.01	99.432585	4.599480	0.47	1.884326	0.633570
0.02	49.442210	3.900805	0.48	1.845306	0.612645
0.03	32.784998	3.489971	0.49	1.808051	0.592250
0.04	24.460955	3.197078	0.50	1.772454	0.572365
0.05	19.470085	2.968879	0.51	1.738415	0.552974
0.06	16.145727	2.781655	0.52	1.705844	0.534060
0.07	13.773601	2.622754	0.53	1.674656	0.515608
0.08	11.996566	2.484620	0.54	1.644773	0.497603
0.09	10.616217	2.362383	0.55	1.616124	0.480031
0.10	9.513508	2.252713	0.56	1.588641	0.462879
0.11	8.612686	2.153236	0.57	1.562263	0.446135
0.12	7.863252	2.062200	0.58	1.536930	0.429787
0.13	7.230242	1.978272	0.59	1.512590	0.413824
0.14	6.688686	1.900417	0.60	1.489192	0.398234
0.15	6.220273	1.827814	0.61	1.466690	0.383008
0.16	5.811269	1.759799	0.62	1.445038	0.368136
0.17	5.451174	1.695831	0.63	1.424197	0.353608
0.18	5.131821	1.635461	0.64	1.404128	0.339417
0.19	4.846763	1.578311	0.65	1.384795	0.325552
0.20	4.590844	1.524064	0.66	1.366164	0.312007
0.21	4.359888	1.472446	0.67	1.348204	0.298773
0.22	4.150482	1.423224	0.68	1.330884	0.285843
0.23	3.959804	1.376194	0.69	1.314177	0.273210
0.24	3.785504	1.331179	0.70	1.298055	0.260867
0.25	3.625610	1.288023	0.71	1.282495	0.248808
0.26	3.478450	1.246587	0.72	1.267473	0.237025
0.27	3.342604	1.206750	0.73	1.252966	0.225514
0.28	3.216852	1.168403	0.74	1.238954	0.214268
0.29	3.100143	1.131448	0.75	1.225417	0.203281
0.30	2.991569	1.095798	0.76	1.212335	0.192549
0.31	2.890336	1.061373	0.77	1.199692	0.182065
0.32	2.795751	1.028101	0.78	1.187471	0.171826
0.33	2.707206	0.995917	0.79	1.175655	0.161825
0.34	2.624163	0.964762	0.80	1.164230	0.152060
0.35	2.546147	0.934581	0.81	1.153181	0.142524
0.36	2.472735	0.905325	0.82	1.142494	0.133214
0.37	2.403550	0.876947	0.83	1.132157	0.124125
0.38	2.338256	0.849405	0.84	1.122158	0.115253
0.39	2.276549	0.822661	0.85	1.112484	0.106595
0.40	2.218160	0.796678	0.86	1.103124	0.098147
0.41	2.162841	0.771422	0.87	1.094069	0.089904
0.42	2.110371	0.746864	0.88	1.085308	0.081864
0.43	2.060549	0.722973	0.89	1.076831	0.074022
0.44	2.013193	0.699722	0.90	1.068629	0.066376
0.45	1.968136	0.677087	0.91	1.060693	0.058923

(Continues)

TABLE 11.1 (*Continued*)

x	$\Gamma(x)$	$\ln\Gamma(x)$	x	$\Gamma(x)$	$\ln\Gamma(x)$
0.92	1.053016	0.051658	1.37	0.889314	−0.117305
0.93	1.045588	0.044579	1.38	0.888537	−0.118179
0.94	1.038403	0.037684	1.39	0.887854	−0.118948
0.95	1.031453	0.030969	1.40	0.887264	−0.119613
0.96	1.024732	0.024431	1.41	0.886765	−0.120176
0.97	1.018232	0.018068	1.42	0.886356	−0.120637
0.98	1.011947	0.011877	1.43	0.886036	−0.120997
0.99	1.005872	0.005855	1.44	0.885805	−0.121258
1	1	0	1.45	0.885661	−0.121421
1.01	0.994326	−0.005690	1.46	0.885604	−0.121485
1.02	0.988844	−0.011218	1.47	0.885633	−0.121452
1.03	0.983550	−0.016587	1.48	0.885747	−0.121324
1.04	0.978438	−0.021798	1.49	0.885945	−0.121100
1.05	0.973504	−0.026853	1.50	0.886227	−0.120782
1.06	0.968744	−0.031755	1.51	0.886592	−0.120371
1.07	0.964152	−0.036506	1.52	0.887039	−0.119867
1.08	0.959725	−0.041108	1.53	0.887568	−0.119271
1.09	0.955459	−0.045563	1.54	0.888178	−0.118583
1.10	0.951351	−0.049872	1.55	0.888868	−0.117806
1.11	0.947396	−0.054039	1.56	0.889639	−0.116939
1.12	0.943590	−0.058063	1.57	0.890490	−0.115984
1.13	0.939931	−0.061948	1.58	0.891420	−0.114940
1.14	0.936416	−0.065695	1.59	0.892428	−0.113809
1.15	0.933041	−0.069306	1.60	0.893515	−0.112592
1.16	0.929803	−0.072782	1.61	0.894681	−0.111288
1.17	0.926700	−0.076126	1.62	0.895924	−0.109900
1.18	0.923728	−0.079338	1.63	0.897244	−0.108427
1.19	0.920885	−0.082420	1.64	0.898642	−0.106871
1.20	0.918169	−0.085374	1.65	0.900117	−0.105231
1.21	0.915576	−0.088201	1.66	0.901668	−0.103508
1.22	0.913106	−0.090903	1.67	0.903296	−0.101704
1.23	0.910755	−0.093482	1.68	0.905001	−0.099819
1.24	0.908521	−0.095937	1.69	0.906782	−0.097853
1.25	0.906402	−0.098272	1.70	0.908639	−0.095808
1.26	0.904397	−0.100487	1.71	0.910572	−0.093683
1.27	0.902503	−0.102583	1.72	0.912581	−0.091479
1.28	0.900718	−0.104563	1.73	0.914665	−0.089197
1.29	0.899042	−0.106426	1.74	0.916826	−0.086838
1.30	0.897471	−0.108175	1.75	0.919063	−0.084401
1.31	0.896004	−0.109810	1.76	0.921375	−0.081888
1.32	0.894640	−0.111333	1.77	0.923763	−0.079300
1.33	0.893378	−0.112745	1.78	0.926227	−0.076636
1.34	0.892216	−0.114048	1.79	0.928767	−0.073897
1.35	0.891151	−0.115241	1.80	0.931384	−0.071084
1.36	0.890185	−0.116326	1.81	0.934076	−0.068197

(*Continues*)

TABLE 11.1 (*Continued*)

x	$\Gamma(x)$	$\ln \Gamma(x)$	x	$\Gamma(x)$	$\ln \Gamma(x)$
1.82	0.936845	−0.065237	1.91	0.965231	−0.035388
1.83	0.939690	−0.062205	1.92	0.968774	−0.031724
1.84	0.942612	−0.059100	1.93	0.972397	−0.027991
1.85	0.945611	−0.055924	1.94	0.976099	−0.024191
1.86	0.948687	−0.052676	1.95	0.979881	−0.020324
1.87	0.951840	−0.049358	1.96	0.983743	−0.016391
1.88	0.955071	−0.045970	1.97	0.987685	−0.012391
1.89	0.958379	−0.042512	1.98	0.991708	−0.008326
1.90	0.961766	−0.038984	1.99	0.995813	−0.004196
			2	1	0

12

Elliptic Integrals and Functions

12.1 Elliptic Integrals

12.1.1 Legendre normal forms

12.1.1.1 An **elliptic integral** is an integral of the form $\int R(x, \sqrt{P(x)})\,dx$, in which R is a rational function of its arguments and $P(x)$ is a third- or fourth-degree polynomial with distinct zeros. Every elliptic integral can be reduced to a sum of integrals expressible in terms of algebraic, trigonometric, inverse trigonometric, logarithmic, and exponential functions (the **elementary functions**), together with one or more of the following three special types of integral:

Elliptic integral of the first kind.

1. $$\int \frac{dx}{\sqrt{(1-x^2)(1-k^2x^2)}} \qquad [k^2 < 1]$$

Elliptic integral of the second kind.

2. $\displaystyle\int \frac{\sqrt{1 - k^2 x^2}}{\sqrt{1 - x^2}}\, dx \qquad [k^2 < 1]$

Elliptic integral of the third kind.

3. $\displaystyle\int \frac{dx}{(1 - nx^2)\sqrt{(1 - x^2)(1 - k^2 x^2)}} \qquad [k^2 < 1]$

These integrals are said to be expressed in the **Legendre normal form.** The number k is called the **modulus** of the elliptic integral, and the number $k' = \sqrt{1 - k^2}$ is called the **complementary modulus** of the elliptic integral. It is usual to set $k = \sin \alpha$, to call α the **modular** angle, and $m = k^2 = \sin^2 \alpha$ the **parameter.** In 12.1.1.1.3, the number n is called the **characteristic parameter** of the elliptic integral of the third kind.

For many purposes it is more convenient to work with the trigonometric forms of these integrals. These are obtained by setting $x = \sin \theta$, and then regarding the integrals as functions of their upper limits by evaluating them over the interval $[0, \varphi]$, thereby defining the **elliptic functions,** or **incomplete elliptic integrals,** $F(\varphi, k)$, $E(\varphi, k)$, and $\Pi(\varphi, k)$ as follows:

Incomplete elliptic integral of the first kind.

4. $\displaystyle F(\varphi, k) = \int_0^\varphi \frac{d\theta}{(1 - k^2 \sin^2 \theta)^{1/2}} = \int_0^{\sin \varphi} \frac{dx}{\sqrt{(1 - x^2)(1 - k^2 x^2)}}$

Incomplete elliptic integral of the second kind.

5. $\displaystyle E(\varphi, k) = \int_0^\varphi (1 - k^2 \sin^2 \theta)^{1/2}\, d\theta = \int_0^{\sin \varphi} \frac{\sqrt{1 - k^2 x^2}}{\sqrt{1 - k^2}}\, dx \qquad [k^2 < 1]$

Incomplete elliptic integral of the third kind.

6. $\displaystyle \Pi(\varphi, n, k) = \int_0^\varphi \frac{d\theta}{(1 - n \sin^2 \theta)(1 - k^2 \sin^2 \theta)^{1/2}}$

$\displaystyle \qquad = \int_0^{\sin \varphi} \frac{dx}{(1 - nx^2)\sqrt{(1 - x^2)(1 - k^2 x^2)}} \qquad [k^2 < 1]$

When the integration in 12.1.1.1.4–5 is extended over the interval $0 \le \theta \le \pi/2$, the integrals are called **complete elliptic integrals.** The notation used for complete elliptic integrals is

7. $\displaystyle \mathbf{K}(k) = F\left(\frac{\pi}{2}, k\right) = \mathbf{K}'(k')$

8. $\mathbf{E}(k) = E\left(\dfrac{\pi}{2}, k\right) = \mathbf{E}'(k')$

9. $\mathbf{K}'(k) = F\left(\dfrac{\pi}{2}, k'\right) = \mathbf{K}(k')$

10. $\mathbf{E}'(k) = E\left(\dfrac{\pi}{2}, k'\right) = \mathbf{E}(k')$

Frequently the modulus k is omitted from this notation with the understanding that

$$\mathbf{K} \equiv \mathbf{K}(k), \qquad \mathbf{K}' \equiv \mathbf{K}'(k), \qquad \mathbf{E} = \mathbf{E}(k), \qquad \mathbf{E}' = \mathbf{E}'(k).$$

12.1.2 Tabulations and trigonometric series representations of complete elliptic integrals

Abbreviated tabulations of complete elliptic integrals of the first and second kinds are given in Tables 12.1 and 12.2, with the respective arguments k^2 and the modular angle $\alpha(k = \sin \alpha)$.

TABLE 12.1 Short Table of Complete Elliptic Integrals of the First and Second Kinds with Argument k^2

$$\mathbf{K}(k) = \int_0^{\pi/2} (1 - k^2 \sin^2 t)^{-1/2} \, dt \qquad \mathbf{K}'(k) = \int_0^{\pi/2} [1 - (1 - k^2) \sin^2 t]^{-1/2} \, dt$$

$$\mathbf{E}(k) = \int_0^{\pi/2} (1 - k^2 \sin^2 t)^{1/2} \, dt \qquad \mathbf{E}'(k) = \int_0^{\pi/2} [1 - (1 - k^2) \sin^2 t]^{1/2} \, dt$$

k^2	$\mathbf{K}(k)$	$\mathbf{K}'(k)$	$\mathbf{E}(k)$	$\mathbf{E}'(k)$
0	1.570796	∞	1.570796	1
0.05	1.591003	2.908337	1.550973	1.060474
0.1	1.612441	2.578092	1.530758	1.104775
0.15	1.635257	2.389016	1.510122	1.143396
0.2	1.659624	2.257205	1.489035	1.17849
0.25	1.68575	2.156516	1.467462	1.211056
0.3	1.713889	2.075363	1.445363	1.241671
0.35	1.744351	2.007598	1.422691	1.270707
0.4	1.777519	1.949568	1.399392	1.298428
0.45	1.813884	1.898925	1.375402	1.325024
0.5	1.854075	1.854075	1.350644	1.350644
0.55	1.898925	1.813884	1.325024	1.375402
0.6	1.949568	1.777519	1.298428	1.399392
0.65	2.007598	1.744351	1.270707	1.422691
0.7	2.075363	1.713889	1.241671	1.445363
0.75	2.156516	1.68575	1.211056	1.467462
0.8	2.257205	1.659624	1.17849	1.489035
0.85	2.389016	1.635257	1.143396	1.510122
0.9	2.578092	1.612441	1.104775	1.530758
0.95	2.908337	1.591003	1.060474	1.550973
1	∞	1.570796	1	1.570796

TABLE 12.2 Short Table of Complete Elliptic Integrals of the First and Second Kinds with Argument Modular Angle α

$$\mathbf{K}(\alpha) = \int_0^{\pi/2} (1 - \sin^2 \alpha \sin^2 t)^{-1/2}\, dt \qquad \mathbf{K}'(\alpha) = \int_0^{\pi/2} [1 - (1 - \sin^2 \alpha)\sin^2 t]^{-1/2}\, dt$$

$$\mathbf{E}(\alpha) = \int_0^{\pi/2} (1 - \sin^2 \alpha \sin^2 t)^{1/2}\, dt \qquad \mathbf{E}'(\alpha) = \int_0^{\pi/2} [1 - (1 - \sin^2 \alpha)\sin^2 t]^{1/2}\, dt$$

α^0	$\mathbf{K}(\alpha)$	$\mathbf{K}'(\alpha)$	$\mathbf{E}(\alpha)$	$\mathbf{E}'(\alpha)$
0	1.570796	∞	1.570796	1
5	1.573792	3.831742	1.567809	1.012664
10	1.582843	3.153385	1.558887	1.040114
15	1.598142	2.768063	1.544150	1.076405
20	1.620026	2.504550	1.523799	1.118378
25	1.648995	2.308787	1.498115	1.163828
30	1.685750	2.156516	1.467462	1.211056
35	1.731245	2.034715	1.432291	1.258680
40	1.786769	1.935581	1.393140	1.305539
45	1.854075	1.854075	1.350644	1.350644
50	1.935581	1.786769	1.305539	1.393140
55	2.034715	1.731245	1.258680	1.432291
60	2.156516	1.685750	1.211056	1.467462
65	2.308787	1.648995	1.163828	1.498115
70	2.504550	1.620026	1.118378	1.523799
75	2.768063	1.598142	1.076405	1.544150
80	3.153385	1.582843	1.040114	1.558887
85	3.831742	1.573792	1.012664	1.567809
90	∞	1.570796	1	1.570796

When an analytical approximation is necessary, the following trigonometric series representations can be used.

1. $\mathbf{K} = \dfrac{\pi}{2}\left\{ 1 + \left(\dfrac{1}{2}\right)^2 k^2 + \left(\dfrac{1\cdot 3}{2\cdot 4}\right)^2 k^4 + \cdots + \left[\dfrac{(2n-1)!!}{2^n n!}\right]^2 k^{2n} + \cdots \right\}$

2. $\mathbf{K} = \dfrac{\pi}{1+k'}\left\{ 1 + \left(\dfrac{1}{2}\right)^2 \left(\dfrac{1-k'}{1+k'}\right)^2 + \left(\dfrac{1\cdot 3}{2\cdot 4}\right)^2 \left(\dfrac{1-k'}{1+k'}\right)^4 + \cdots \right.$

    $\left. + \left[\dfrac{(2n-1)!!}{2^n n!}\right]^2 \left(\dfrac{1-k'}{1+k'}\right)^{2n} + \cdots \right\}$

3. $\mathbf{K} = \ln\left(\dfrac{4}{k'}\right) + \left(\dfrac{1}{2}\right)^2 \left(\ln\left(\dfrac{4}{k'}\right) - \dfrac{2}{1\cdot 2}\right) k'^2$

    $+ \left(\dfrac{1\cdot 3}{2\cdot 4}\right)^2 \left(\ln\left(\dfrac{4}{k'}\right) - \dfrac{2}{1\cdot 2} - \dfrac{2}{2\cdot 3}\right) k'^4 + \cdots$

    $+ \left(\dfrac{1\cdot 3\cdot 5}{2\cdot 4\cdot 6}\right)^2 \left(\ln\left(\dfrac{4}{k'}\right) - \dfrac{2}{1\cdot 2} - \dfrac{2}{3\cdot 4} - \dfrac{2}{5\cdot 6}\right) k'^6 + \cdots$

4. $E = \dfrac{\pi}{2}\left\{1 - \dfrac{1}{2^2}k^2 - \dfrac{1^2 \cdot 3}{2^2 \cdot 4}k^4 - \cdots - \left[\dfrac{(2n-1)!!}{2^n n!}\right]^2 \dfrac{k^{2n}}{2n-1} - \cdots\right\}$

5. $E = \dfrac{(1+k')\pi}{4}\left\{1 + \dfrac{1}{2^2}\left(\dfrac{1-k'}{1+k'}\right)^2 + \dfrac{1^2}{2^2 \cdot 4^2}\left(\dfrac{1-k'}{1+k'}\right)^4 + \cdots\right.$

   $\left. + \left[\dfrac{(2n-3)!!}{2^n n!}\right]^2\left(\dfrac{1-k'}{1+k'}\right)^{2n} + \cdots\right\}$

6. $E = 1 + \dfrac{1}{2}\left(\ln\left(\dfrac{4}{k'}\right) - \dfrac{2}{1 \cdot 2}\right)k'^2 + \dfrac{1^2 \cdot 3}{2^2 \cdot 4}\left(\ln\left(\dfrac{4}{k'}\right) - \dfrac{2}{1 \cdot 2} - \dfrac{1}{3 \cdot 4}\right)k'^4$

   $+ \dfrac{1^2 \cdot 3^2 \cdot 5}{2^2 \cdot 4^2 \cdot 6}\left(\ln\left(\dfrac{4}{k'}\right) - \dfrac{2}{1 \cdot 2} - \dfrac{2}{3 \cdot 4} - \dfrac{2}{5 \cdot 6}\right)k'^6 + \cdots$

$$[(2n-1)!! = 1 \cdot 3 \cdot 5 \cdots (2n-1)]$$

12.1.3 Tabulations and trigonometric series for $E(\varphi, k)$ and $F(\varphi, k)$

Abbreviated tabulations of incomplete elliptic integrals of the first and second kinds are given in Table 12.3, with the argument being the modular angle α $(k = \sin\alpha)$. When an analytical approximation is necessary the following trigonometric series representations can be used. For small values of φ and k:

1. $F(\varphi, k) = \dfrac{2}{\pi}K\varphi - \sin\varphi\cos\varphi\left(a_0 + \dfrac{2}{3}a_1\sin^2\varphi + \dfrac{2 \cdot 4}{3 \cdot 5}a_2\sin^4\varphi + \cdots\right),$

   where

   $$a_0 = \dfrac{2}{\pi}K - 1, \quad a_n = a_{n-1} - \left[\dfrac{(2n-1)!!}{2^n n!}\right]^2 k^{2n}.$$

2. $E(\varphi, k) = \dfrac{2}{\pi}E\varphi + \sin\varphi\cos\varphi\left(b_0 + \dfrac{2}{3}b_1\sin^2\varphi + \dfrac{2 \cdot 4}{3 \cdot 5}b_2\sin^4\varphi + \cdots\right),$

   where

   $$b_0 = 1 - \dfrac{2}{\pi}K, \quad b_n = b_{n-1} - \left[\dfrac{(2n-1)!!}{2^n n!}\right]^2 \dfrac{k^{2n}}{2n-1}.$$

For k close to 1:

3. $F(\varphi, k) = \dfrac{2}{\pi}K'\ln\tan\left(\dfrac{\varphi}{2} + \dfrac{\pi}{4}\right)$

   $- \dfrac{\tan\varphi}{\cos\varphi}\left(a'_0 - \dfrac{2}{3}a'_1\tan^2\varphi + \dfrac{2 \cdot 4}{3 \cdot 5}a'_2\sin^4\varphi - \cdots\right),$

   where

   $$a'_0 = \dfrac{2}{\pi}K' - 1, \quad a'_n = a_{n-1} - \left[\dfrac{(2n-1)!!}{2^n n!}\right]^2 k'^{2n}.$$

4. $E(\varphi, k) = \dfrac{2}{\pi}(\mathbf{K}' - \mathbf{E}') \ln \tan\left(\dfrac{\varphi}{2} + \dfrac{\pi}{4}\right) + \dfrac{\tan \varphi}{\cos \varphi}\left(b_1' - \dfrac{2}{3}b_2' \tan^2 \varphi\right.$

$$\left. + \dfrac{2 \cdot 4}{3 \cdot 5}b_3' \tan^4 \varphi - \cdots \right) + \dfrac{1}{\sin \varphi}[1 - \cos \varphi\sqrt{1 - k^2 \sin \varphi}],$$

where

$$b_0' = \dfrac{2}{\pi}(\mathbf{K}' - \mathbf{E}'), \quad b_n' = b_{n-1}' - \left[\dfrac{(2n-3)!!}{2^{n-1}(n-1)!}\right]^2 \left(\dfrac{2n-1}{2n}\right)k'^{2n}.$$

TABLE 12.3   Short Table of Incomplete Elliptic Integrals of the First and Second Kinds with Argument Modular Angle α

$$F(\varphi, \alpha) = \int_0^\varphi (1 - \sin^2 \alpha \sin^2 t)^{-1/2}\, dt \qquad [k = \sin \alpha]$$

(φ)	$0°$	$10°$	$20°$	$30°$	$40°$	$50°$	$60°$	$70°$	$80°$	$90°$
$0°$	0	0	0	0	0	0	0	0	0	0
$10°$	0.174533	0.174559	0.174636	0.174754	0.174899	0.175054	0.175200	0.175320	0.175399	0.175426
$20°$	0.349066	0.349275	0.349880	0.350819	0.351989	0.353257	0.354472	0.355480	0.356146	0.356379
$30°$	0.523599	0.524284	0.526284	0.529429	0.533427	0.537868	0.542229	0.545932	0.548425	0.549306
$40°$	0.698132	0.699692	0.704287	0.711647	0.721262	0.732308	0.743581	0.753521	0.760426	0.762910
$50°$	0.872665	0.875555	0.884162	0.898245	0.917255	0.940076	0.964652	0.987623	1.004439	1.010683
$60°$	1.047198	1.051879	1.065969	1.089551	1.122557	1.164316	1.212597	1.261860	1.301353	1.316958
$70°$	1.221730	1.228610	1.249526	1.285301	1.337228	1.406769	1.494411	1.595906	1.691815	1.735415
$80°$	1.396263	1.405645	1.434416	1.484555	1.559734	1.665965	1.812530	2.011928	2.265272	2.436246
$90°$	1.570796	1.582843	1.620026	1.685750	1.786769	1.935581	2.156516	2.504549	3.153384	∞

$$E(\varphi, \alpha) = \int_0^\varphi (1 - \sin^2 \alpha \sin^2 t)^{1/2}\, dt \qquad [k = \sin \alpha]$$

(φ)	$0°$	$10°$	$20°$	$30°$	$40°$	$50°$	$60°$	$70°$	$80°$	$90°$
$0°$	0	0	0	0	0	0	0	0	0	0
$10°$	0.174533	0.174506	0.174430	0.174312	0.174168	0.174015	0.173870	0.173752	0.173675	0.173648
$20°$	0.349066	0.348857	0.348255	0.347329	0.346186	0.344963	0.343806	0.342858	0.342236	0.342020
$30°$	0.523599	0.522915	0.520938	0.517882	0.514089	0.509995	0.506092	0.502868	0.500742	0.500000
$40°$	0.698132	0.696578	0.692070	0.685060	0.676282	0.666705	0.657463	0.649737	0.644593	0.642788
$50°$	0.872665	0.869790	0.861421	0.848317	0.831732	0.813383	0.795380	0.780066	0.769713	0.766044
$60°$	1.047198	1.042550	1.028972	1.007556	0.980134	0.949298	0.918393	0.891436	0.872755	0.866025
$70°$	1.221730	1.214913	1.194925	1.163177	1.122054	1.074998	1.026637	0.982976	0.951438	0.939693
$80°$	1.396263	1.386979	1.359682	1.316058	1.258967	1.192553	1.122486	1.056482	1.005433	0.984808
$90°$	1.570796	1.558887	1.523799	1.467462	1.393140	1.305539	1.211056	1.118378	1.040114	1.000000

12.2 Jacobian Elliptic Functions

12.2.1 The functions sn u, cn u, and dn u

12.2.1.1   The **Jacobian elliptic functions** are defined in terms of the elliptic integral

1. $u = \displaystyle\int_0^\varphi \frac{d\theta}{(1 - k^2 \sin^2 \theta)^{1/2}} = F(\varphi, k),$

in which φ is called the **amplitude** of u and is written

2. $\varphi = \mathrm{am}\, u.$

The Jacobian elliptic functions sn u, cn u, and dn u are defined as

3. $\mathrm{sn}\, u = \sin \varphi$

4. $\mathrm{cn}\, u = \cos \varphi = (1 - \mathrm{sn}^2\, u)^{1/2}$

5. $\mathrm{dn}\, u = (1 - k^2 \sin^2 \varphi)^{1/2} = \Delta(\varphi)$ (the **delta amplitude**)

     In terms of the Legendre normal form of 12.2.1.1.1,

$$u = \int_0^x \frac{dv}{\sqrt{(1 - v^2)(1 - k^2 v^2)}},$$

we have

6. $x = \sin \varphi = \mathrm{sn}\, u,$

7. $\varphi = \mathrm{am}\, u = \arcsin(\mathrm{sn}\, u) = \arcsin x.$

12.2.2 Basic results

12.2.2.1 Even and odd properties and special values.

1. $\mathrm{am}(-u) = -\,\mathrm{am}\, u$ (**odd** function)
2. $\mathrm{sn}(-u) = -\,\mathrm{sn}\, u$ (odd function)
3. $\mathrm{cn}(-u) = \mathrm{cn}\, u$ (**even** function)
4. $\mathrm{dn}(-u) = \mathrm{dn}\, u$ (even function)
5. $\mathrm{am}\, 0 = 0$
6. $\mathrm{sn}\, 0 = 0$
7. $\mathrm{cn}\, 0 = 1$
8. $\mathrm{dn}\, 0 = 1$

12.2.2.2 Identities.

1. $\mathrm{sn}^2\, u + \mathrm{cn}^2\, u = 1$
2. $\mathrm{dn}^2\, u + k^2 \mathrm{sn}^2\, u = 1$
3. $\mathrm{dn}^2\, u - k^2 \mathrm{cn}^2\, u = 1 - k^2$

12.2.2.3 Addition formulas for double and half-arguments.

1. $\text{sn}(u \pm v) = \dfrac{\text{sn}\, u \,\text{cn}\, v \,\text{dn}\, v \pm \text{cn}\, u \,\text{sn}\, v \,\text{dn}\, u}{1 - k^2 \,\text{sn}^2 u \,\text{sn}^2 v}$

2. $\text{cn}(u \pm v) = \dfrac{\text{cn}\, u \,\text{cn}\, v \mp \text{sn}\, u \,\text{dn}\, u \,\text{sn}\, v \,\text{dn}\, v}{1 - k^2 \,\text{sn}^2 u \,\text{sn}^2 v}$

3. $\text{dn}(u \pm v) = \dfrac{\text{dn}\, u \,\text{dn}\, v \mp k^2 \,\text{sn}\, u \,\text{cn}\, u \,\text{sn}\, v \,\text{cn}\, v}{1 - k^2 \,\text{sn}^2 u \,\text{sn}^2 v}$

4. $\text{sn}(2u) = \dfrac{2 \,\text{sn}\, u \,\text{cn}\, u \,\text{dn}\, u}{1 - k^2 \,\text{sn}^4 u}$

5. $\text{cn}(2u) = \dfrac{\text{cn}^2 u - \text{sn}^2 u \,\text{dn}^2 u}{1 - k^2 \,\text{sn}^4 u}$

6. $\text{dn}(2u) = \dfrac{\text{dn}^2 u - k^2 \,\text{sn}^2 u \,\text{cn}^2 u}{1 - k^2 \,\text{sn}^4 u}$

7. $\text{sn}\left(\dfrac{1}{2}u\right) = \left(\dfrac{1 - \text{cn}\, u}{1 + \text{dn}\, u}\right)^{1/2}$

8. $\text{cn}\left(\dfrac{1}{2}u\right) = \left(\dfrac{\text{cn}\, u + \text{dn}\, u}{1 + \text{dn}\, u}\right)^{1/2}$

9. $\text{dn}\left(\dfrac{1}{2}u\right) = \left(\dfrac{\text{cn}\, u + \text{dn}\, u}{1 + \text{cn}\, u}\right)^{1/2}$

12.2.2.4 Series expansions of sn u, cn u, dn u, and am u in powers of u.   Using the notation $pq(u \,|\, k^2)$ to signify that pq is a function of u and k^2 we have:

1. $\text{sn}(u \,|\, k^2) = u - \dfrac{1}{3!}(1 + k^2)u^3 + \dfrac{1}{5!}(1 + 14k^2 + k^4)u^5 - \cdots$

2. $\text{cn}(u \,|\, k^2) = 1 - \dfrac{1}{2!}k^2 u^2 + \dfrac{1}{4!}(1 + 4k^2)u^4 - \dfrac{1}{6!}(1 + 44k^2 + 16k^4)u^6 + \cdots$

3. $\text{dn}(u \,|\, k^2) = 1 - \dfrac{1}{2!}k^2 u^2 + \dfrac{1}{4!}k^2(4 + k^2)u^4 - \dfrac{1}{6!}k^2(16 + 44k^2 + k^4)u^6 + \cdots$

4. $\text{am}(u \,|\, k^2) = u - \dfrac{1}{3!}k^2 u^3 + \dfrac{1}{5!}k^2(4 + k^2)u^5 - \dfrac{1}{7!}k^2(16 + 44k^2 + k^4)u^7 + \cdots$

5. $\text{sn}(u \,|\, 0) = \sin u$

6. $\text{cn}(u \,|\, 0) = \cos u$

7. $\text{dn}(u \,|\, 0) = 1$

8. $\text{sn}(u \,|\, 1) = \tanh u$

9. $cn(u \mid 1) = \operatorname{sech} u$

10. $dn(u \mid 1) = \operatorname{sech} u$

12.3 Derivatives and Integrals

12.3.1 Derivatives of sn *u*, cn *u*, and dn *u*

12.3.1.1

1. $\dfrac{d}{du}[\operatorname{sn} u] = \operatorname{cn} u \, \operatorname{dn} u$

2. $\dfrac{d}{du}[\operatorname{cn} u] = -\operatorname{sn} u \, \operatorname{dn} u$

3. $\dfrac{d}{du}[\operatorname{dn} u] = -k^2 \operatorname{cn} u \, \operatorname{dn} u$

12.3.2 Integrals involving sn *u*, cn *u*, and dn *u*

12.3.2.1

1. $\displaystyle\int \operatorname{sn} u \, du = \dfrac{1}{k}(\operatorname{dn} u - k \operatorname{cn} u)$

2. $\displaystyle\int \operatorname{cn} u \, du = \dfrac{1}{k}\arccos(\operatorname{dn} u)$

3. $\displaystyle\int \operatorname{dn} u \, du = \arcsin(\operatorname{sn} u) = \operatorname{am} u$

4. $\displaystyle\int \dfrac{du}{\operatorname{sn} u} = \ln\left|\dfrac{\operatorname{sn} u}{\operatorname{cn} u + \operatorname{dn} u}\right|$

5. $\displaystyle\int \dfrac{du}{\operatorname{cn} u} = \left(\dfrac{1}{1 - k^2}\right)\ln\left|\dfrac{(1 - k^2)\operatorname{sn} u + \operatorname{dn} u}{\operatorname{cn} u}\right|$

6. $\displaystyle\int \dfrac{du}{\operatorname{dn} u} = \left(\dfrac{1}{1 - k^2}\right)\arctan\left|\dfrac{(1 - k^2)\operatorname{sn} u - \operatorname{cn} u}{(1 - k^2)\operatorname{sn} u + \operatorname{cn} u}\right|$

12.4 Inverse Jacobian Elliptic Functions

12.4.1 Definitions

12.4.1.1 $\operatorname{sn}^{-1} u$, $\operatorname{cn}^{-1} u$, and $\operatorname{dn}^{-1} u$.   Each Jacobian elliptic function is the inverse of an appropriate elliptic integral. Since $\varphi = \operatorname{am} u$ and

1. $u = \displaystyle\int_0^{\varphi} \dfrac{d\theta}{(1 - k^2 \sin^2 \theta)^{1/2}} \qquad [0 < \varphi \leq \pi/2],$

it follows that

2. $\text{am}^{-1}\varphi = \int_0^\varphi \dfrac{d\theta}{(1 - k^2 \sin^2 \theta)^{1/2}} = F(\varphi, k).$

Analogously,

3. $\text{sn}^{-1} u = \int_0^u \dfrac{dx}{\sqrt{(1 - x^2)(1 - k^2 x^2)}} = F(\arcsin x, k) \qquad [0 < x \le 1]$

4. $\text{cn}^{-1} u = \int_u^1 \dfrac{dx}{\sqrt{(1 - x^2)(1 - k^2 + k^2 x^2)}} = F(\arcsin \sqrt{1 - x^2}, k)$

$$[0 \le x < 1]$$

5. $\text{dn}^{-1} u = \int_u^1 \dfrac{dx}{\sqrt{(1 - x^2)(x^2 + k^2 - 1)}} = F(\arcsin \sqrt{(1 - x^2)/k^2}, k)$

$$[1 - k^2 \le x < 1]$$

12.4.1.2 Special values.

1. $\text{sn}^{-1}(\varphi, 1) = \ln|\sec\varphi + \tan\varphi|$
2. $\text{cn}^{-1}(u, 1) = \ln|(1 + u)^{1/2}(1 - u)^{-1/2}|$
3. $\text{dn}^{-1}(u, 1) = \ln|[1 + (1 - u^2)^{1/2}]/u|$
4. $\text{sn}^{-1}(u, 0) = \arcsin u$
5. $\text{cn}^{-1}(u, 0) = \arccos u$

12.4.1.3 Integrals of $\text{sn}^{-1} u$, $\text{cn}^{-1} u$, and $\text{dn}^{-1} u$.

1. $\displaystyle\int \text{sn}^{-1} u \, du = u \, \text{sn}^{-1} u + \dfrac{1}{k} \cosh\left[\dfrac{(1 - k^2 u^2)^{1/2}}{1 - k^2}\right]$

2. $\displaystyle\int \text{cn}^{-1} u \, du = u \, \text{cn}^{-1} u - \dfrac{1}{k} \text{arccosh}[(1 - k^2 + k^2 u^2)^{1/2}]$

3. $\displaystyle\int \text{dn}^{-1} u = u \, \text{dn}^{-1} u - \arcsin\left[\dfrac{(1 - u^2)^{1/2}}{k}\right]$

13

Probability Integrals and the Error Function

13.1 Normal Distribution

13.1.1 Definitions

13.1.1.1   The **normal** or **Gaussian probability density function,**

1. $f(x) = \dfrac{1}{\sigma\sqrt{2\pi}}\exp\left\{-\dfrac{1}{2}\left(\dfrac{x-\mu}{\sigma}\right)^2\right\},$

is symmetrical about $x = \mu$ and such that

2. $\displaystyle\int_{-\infty}^{\infty} f(x)\,dx = \dfrac{1}{\sigma\sqrt{2\pi}}\int_{-\infty}^{\infty}\exp\left\{-\dfrac{1}{2}\left(\dfrac{x-\mu}{\sigma}\right)^2\right\}dx = 1.$

The distribution has a **mean** of

3. $\mu = \displaystyle\int_{-\infty}^{\infty} x f(x)\,dx = \dfrac{1}{\sigma\sqrt{2\pi}}\int_{-\infty}^{\infty} x\exp\left\{-\dfrac{1}{2}\left(\dfrac{x-\mu}{\sigma}\right)^2\right\}dx$

and a **variance** of

4. $\sigma^2 = \dfrac{1}{\sigma\sqrt{2\pi}} \displaystyle\int_{-\infty}^{\infty} (x-\mu)^2 \exp\left\{ -\dfrac{1}{2}\left(\dfrac{x-\mu}{\sigma} \right)^2 \right\} dx.$

When the normal probability density $f(x)$ is standardized with zero mean ($\mu = 0$) and unit variance ($\sigma = 1$) the **normal probability distribution** $P(x)$, also denoted by $\Phi(x)$, is defined as

5. $P(x) = \dfrac{1}{\sqrt{2\pi}} \displaystyle\int_{-\infty}^{x} e^{-t^2/2}\, dt.$

Related functions used in statistics are

6. $Q(x) = \dfrac{1}{\sqrt{2\pi}} \displaystyle\int_{x}^{\infty} e^{-t^2/2}\, dt = 1 - P(x) = P(-x)$

7. $A(x) = \dfrac{1}{\sqrt{2\pi}} \displaystyle\int_{-x}^{x} e^{-t^2/2}\, dt = 2P(x) - 1.$

Special values of $P(x)$, $Q(x)$ and $A(x)$ are

$$P(-\infty) = 0, \quad P(0) = 0.5, \quad P(\infty) = 1, \quad Q(-\infty) = 1, \quad Q(0) = 0.5,$$
$$Q(\infty) = 0, \quad A(0) = 0, \quad A(\infty) = 1.$$

An abbreviated tabulation of $P(x)$, $Q(x)$, and $A(x)$ is given in Table 13.1.

13.1.2 Power series representations ($x \geq 0$)

13.1.2.1

1. $P(x) = \dfrac{1}{2} + \dfrac{1}{\sqrt{2\pi}}\left\{ x - \dfrac{x^3}{6} + \dfrac{x^5}{40} - \dfrac{x^7}{336} + \dfrac{x^9}{3456} - \cdots \right\}$

$\qquad = \dfrac{1}{2} + \dfrac{1}{\sqrt{2\pi}} \displaystyle\sum_{k=0}^{\infty} \dfrac{(-1)^k x^{2k+1}}{k!\, 2^k (2k+1)} \qquad [x \geq 0]$

2. $Q(x) = \dfrac{1}{2} - \dfrac{1}{\sqrt{2\pi}}\left\{ x - \dfrac{x^3}{6} + \dfrac{x^5}{40} - \dfrac{x^7}{336} + \dfrac{x^9}{3456} - \cdots \right\}$

$\qquad = \dfrac{1}{2} - \dfrac{1}{\sqrt{2\pi}} \displaystyle\sum_{k=0}^{\infty} \dfrac{(-1)^k x^{2k+1}}{k!\, 2^k (2k+1)} \qquad [x \geq 0]$

3. $A(x) = \sqrt{\dfrac{2}{\pi}}\left\{ x - \dfrac{x^3}{6} + \dfrac{x^5}{40} - \dfrac{x^7}{336} + \dfrac{x^9}{3456} - \cdots \right\}$

$\qquad = \sqrt{\dfrac{2}{\pi}} \displaystyle\sum_{k=0}^{\infty} \dfrac{(-1)^k x^{2k+1}}{k!\, 2^k (2k+1)} \qquad [x \geq 0]$

TABLE 13.1 Abbreviated Tabulation of $P(x)$, $Q(x)$, $A(x)$, and erf x

x	$P(x)$	$Q(x)$	$A(x)$	erf x
0	0.5	0.5	0	0
0.1	0.539828	0.460172	0.079656	0.112463
0.2	0.579260	0.420740	0.158519	0.222702
0.3	0.617911	0.382089	0.235823	0.328627
0.4	0.655422	0.344578	0.310843	0.428392
0.5	0.691462	0.308538	0.382925	0.520500
0.6	0.725747	0.274253	0.451494	0.603856
0.7	0.758036	0.241964	0.516073	0.677801
0.8	0.788145	0.211855	0.576289	0.742101
0.9	0.815940	0.184060	0.631880	0.796908
1.0	0.841345	0.158655	0.682689	0.842701
1.1	0.864334	0.135666	0.728668	0.880205
1.2	0.884930	0.115070	0.769861	0.910314
1.3	0.903199	0.096801	0.806399	0.934008
1.4	0.919243	0.080757	0.838487	0.952285
1.5	0.933193	0.066807	0.866386	0.966105
1.6	0.945201	0.054799	0.890401	0.976348
1.7	0.955435	0.044565	0.910869	0.983790
1.8	0.964070	0.035930	0.928139	0.989090
1.9	0.971284	0.028716	0.942567	0.992790
2.0	0.977250	0.022750	0.954500	0.995322
2.1	0.982136	0.017864	0.964271	0.997020
2.2	0.986097	0.013903	0.972193	0.998137
2.3	0.989276	0.010724	0.978552	0.998857
2.4	0.991802	0.008198	0.983605	0.999311
2.5	0.993790	0.006210	0.987581	0.999593
2.6	0.995339	0.004661	0.990678	0.999764
2.7	0.996533	0.003467	0.993066	0.999866
2.8	0.997445	0.002555	0.994890	0.999925
2.9	0.998134	0.001866	0.996268	0.999959
3.0	0.998650	0.001350	0.997300	0.999978
∞	1	0	1	1

13.1.3 Asymptotic expansions ($x \gg 0$)

13.1.3.1

1. $$P(x) \sim 1 - \frac{1}{\sqrt{2\pi}} \frac{e^{-x^2/2}}{x} \left[1 - \frac{1}{x^2} + \frac{1 \cdot 3}{x^4} - \frac{1 \cdot 3 \cdot 5}{x^6} + \cdots \right.$$

$$\left. + \frac{(-1)^n 1 \cdot 3 \cdot 5 \cdots (2n-1)}{x^{2n}} \right] + R_n,$$

where the remainder (error) term R_n is such that $|R_n|$ is less than the magnitude of the first term to be neglected in the asymptotic expansion and of the same sign for $x \gg 0$.

2. $Q(x) \sim 1 - \dfrac{1}{\sqrt{2\pi}} \dfrac{e^{-x^2/2}}{x} \left[1 - \dfrac{1}{x^2} + \dfrac{1\cdot 3}{x^4} - \dfrac{1\cdot 3\cdot 5}{x^6} + \cdots \right.$

$$\left. + \dfrac{(-1)^n 1\cdot 3\cdot 5 \cdots (2n-1)}{x^{2n}} \right] + R_n,$$

where the remainder (error) term R_n is such that $|R_n|$ is less than the magnitude of the first term to be neglected in the asymptotic expansion and of the same sign for $x \gg 0$.

3. $A(x) \sim 1 - \sqrt{\dfrac{2}{\pi}} \dfrac{e^{-x^2/2}}{x} \left[1 - \dfrac{1}{x^2} + \dfrac{1\cdot 3}{x^4} - \dfrac{1\cdot 3\cdot 5}{x^6} + \cdots \right.$

$$\left. + \dfrac{(-1)^n 1\cdot 3\cdot 5 \cdots (2n-1)}{x^{2n}} \right] + R_n,$$

where the remainder (error) term R_n is such that $|R_n|$ is less than the magnitude of the first term to be neglected in the asymptotic expansion and of the same sign for $x \gg 0$.

13.2 The Error Function

13.2.1 Definitions

13.2.1.1   The **error function** erf x occurs in probability theory and statistics and diffusion problems, and is defined as

1. $\operatorname{erf} x = \dfrac{2}{\sqrt{\pi}} \displaystyle\int_0^x e^{-t^2}\, dt,$

and it obeys the **symmetry relation**

2. $\operatorname{erf}(-x) = -\operatorname{erf} x.$

Special values of erf x are

3. $\operatorname{erf} 0 = 0$ and $\operatorname{erf} \infty = 1.$

The **complementary error function** erfc x is defined as

4. $\operatorname{erfc} x = 1 - \operatorname{erf} x.$

13.2.2 Power series representation

13.2.2.1

1. $\operatorname{erf} x = \dfrac{2}{\sqrt{\pi}} \left\{ x - \dfrac{x^3}{3} + \dfrac{x^5}{10} - \dfrac{x^7}{42} + \dfrac{x^9}{216} - \cdots \right\}$

$$= \dfrac{2}{\sqrt{\pi}} \sum_{k=0}^{\infty} \dfrac{(-1)^k x^{2k+1}}{k!(2k+1)}.$$

An abbreviated tabulation of erf x is given in Table 13.1.

13.2.3 Asymptotic expansion ($x \gg 0$)

13.2.3.1

1. $\quad \text{erf}\, x \sim 1 - \dfrac{e^{-x^2}}{x\sqrt{\pi}}\left\{ 1 - \dfrac{1}{2x^2} + \dfrac{1\cdot 3}{2^2 x^4} - \dfrac{1\cdot 3\cdot 5}{2^3 x^6} + \cdots + \dfrac{(-1)^n (2n)!}{n!\, 2^{2n} x^{2n}} \right\} + R_n$

where the remainder (error) term R_n is less than the magnitude of the first term to be neglected in the asymptotic expansion and of the same sign, for $x \gg 0$.

13.2.4 Connection between $P(x)$ and erf x

13.2.4.1

1. $\quad P(x) = \dfrac{1}{2}\left[1 + \text{erf}\, \dfrac{x}{\sqrt{2}} \right].$

13.2.5 Integrals expressible in terms of erf x

13.2.5.1

1. $\quad \displaystyle\int_0^\infty e^{-t^2}\, \dfrac{\sin 2tx}{t}\, dt = \dfrac{1}{2}\pi\, \text{erf}\, x$

2. $\quad \displaystyle\int_0^\infty e^{-t^2} \sin 2tx\, dx = \dfrac{1}{2}\sqrt{\pi}\, e^{-x^2}\, \text{erf}\, x$

13.2.6 Derivatives of erf x

13.2.6.1

1. $\quad \dfrac{d}{dx}[\text{erf}\, x] = \dfrac{2}{\sqrt{\pi}} e^{-x^2}$

2. $\quad \dfrac{d^2}{dx^2}[\text{erf}\, x] = -\dfrac{4}{\sqrt{\pi}} x e^{-x^2}$

3. $\quad \dfrac{d^3}{dx^3}[\text{erf}\, x] = -\dfrac{1}{\sqrt{\pi}}(8x^2 - 4)e^{-x^2}$

4. $\quad \dfrac{d^4}{dx^4}[\text{erf}\, x] = -\dfrac{8}{\sqrt{\pi}}(3 - 2x^2)x e^{-x^2}$

13.2.7 Integrals of erfc x

13.2.7.1 The *n*'th repeated integral of erfc x arises in the study of diffusion and heat conduction and is denoted by $i^n\, \text{erfc}\, x$, with

1. $\quad i^n\, \text{erfc}\, x = \displaystyle\int_x^\infty i^{n-1}\, \text{erfc}\, t\, dt \qquad [n = 0, 1, 2, \ldots],$

2. $i^{-1}\,\mathrm{erfc}\,x = \dfrac{2}{\sqrt{\pi}}e^{-x^2}$ and $i^0\,\mathrm{erfc}\,x = \mathrm{erfc}\,x$.

In particular,

3. $i\,\mathrm{erfc}\,x = -\dfrac{1}{\pi}e^{-x^2} - x\,\mathrm{erfc}\,x$

4. $i^2\,\mathrm{erfc}\,x = -\frac{1}{4}(\mathrm{erfc}\,x - 2xi\,\mathrm{erfc}\,x)$

and $i^n\,\mathrm{erfc}\,x$ satisfies the **recurrence relation**

5. $2n\,i^n\,\mathrm{erfc}\,x = -i^{n-2}\,\mathrm{erfc}\,x + 2x\,i^{n-1}\,\mathrm{erfc}\,x$ $[n = 1, 2, 3, \ldots]$.

13.2.8 Integral and power series representation of i^n erfc x

13.2.8.1

1. $i^n\,\mathrm{erfc}\,x = \dfrac{2}{\sqrt{\pi}} \displaystyle\int_x^\infty \dfrac{(t-x)^n}{n!}e^{-t^2}\,dt$

2. $i^n\,\mathrm{erfc}\,x = \displaystyle\sum_{k=0}^\infty \dfrac{(-1)^k x^k}{2^{n-k}k!\,\Gamma\!\left(1 + \dfrac{n-k}{2}\right)}$

13.2.9 Value of i^n erfc x at zero

13.2.9.1

1. $i^n\,\mathrm{erfc}\,0 = \dfrac{1}{2^n\,\Gamma(1 + n/2)}$

2. $i^0\,\mathrm{erfc}\,0 = 1$, $i\,\mathrm{erfc}\,0 = 1/\sqrt{\pi}$, $i^2\,\mathrm{erfc}\,0 = 1/4$

$i^3\,\mathrm{erfc}\,0 = 1/(6\sqrt{\pi})$, $i^4\,\mathrm{erfc}\,0 = 1/32$, $i^5\,\mathrm{erfc}\,0 = 1/(60\sqrt{\pi})$

14

Fresnel Integrals

14.1 Definitions, Series Representations, and Values at Infinity

14.1.1 The Fresnel integrals

14.1.1.1 The Fresnel integrals arise in diffraction problems and they are defined as

1. $C(x) = \int_0^x \cos\left(\frac{\pi}{2}t^2\right) dt$

2. $S(x) = \int_0^x \sin\left(\frac{\pi}{2}t^2\right) dt$

14.1.2 Series representations

14.1.2.1

1. $C(x) = \sum_{n=0}^{\infty} \frac{(-1)^n (\pi/2)^{2n}}{(2n)!(4n+1)} x^{4n+1}$

2. $C(x) = \cos\left(\dfrac{\pi}{2}x^2\right) \displaystyle\sum_{n=0}^{\infty} \dfrac{(-1)^n \pi^{2n}}{1 \cdot 3 \cdot 5 \cdots (4n+1)} x^{4n+1}$

$\quad + \sin\left(\dfrac{\pi}{2}x^2\right) \displaystyle\sum_{n=0}^{\infty} \dfrac{(-1)^n \pi^{2n+1}}{1 \cdot 3 \cdot 5 \cdots (4n+3)} x^{4n+3}$

3. $S(x) = \displaystyle\sum_{n=0}^{\infty} \dfrac{(-1)^n (\pi/2)^{2n+1}}{(2n+1)!(4n+3)} x^{4n+3}$

4. $S(x) = -\cos\left(\dfrac{\pi}{2}x^2\right) \displaystyle\sum_{n=0}^{\infty} \dfrac{(-1)^n \pi^{2n+1}}{1 \cdot 3 \cdot 5 \cdots (4n+3)} x^{4n+3}$

$\quad + \sin\left(\dfrac{\pi}{2}x^2\right) \displaystyle\sum_{n=0}^{\infty} \dfrac{(-1)^n \pi^{2n}}{1 \cdot 3 \cdot 5 \cdots (4n+1)} x^{4n+1}$

14.1.3 Limiting values as $x \to \infty$

14.1.3.1

1. $\displaystyle\lim_{x \to \infty} C(x) = \dfrac{1}{2}, \qquad \lim_{x \to \infty} S(x) = \dfrac{1}{2}$

15

Definite Integrals

15.1 Integrands Involving Powers of *x*

15.1.1

1. $\displaystyle\int_1^\infty \frac{dx}{x^n} = \frac{1}{n-1}$ $[n > 1]$

2. $\displaystyle\int_0^\infty \frac{x^{p-1}dx}{1+x} = \frac{\pi}{\sin p\pi}$ $[0 < p < 1]$

3. $\displaystyle\int_0^1 \frac{x^{p-1} - x^{-p}}{1+x}\, dx = \pi \operatorname{cosec} p\pi$ $[p^2 < 1]$

4. $\displaystyle\int_0^1 \frac{x^p - x^{-p}}{1+x}\, dx = \frac{1}{p} - \pi \operatorname{cosec} p\pi$ $[p^2 < 1]$

5. $\displaystyle\int_0^1 \frac{x^{p-1} - x^{-p}}{1-x}\, dx = \pi \cot p\pi$ $[p^2 < 1]$

6. $\displaystyle\int_0^1 \frac{x^p - x^{-p}}{x - 1}\, dx = \frac{1}{p} - \pi \cot p\pi$ $[p^2 < 1]$

7. $\displaystyle\int_0^\infty \frac{x^{p-1} - x^{q-1}}{1 - x}\, dx = \pi(\cot p\pi - \cot q\pi)$ $[p > 0, q > 0]$

8. $\displaystyle\int_0^1 \frac{x^p - x^{-p}}{1 - x^2}\, dx = \frac{\pi}{2} \cot \frac{p\pi}{2} - \frac{1}{p}$ $[p^2 < 1]$

9. $\displaystyle\int_0^1 \frac{x^p - x^{-p}}{1 + x^2}\, dx = \frac{1}{p} - \frac{\pi}{2} \csc \frac{p\pi}{2}$ $[p^2 < 1]$

10. $\displaystyle\int_0^\infty \frac{x^{p-1} dx}{1 - x^q} = \frac{\pi}{q} \cot \frac{p\pi}{q}$ $[p < q]$

11. $\displaystyle\int_0^\infty \frac{x^{p-1} dx}{1 - x^q} = \frac{\pi}{q} \csc \frac{p\pi}{q}$ $[0 < p < q]$

12. $\displaystyle\int_0^\infty \frac{x^{p-1} dx}{(1 + x^q)^2} = \frac{(p - q)\pi}{q^2} \csc \frac{(p - q)\pi}{q}$ $[p < 2q]$

13. $\displaystyle\int_0^\infty \frac{x^{p+1} dx}{(1 + x^2)^2} = \frac{p\pi}{4} \csc \frac{p\pi}{2}$ $[|p| < 2]$

14. $\displaystyle\int_0^\infty \frac{x^{2n+1} dx}{\sqrt{1 - x^2}} = \frac{(2n)!!}{(2n + 1)!!}$

15. $\displaystyle\int_0^\infty \frac{a\, dx}{a^2 + x^2} = \begin{cases} \dfrac{\pi}{2} \operatorname{sign} a, & a \neq 0 \\[2mm] 0, & a = 0 \end{cases}$

16. $\displaystyle\int_0^\infty \frac{dx}{(a^2 + x^2)(b^2 + x^2)} = \frac{\pi}{2ab(a + b)}$

17. $\displaystyle\int_0^\infty \frac{x^{p-1}\, dx}{(a^2 + x^2)(b^2 + x^2)} = \frac{\pi}{2} \frac{b^{p-2} - a^{p-2}}{a^2 - b^2} \csc \frac{p\pi}{2}$

$[a > 0, b > 0, 0 < p < 4]$

18. $\displaystyle\int_0^\infty \frac{x^{p-1} dx}{(a^2 + x^2)(b^2 - x^2)} = \frac{\pi}{2} \frac{a^{p-2} + b^{p-2} \cos \dfrac{p\pi}{2}}{a^2 + b^2} \csc \frac{p\pi}{2}$

$[a > 0, b > 0, 0 < p < 4]$

19. $\displaystyle\int_0^\infty \frac{dx}{(ax^2 + 2bx + c)^n} = \frac{(-1)^{n-1}}{(n - 1)!} \frac{\partial^{n-1}}{\partial c^{n-1}} \left[\frac{1}{\sqrt{ac - b^2}} \operatorname{arccot} \frac{b}{\sqrt{ac - b^2}} \right]$

$[a > 0, ac > b^2]$

20. $\displaystyle\int_{-\infty}^{\infty} \frac{dx}{(ax^2 + 2bx + c)^n} = \frac{(2n-3)!!\pi a^{n-1}}{(2n-2)!!(ac - b^2)^{n-(1/2)}}$ $\qquad [a > 0, ac > b^2]$

21. $\displaystyle\int_0^{\infty} \frac{x\, dx}{(ax^2 + 2bx + c)^n}$

$$= \frac{(-1)^n}{(n-1)!} \frac{\partial^{n-2}}{\partial c^{n-2}} \left[\frac{1}{2(ac - b^2)} - \frac{b}{2(ac - b^2)^{3/2}} \operatorname{arccot} \frac{b}{\sqrt{ac - b^2}} \right]$$

$$[ac > b^2]$$

$$= \frac{(-1)^n}{(n-1)!} \frac{\partial^{n-2}}{\partial c^{n-2}} \left[\frac{1}{2(ac - b^2)} + \frac{b}{4(b^2 - ac)^{3/2}} \ln\left(\frac{b + \sqrt{b^2 - ac}}{b - \sqrt{b^2 - ac}} \right) \right]$$

$$[b^2 > ac > 0]$$

$$= \frac{a^{n-2}}{2(n-1)(2n-1)b^{2n-2}} \qquad [ac = b^2], a > 0, b > 0, n \geq 2$$

22. $\displaystyle\int_{-\infty}^{\infty} \frac{x\, dx}{(ax^2 + 2bx + c)^n} = -\frac{(2n-3)!!\pi ba^{n-2}}{(2n-2)!!(ac - b^2)^{(2n-1)/2}}$

$$[ac > b^2, a > 0, n \geq 2]$$

23. $\displaystyle\int_0^{\infty} \left(\sqrt{x^2 + a^2} - x \right)^n dx = \frac{na^{n+1}}{n^2 - 1}$ $\qquad [n \geq 2]$

24. $\displaystyle\int_0^{\infty} \frac{dx}{(x + \sqrt{x^2 + a^2})^n} = \frac{n}{a^{n-1}(n^2 - 1)}$ $\qquad [n \geq 2]$

15.2 Integrands Involving Trigonometric Functions

15.2.1

1. $\displaystyle\int_0^{\pi} \sin mx \sin nx\, dx = 0$ $\qquad [m, n \text{ integers}, m \neq n]$

2. $\displaystyle\int_0^{\pi} \cos mx \cos nx\, dx = 0$ $\qquad [m, n \text{ integers}, m \neq n]$

3. $\displaystyle\int_0^{\pi} \sin mx \cos nx\, dx = \frac{m}{m^2 - n^2}[1 - (-1)^{m+n}]$ $\qquad [m, n \text{ integers}, m \neq n]$

4. $\displaystyle\int_0^{2\pi} \sin mx \sin nx\, dx = 0$ $\qquad [m, n \text{ integers}, m \neq n]$

5. $\displaystyle\int_0^{2\pi} \sin^2 nx\, dx = \pi$ $\qquad [n \neq 0 \text{ integral}]$

6. $\displaystyle\int_0^{2\pi} \sin mx \cos nx\, dx = 0$ [m, n integers]

7. $\displaystyle\int_0^{2\pi} \cos mx \cos nx\, dx = \begin{cases} 0, & m \neq n \\ \pi, & m = n \neq 0 \\ 2\pi, & m = n = 0 \end{cases}$ [m, n integers]

8. $\displaystyle\int_0^{\pi} \frac{\sin nx}{\sin x}\, dx = \begin{cases} 0, & n \text{ even} \\ \pi, & n \text{ odd} \end{cases}$

9. $\displaystyle\int_0^{\pi/2} \frac{\sin(2n-1)x}{\sin x}\, dx = \frac{\pi}{2}$

10. $\displaystyle\int_0^{\pi/2} \sin^{2m} x\, dx = \int_0^{\pi/2} \cos^{2m} x\, dx = \frac{(2m-1)!!}{(2m)!!}\frac{\pi}{2}$

11. $\displaystyle\int_0^{\pi/2} \sin^{2m+1} x\, dx = \int_0^{\pi/2} \cos^{2m+1} x\, dx = \frac{(2m)!!}{(2m+1)!!}$

12. $\displaystyle\int_0^{\infty} \frac{\sin ax}{\sqrt{x}}\, dx = \int_0^{\infty} \frac{\cos ax}{\sqrt{x}}\, dx = \sqrt{\frac{\pi}{2a}}$ [$a > 0$]

13. $\displaystyle\int_0^{\infty} \frac{\sin ax}{x}\, dx = \begin{cases} \dfrac{\pi}{2}\,\text{sign } a, & a > 0 \\ 0, & a = 0 \end{cases}$

14. $\displaystyle\int_0^{\infty} \frac{\sin x \cos ax}{x}\, dx = \begin{cases} 0, & a^2 > 0 \\ \dfrac{\pi}{4}, & a = \pm 1 \\ \dfrac{\pi}{2}, & a^2 < 1 \end{cases}$

15. $\displaystyle\int_0^{2\pi} \frac{dx}{1 + a \cos x} = \frac{2\pi}{\sqrt{1 - a^2}}$ [$a^2 < 1$]

16. $\displaystyle\int_0^{\pi/2} \frac{dx}{1 + a \cos x} = \frac{\arccos a}{\sqrt{1 - a^2}}$ [$a^2 < 1$]

17. $\displaystyle\int_0^{\infty} \frac{\cos ax}{b^2 + x^2}\, dx = \begin{cases} \dfrac{\pi}{2b}e^{-ab}, & a > 0, b > 0 \\ \dfrac{\pi}{2b}e^{ab}, & a < 0, b > 0 \end{cases}$

18. $\displaystyle\int_0^{\infty} \frac{\cos ax}{b^2 - x^2}\, dx = \frac{\pi}{2b}\sin ab$ [$a > 0, b > 0$]

19. $\displaystyle\int_0^\infty \frac{x \sin ax}{b^2 + x^2}\, dx = \pi e^{-ab} \qquad [a > 0, b > 0]$

20. $\displaystyle\int_0^\infty \frac{x \sin ax}{b^2 - x^2}\, dx = -\frac{\pi}{2} \cos ab \qquad [a > 0]$

21. $\displaystyle\int_0^\infty \frac{\sin^2 x}{x^2}\, dx = \frac{\pi}{2}$

22. $\displaystyle\int_0^\infty \frac{\sin^2 ax}{x^2 + b^2}\, dx = \frac{\pi}{4b}(1 - e^{-2ab}) \qquad [a > 0, b > 0]$

23. $\displaystyle\int_0^\infty \frac{\cos^2 ax}{x^2 + b^2}\, dx = \frac{\pi}{4b}(1 + e^{-2ab}) \qquad [a > 0, b > 0]$

24. $\displaystyle\int_0^\infty \frac{\cos ax}{b^4 + x^4}\, dx = \frac{\pi\sqrt{2}}{4b^3} \exp\!\left(\frac{-ab}{\sqrt{2}}\right)\!\left(\cos\frac{ab}{\sqrt{2}} + \sin\frac{ab}{\sqrt{2}}\right) \qquad [a > 0, b > 0]$

25. $\displaystyle\int_0^\infty \frac{\cos ax}{b^4 - x^4}\, dx = \frac{\pi}{4b^3}(e^{-ab} + \sin ab) \qquad [a > 0, b > 0]$

26. $\displaystyle\int_0^{\pi/2} \frac{\cos^2 x\, dx}{1 - 2a\cos 2x + a^2} = \begin{cases} \dfrac{\pi}{4(1 - a)}, & a^2 < 1 \\[2ex] \dfrac{\pi}{4(a - 1)}, & a^2 > 1 \end{cases}$

27. $\displaystyle\int_0^\infty \frac{\cos ax\, dx}{(b^2 + x^2)(c^2 + x^2)} = \frac{\pi(be^{-ac} - ce^{-ab})}{2bc(b^2 - c^2)} \qquad [a > 0, b > 0, c > 0]$

28. $\displaystyle\int_0^\infty \frac{x \sin ax\, dx}{(b^2 + x^2)(c^2 + x^2)} = \frac{\pi(e^{-ab} - e^{-ac})}{2(c^2 - b^2)} \qquad [a > 0, b > 0, c > 0]$

29. $\displaystyle\int_0^\infty \frac{\cos ax\, dx}{(b^2 + x^2)^2} = \frac{\pi}{4b^3}(1 + ab)e^{-ab} \qquad [a > 0, b > 0]$

30. $\displaystyle\int_0^\infty \frac{x \sin ax\, dx}{(b^2 + x^2)^2} = \frac{\pi}{4b}ae^{-ab} \qquad [a > 0, b > 0]$

31. $\displaystyle\int_0^\pi \frac{\cos nx\, dx}{1 - 2a\cos x + a^2} = \begin{cases} \dfrac{\pi a^n}{1 - a^2}, & a^2 < 1 \\[2ex] \dfrac{\pi}{(a^2 - 1)a^n}, & a^2 > 1 \end{cases} \qquad [n \geq 0 \text{ integral}]$

32. $\displaystyle\int_0^{\pi/2} \frac{\sin^2 x\, dx}{1 - 2a\cos 2x + a^2} = \frac{\pi}{4(1 + a)} \qquad [a^2 < 1]$

33. $\displaystyle\int_0^\infty \sin ax^2\, dx = \int_0^\infty \cos ax^2\, dx = \frac{1}{2}\sqrt{\frac{\pi}{2a}} \qquad \text{(see 14.1.1.1)}$

34. $\displaystyle\int_0^\infty \sin ax^2 \cos 2bx\, dx = \frac{1}{2}\sqrt{\frac{\pi}{2a}}\left[\cos\frac{b^2}{a} - \sin\frac{b^2}{a}\right]$ $[a > 0, b > 0]$

35. $\displaystyle\int_0^\infty \cos ax^2 \cos 2bx\, dx = \frac{1}{2}\sqrt{\frac{\pi}{2a}}\left(\cos\frac{b^2}{a} + \sin\frac{b^2}{a}\right)$ $[a > 0, b > 0]$

36. $\displaystyle\int_0^\pi \cos(x \sin\theta)\, d\theta = \pi J_0(x)$

37. $\displaystyle\int_0^\pi \cos(n\theta - x\sin\theta)\, d\theta = \pi J_n(x)$

Integrands Involving the Exponential Function

15.3.1

1. $\displaystyle\int_0^\infty e^{-px}\, dx = \frac{1}{p}$ $[p > 0]$

2. $\displaystyle\int_0^\infty x^n e^{-px}\, dx = n!/p^{n+1}$ $[p > 0, n \text{ integral}]$

3. $\displaystyle\int_0^u xe^{-px}\, dx = \frac{1}{p^2} - \frac{1}{p^2}e^{-pu}(1 + pu)$ $[u > 0]$

4. $\displaystyle\int_0^u x^2 e^{-px}\, dx = \frac{2}{p^3} - \frac{1}{p^3}e^{-pu}(2 + 2pu - p^2u^2)$ $[u > 0]$

5. $\displaystyle\int_0^u x^3 e^{-px}\, dx = \frac{6}{p^4} - \frac{1}{p^4}e^{-pu}(6 + 6pu + 3p^2u^2 + p^3u^3)$ $[u > 0]$

6. $\displaystyle\int_0^1 \frac{xe^x\, dx}{(1+x)^2} = \frac{1}{2}e - 1$

7. $\displaystyle\int_0^\infty x^{\nu-1}e^{-\mu x}\, dx = \frac{1}{\mu^\nu}\Gamma(\nu)$ $[\mu > 0, \nu > 0]$

8. $\displaystyle\int_0^\infty \frac{dx}{1 + e^{px}} = \frac{\ln 2}{p}$ $[p > 0]$

9. $\displaystyle\int_0^u \frac{e^{-qx}}{\sqrt{x}}\, dx = \sqrt{\frac{\pi}{q}}P(\sqrt{qu})$ $[q > 0]$ (see 13.1.1.1.5)

10. $\displaystyle\int_0^\infty \frac{e^{-qx}}{\sqrt{x}}\, dx = \sqrt{\frac{\pi}{q}}$ $[q > 0]$

11. $\displaystyle\int_{-\infty}^\infty \frac{e^{-px}}{1 + e^{-qx}}\, dx = \frac{\pi}{q}\csc\frac{p\pi}{q}$ $[q > p > 0 \text{ or } 0 > p > q]$

12. $\displaystyle\int_{-\infty}^{\infty} \frac{e^{-px}}{b - e^{-x}}\, dx = \pi b^{p-1} \cot p\pi$ $[b > 0, 0 < p < 1]$

13. $\displaystyle\int_{-\infty}^{\infty} \frac{e^{-px}}{b + e^{-x}}\, dx = \pi b^{p-1} \operatorname{cosec} p\pi$ $[b > 0, 0 < p < 1]$

14. $\displaystyle\int_{1}^{\infty} \frac{e^{-px}}{\sqrt{x - 1}}\, dx = \sqrt{\frac{\pi}{p}} e^{-p}$ $[p > 0]$

15. $\displaystyle\int_{0}^{\infty} \frac{e^{-px}}{\sqrt{x + \beta}}\, dx = \sqrt{\frac{\pi}{p}} e^{\beta p}[1 - P(\beta p)]$ $[p > 0, \beta > 0]$ (see 13.1.1.1.5)

16. $\displaystyle\int_{u}^{\infty} \frac{e^{-px}}{x\sqrt{x - u}}\, dx = \frac{\pi}{\sqrt{u}}[1 - P(\sqrt{pu})]$ $[p \geq 0, u > 0]$ (see 13.1.1.1.5)

17. $\displaystyle\int_{0}^{\infty} \frac{xe^{-x}}{e^x - 1}\, dx = \frac{\pi^2}{6} - 1$

18. $\displaystyle\int_{0}^{\infty} \frac{xe^{-2x}}{e^{-x} + 1}\, dx = 1 - \frac{\pi^2}{12}$

19. $\displaystyle\int_{0}^{\infty} \frac{xe^{-3x}}{e^{-x} + 1}\, dx = \frac{\pi^2}{12} - \frac{3}{4}$

20. $\displaystyle\int_{0}^{\infty} \frac{x\, dx}{\sqrt{e^x - 1}} = 2\pi \ln 2$

21. $\displaystyle\int_{0}^{\infty} \frac{x^2\, dx}{\sqrt{e^x - 1}} = 4\pi \left[(\ln 2)^2 + \frac{\pi^2}{12}\right]$

22. $\displaystyle\int_{0}^{\infty} \frac{xe^{-x}}{\sqrt{e^x - 1}}\, dx = \frac{\pi}{2}(2 \ln 2 - 1)$

23. $\displaystyle\int_{0}^{1} \left(\frac{1}{\ln x} + \frac{1}{1 - x}\right) dx = \gamma$ (see 11.1.1.1.7)

24. $\displaystyle\int_{0}^{\infty} \frac{1}{x}\left(\frac{1}{1 + x^2} - e^{-x}\right) dx = \gamma$ (see 11.1.1.1.7)

25. $\displaystyle\int_{0}^{\infty} \left(\frac{1}{e^x - 1} - \frac{1}{xe^x}\right) dx = \gamma$ (see 11.1.1.1.7)

26. $\displaystyle\int_{0}^{\infty} x^{2n} e^{-px^2}\, dx = \frac{(2n - 1)!!}{2(2p)^n}\sqrt{\frac{\pi}{p}}$

                         $[p > 0, n = 0, 1, 2, \ldots, (2n - 1)!! = 1 \cdot 3 \cdot 5 \cdots (2n - 1)]$

27. $\displaystyle\int_{0}^{\infty} x^{2n+1} e^{-px^2}\, dx = \frac{n!}{2p^{n+1}}$ $[p > 0, n = 0, 1, 2, \ldots]$

28. $\displaystyle \int_0^x e^{-q^2 t^2}\, dt = \frac{\sqrt{\pi}}{2q} P(qx)$ $[q > 0]$ (see 13.1.1.1.5)

29. $\displaystyle \int_0^\infty e^{-q^2 t^2}\, dt = \frac{\sqrt{\pi}}{2q}$ $[q > 0]$

30. $\displaystyle \int_{-\infty}^\infty \exp(-p^2 x^2 \pm qx)\, dx = \exp\!\left(\frac{q^2}{4p^2}\right) \frac{\sqrt{\pi}}{p}$ $[p > 0]$

15.4 Integrands Involving the Hyperbolic Function

15.4.1

1. $\displaystyle \int_0^\infty \frac{dx}{\cosh ax} = \frac{\pi}{2a}$ $[a > 0]$

2. $\displaystyle \int_0^\infty \frac{\sinh ax}{\sinh bx}\, dx = \frac{\pi}{2b} \tan \frac{a\pi}{2b}$ $[b > |a|]$

3. $\displaystyle \int_0^\infty \frac{dx}{a + b \sinh x} = \frac{1}{\sqrt{a^2 + b^2}} \ln\!\left[\frac{a + b + \sqrt{a^2 + b^2}}{a + b - \sqrt{a^2 + b^2}}\right]$ $[ab \neq 0]$

4. $\displaystyle \int_0^\infty \frac{dx}{a + b \cosh x} = \frac{2}{\sqrt{b^2 - a^2}} \arctan \frac{\sqrt{b^2 - a^2}}{a + b}$ $[b^2 > a^2]$

   $\displaystyle \qquad = \frac{1}{\sqrt{a^2 - b^2}} \ln\!\left[\frac{a + b + \sqrt{a^2 - b^2}}{a + b - \sqrt{a^2 - b^2}}\right]$ $[a^2 > b^2]$

5. $\displaystyle \int_0^\infty \frac{dx}{a \sinh x + b \cosh x} = \frac{2}{\sqrt{b^2 - a^2}} \arctan \frac{\sqrt{b^2 - a^2}}{a + b}$ $[b^2 > a^2]$

   $\displaystyle \qquad = \frac{1}{\sqrt{a^2 - b^2}} \ln\!\left[\frac{a + b + \sqrt{a^2 - b^2}}{a + b - \sqrt{a^2 - b^2}}\right]$ $[a^2 > b^2]$

15.5 Integrands Involving the Logarithmic Function

15.5.1

1. $\displaystyle \int_0^1 \frac{\ln x}{1 - x}\, dx = -\frac{\pi^2}{6}$

2. $\displaystyle \int_0^1 \frac{\ln x}{1 + x}\, dx = -\frac{\pi^2}{12}$

3. $\displaystyle \int_0^1 \frac{\ln(1 + x)}{1 + x^2}\, dx = \frac{\pi}{8} \ln 2$

4. $\displaystyle\int_0^1 (\ln x)^p \, dx = (-1)^p p!$ $[p = 0, 1, 2, \ldots]$

5. $\displaystyle\int_0^1 x \ln(1 + x) \, dx = \frac{1}{4}$

6. $\displaystyle\int_0^{\pi/2} \ln(\sin x) \, dx = \int_0^{\pi/2} \ln(\cos x) \, dx = -\frac{\pi}{2} \ln 2$

16

Different Forms of Fourier Series

16.1 Fourier Series for $f(x)$ on $-\pi \leq x \leq \pi$

16.1.1 The Fourier series

16.1.1.1

1. $\dfrac{1}{2}a_0 + \displaystyle\sum_{n=1}^{\infty}(a_n \cos nx + b_n \sin nx)$

2. *Fourier coefficients*

$$a_0 = \frac{1}{\pi}\int_{-\pi}^{\pi} f(x)\,dx, \quad a_n = \frac{1}{\pi}\int_{-\pi}^{\pi} f(x)\cos nx\,dx \qquad [n = 1, 2, \ldots]$$

$$b_n = \frac{1}{\pi}\int_{-\pi}^{\pi} f(x)\sin nx\,dx \qquad [n = 1, 2, \ldots]$$

3. *Parseval relation*

$$\frac{1}{\pi}\int_{-\pi}^{\pi}[f(x)]^2\,dx = \frac{1}{2}a_0^2 + \sum_{n=1}^{\infty}\left(a_n^2 + b_n^2\right)$$

If $f(x)$ is periodic with period 2π the Fourier series represents $f(x)$ for all x and these integrals may be evaluated over any interval of length 2π.

16.2 Fourier Series for $f(x)$ on $-L \leq x \leq L$

16.2.1 The Fourier series

16.2.1.1

1. $$\frac{1}{2}a_0 + \sum_{n=1}^{\infty}\left[a_n\cos\left(\frac{n\pi x}{L}\right) + b_n\sin\left(\frac{n\pi x}{L}\right)\right]$$

2. *Fourier coefficients*

$$a_0 = \frac{1}{L}\int_{-L}^{L} f(x)\,dx, \quad a_n = \frac{1}{L}\int_{-L}^{L} f(x)\cos\left(\frac{n\pi x}{L}\right)dx$$

$$[n = 1, 2, \ldots]$$

$$b_n = \frac{1}{L}\int_{-L}^{L} f(x)\sin\left(\frac{n\pi x}{L}\right)dx \quad [n = 1, 2, \ldots]$$

3. *Parseval relation*

$$\frac{1}{L}\int_{-L}^{L}[f(x)]^2\,dx = \frac{1}{2}a_0^2 + \sum_{n=1}^{\infty}\left(a_n^2 + b_n^2\right)$$

If $f(x)$ is periodic with period $2L$ the Fourier series represents $f(x)$ for all x and these integrals may be evaluated over any interval of length $2L$.

16.3 Fourier Series for $f(x)$ on $a \leq x \leq b$

16.3.1 The Fourier series

16.3.1.1

1. $$\frac{1}{2}a_0 + \sum_{n=1}^{\infty}\left[a_n\cos\left(\frac{2n\pi x}{b-a}\right) + b_n\sin\left(\frac{2n\pi x}{b-a}\right)\right]$$

2. *Fourier coefficients*

$$a_0 = \frac{2}{b-a}\int_{a}^{b} f(x)\,dx, \quad a_n = \frac{2}{b-a}\int_{a}^{b} f(x)\cos\left(\frac{2n\pi x}{b-a}\right)dx$$

$$[n = 1, 2, \ldots]$$

$$b_n = \frac{2}{b-a}\int_{a}^{b} f(x)\sin\left(\frac{2n\pi x}{b-a}\right)dx \quad [n = 1, 2, \ldots]$$

3. *Parseval relation*

$$\frac{2}{b-a}\int_{a}^{b}[f(x)]^2\,dx = \frac{1}{2}a_0^2 + \sum_{n=1}^{\infty}\left(a_n^2 + b_n^2\right)$$

If $f(x)$ is periodic with period $(b - a)$ the Fourier series represents $f(x)$ for all x and these integrals may be evaluated over any interval of length $(b - a)$.

16.4 Half-Range Fourier Cosine Series for $f(x)$ on $0 \leq x \leq \pi$

16.4.1 The Fourier series

16.4.1.1

1. $\dfrac{1}{2}a_0 + \displaystyle\sum_{n=1}^{\infty} a_n \cos nx$

2. *Fourier coefficients*

$$a_0 = \frac{2}{\pi} \int_0^{\pi} f(x)\, dx, \quad a_n = \frac{2}{\pi} \int_0^{\pi} f(x)\cos nx\, dx \qquad [n = 1, 2, \ldots]$$

3. *Parseval relation*

$$\frac{2}{\pi} \int_0^{\pi} [f(x)]^2\, dx = \frac{1}{2}a_0^2 + \sum_{n=1}^{\infty} a_n^2$$

If $f(x)$ is an even function, or it is extended to the interval $-\pi \leq x \leq 0$ as an even function so that $f(x) = f(-x)$, the Fourier series represents $f(x)$ on the interval $-\pi \leq x \leq \pi$.

16.5 Half-Range Fourier Cosine Series for $f(x)$ on $0 \leq x \leq L$

16.5.1 The Fourier series

16.5.1.1

1. $\dfrac{1}{2}a_0 + \displaystyle\sum_{n=1}^{\infty} a_n \cos\left(\dfrac{n\pi x}{L}\right)$

2. *Fourier coefficients*

$$a_0 = \frac{2}{L} \int_0^{L} f(x)\, dx, \quad a_n = \frac{2}{L} \int_0^{L} f(x)\cos\left(\frac{n\pi x}{L}\right) dx$$

$$[n = 1, 2, \ldots]$$

3. *Parseval relation*

$$\frac{2}{L} \int_0^{L} [f(x)]^2\, dx = \frac{1}{2}a_0^2 + \sum_{n=1}^{\infty} a_n^2$$

If $f(x)$ is an even function, or it is extended to the interval $-L \leq x \leq 0$ as an even function so that $f(-x) = f(x)$, the Fourier series represents $f(x)$ on the interval $-L \leq x \leq L$.

16.6 Half-Range Fourier Sine Series for $f(x)$ on $0 \leq x \leq \pi$

16.6.1 The Fourier series

16.6.1.1

1. $\displaystyle\sum_{n=1}^{\infty} b_n \sin nx$

2. **Fourier coefficients**

$$b_n = \frac{2}{\pi} \int_0^{\pi} f(x) \sin nx \, dx \qquad [n = 1, 2, \ldots]$$

3. **Parseval relation**

$$\frac{2}{\pi} \int_0^{\pi} [f(x)]^2 \, dx = \sum_{n=1}^{\infty} b_n^2$$

If $f(x)$ is an odd function, or it is extended to the interval $-\pi \leq x \leq 0$ as an odd function so that $f(-x) = -f(x)$, the Fourier series represents $f(x)$ on the interval $-\pi \leq x \leq \pi$.

16.7 Half-Range Fourier Sine Series for $f(x)$ on $0 \leq x \leq L$

16.7.1 The Fourier series

16.7.1.1

1. $\displaystyle\sum_{n=1}^{\infty} b_n \sin\left(\frac{n\pi x}{L}\right)$

2. **Fourier coefficients**

$$b_n = \frac{2}{L} \int_0^{L} f(x) \sin\left(\frac{n\pi x}{L}\right) dx \qquad [n = 1, 2, \ldots]$$

3. **Parseval relation**

$$\frac{2}{L} \int_0^{L} [f(x)]^2 \, dx = \sum_{n=1}^{\infty} b_n^2$$

If $f(x)$ is an odd function, or it is extended to the interval $-L \leq x \leq 0$ as an odd function so that $f(-x) = -f(x)$, the Fourier series represents $f(x)$ on the interval $-L \leq x \leq L$.

16.8 Complex (Exponential) Fourier Series for $f(x)$ on $-\pi \leq x \leq \pi$

16.8.1 The Fourier series

16.8.1.1

1. $\displaystyle\lim_{m \to \infty} \sum_{n=-m}^{m} c_n e^{inx}$

2. ***Fourier coefficients***

$$c_n = \frac{1}{2\pi} \int_{-\pi}^{\pi} f(x) e^{-inx} \, dx \qquad [n = 0, \pm 1, \pm 2, \ldots]$$

3. ***Parseval relation***

$$\frac{1}{2\pi} \int_{-\pi}^{\pi} |f(x)|^2 \, dx = \lim_{m \to \infty} \sum_{n=-m}^{m} |c_n|^2$$

If $f(x)$ is periodic with period 2π the Fourier series represents $f(x)$ for all x and these integrals may be evaluated over any interval of length 2π.

16.9 Complex (Exponential) Fourier Series for $f(x)$ on $-L \leq x \leq L$

16.9.1 The Fourier series

16.9.1.1

1. $\lim\limits_{m \to \infty} \sum\limits_{n=-m}^{m} c_n \exp[(in\pi x)/L]$

2. ***Fourier coefficients***

$$c_n = \frac{1}{2L} \int_{-L}^{L} f(x) \exp[-(in\pi x)/L] \, dx \qquad [n = 0, \pm 1, \pm 2, \ldots]$$

3. ***Parseval relation***

$$\frac{1}{2L} \int_{-L}^{L} |f(x)|^2 \, dx = \lim_{m \to \infty} \sum_{n=-m}^{m} |c_n|^2.$$

If $f(x)$ is periodic with period $2L$ the Fourier series represents $f(x)$ for all x and these integrals may be evaluated over any interval of length $2L$.

16.10 Representative Examples of Fourier Series

16.10.1

1. $f(x) = x \qquad [-\pi \leq x \leq \pi]$

Fourier series.

$$2 \sum_{n=1}^{\infty} \frac{(-1)^{n+1}}{n} \sin nx$$

Converges pointwise to $f(x)$ for $-\pi < x < \pi$.

Fourier coefficients.

$a_n = 0 \qquad [n = 0, 1, 2, \ldots]$

$b_n = \dfrac{2(-1)^{n+1}}{n} \qquad [n = 1, 2, \ldots]$

Parseval relation.

$$\frac{1}{\pi} \int_{-\pi}^{\pi} x^2 \, dx = \sum_{n=1}^{\infty} b_n^2 \quad \text{or} \quad \frac{\pi^2}{6} = \sum_{n=1}^{\infty} \frac{1}{n^2}$$

2. $f(x) = |x| \qquad [-\pi \le x \le \pi]$

Fourier series.

$$\frac{\pi}{2} - \frac{4}{\pi} \sum_{n=1}^{\infty} \frac{\cos(2n-1)x}{(2n-1)^2}$$

Converges pointwise to $f(x)$ for $-\pi \le x \le \pi$.

Fourier coefficients.

$$a_0 = \pi, \quad a_{2n} = 0, \quad a_{2n-1} = \frac{-4}{\pi(2n-1)^2} \qquad [n = 1, 2, \ldots]$$

$$b_n = 0 \qquad [n = 1, 2, \ldots]$$

Parseval relation.

$$\frac{1}{\pi} \int_{-\pi}^{\pi} |x|^2 \, dx = \frac{1}{2}a_0^2 + \sum_{n=1}^{\infty} a_n^2 \quad \text{or} \quad \frac{\pi^4}{96} = \sum_{n=1}^{\infty} \frac{1}{(2n-1)^4}$$

3. $f(x) = \begin{cases} -1, & -\pi \le x < 0 \\ 1, & 0 < x \le \pi \end{cases}$

Fourier series.

$$\frac{4}{\pi} \sum_{n=1}^{\infty} \frac{\sin(2n-1)x}{2n-1}$$

Converges pointwise to $f(x)$ for $-\pi < x < \pi$ and $x \ne 0$.

Fourier coefficients.

$$a_n = 0 \qquad [n = 0, 1, 2, \ldots]$$

$$b_{2n} = 0, \quad b_{2n-1} = \frac{4}{\pi(2n-1)} \qquad [n = 1, 2, \ldots]$$

Parseval relation.

$$\frac{1}{\pi} \int_{-\pi}^{0} (-1)^2 \, dx + \frac{1}{\pi} \int_{0}^{\pi} 1^2 \, dx = \sum_{n=1}^{\infty} b_n^2 \quad \text{or} \quad \frac{\pi^2}{8} = \sum_{n=1}^{\infty} \frac{1}{(2n-1)^2}$$

4. $f(x) = |\sin x| \qquad [-\pi \le x \le \pi]$

Fourier series.

$$\frac{2}{\pi} - \frac{4}{\pi} \sum_{n=1}^{\infty} \frac{\cos 2nx}{4n^2 - 1}$$

Converges pointwise to $f(x)$ for $-\pi \le x \le \pi$.

Fourier coefficients.

$$a_0 = \frac{4}{\pi}, \quad a_{2n} = \frac{-4}{\pi(4n^2 - 1)}, \quad a_{2n-1} = 0 \qquad [n = 1, 2, \ldots]$$

Parseval relation.

$$\frac{1}{\pi} \int_{-\pi}^{\pi} (\sin x)^2 \, dx = \frac{1}{2} a_0^2 + \sum_{n=1}^{\infty} a_n^2 \quad \text{or} \quad \frac{\pi^2}{16} = \frac{1}{2} + \sum_{n=1}^{\infty} \frac{1}{(4n^2 - 1)^2}$$

5. $f(x) = x^2 \qquad [-\pi \le x \le \pi]$

Fourier series.

$$\frac{\pi^2}{3} + 4 \sum_{n=1}^{\infty} \frac{(-1)^n}{n^2} \cos nx$$

Converges pointwise to $f(x)$ for $-\pi \le x \le \pi$.

Fourier coefficients.

$$a_0 = \frac{2\pi^2}{3}, \quad a_n = \frac{4(-1)^n}{n^2} \qquad [n = 1, 2, \ldots]$$
$$b_n = 0 \qquad [n = 1, 2, \ldots]$$

Parseval relation.

$$\frac{1}{\pi} \int_{-\pi}^{\pi} x^4 \, dx = \frac{1}{2} a_0^2 + \sum_{n=1}^{\infty} a_n^2 \quad \text{or} \quad \frac{\pi^4}{90} = \sum_{n=1}^{\infty} \frac{1}{n^4}$$

6. $f(x) = x^2 \qquad [0 \le x \le 2\pi]$

Fourier series.

$$\frac{4\pi^2}{3} + 4 \sum_{n=1}^{\infty} \left(\frac{1}{n^2} \cos nx - \frac{\pi}{n} \sin nx \right)$$

Converges pointwise to $f(x)$ for $0 < x < 2\pi$.

Fourier coefficients.

$$a_0 = \frac{8\pi^2}{3}, \quad a_n = \frac{4}{n^2} \qquad [n = 1, 2, \ldots]$$
$$b_n = -\frac{4\pi}{n} \qquad [n = 1, 2, \ldots]$$

Parseval relation.

$$\frac{1}{\pi} \int_0^{2\pi} x^4 \, dx = \frac{1}{2} a_0^2 + \sum_{n=1}^{\infty} \left(a_n^2 + b_n^2 \right) \quad \text{or}$$

$$\frac{128\pi^4}{45} = 16 \sum_{n=1}^{\infty} \frac{1}{n^4} + 16\pi^2 \sum_{n=1}^{\infty} \frac{1}{n^2},$$

but from 16.10.1.5 above $\sum_{n=1}^{\infty} 1/n^4 = \pi^2/90$, so it follows that

$$\frac{\pi^2}{6} = \sum_{n=1}^{\infty} \frac{1}{n^2}.$$

7. $f(x) = x - x^2 \qquad [0 \leq x \leq 1]$

Half-range Fourier cosine series.

$$\frac{1}{6} - \frac{1}{\pi^2} \sum_{n=1}^{\infty} \frac{\cos 2n\pi x}{n^2}$$

Converges pointwise to $f(x)$ for $0 \leq x \leq 1$.

Half-range Fourier cosine series coefficients.

$$a_0 = \frac{1}{3}, \quad a_n = \frac{-1}{\pi^2 n^2} \qquad [n = 1, 2, \ldots]$$

Parseval relation.

$$2 \int_0^1 (x - x^2)^2 \, dx = \frac{1}{2} a_0^2 + \sum_{n=1}^{\infty} a_n^2 \quad \text{or} \quad \frac{\pi^4}{90} = \sum_{n=1}^{\infty} \frac{1}{n^4}$$

8. $f(x) = x - x^2 \qquad [0 \leq x \leq 1]$

Half-range Fourier sine series.

$$\frac{8}{\pi^2} \sum_{n=1}^{\infty} \frac{\sin(2n - 1)\pi x}{(2n - 1)^3}$$

Converges pointwise to $f(x)$ for $0 \leq x \leq 1$.

Half-range Fourier sine series coefficients.

$$b_{2n} = 0 \qquad [n = 1, 2, \ldots]$$

$$b_{2n-1} = \frac{8}{\pi^2 (2n - 1)^3} \qquad [n = 1, 2, \ldots]$$

Parseval relation.

$$2 \int_0^1 (x - x^2)^2 \, dx = \sum_{n=1}^{\infty} b_n^2 \quad \text{or} \quad \frac{\pi^6}{960} = \sum_{n=1}^{\infty} \frac{1}{(2n-1)^6}$$

9. $f(x) = e^{ax}$ $[-\pi \le x \le \pi]$

Complex Fourier series.

$$\frac{\sinh \pi a}{\pi} \left[\lim_{m \to \infty} \sum_{n=-m}^{m} \left(\frac{1}{a - in} \right) e^{inx} \right]$$

Converges pointwise to $f(x)$ for $-\pi < x < \pi$.

Complex Fourier series coefficients.

$$c_n = (-1)^n \frac{\sinh \pi a}{\pi} \left(\frac{1}{a - in} \right)$$

The complex Fourier series reduces to the real Fourier series

$$\frac{\sinh \pi a}{\pi a} + \frac{2 \sinh \pi a}{\pi} \sum_{n=1}^{\infty} (-1)^n \left[\left(\frac{a}{a^2 + n^2} \right) \cos nx - \left(\frac{n}{a^2 + n^2} \right) \sin nx \right],$$

which like the complex series converges pointwise to $f(x)$ for $-\pi < x < \pi$.

Parseval relation.

$$\frac{1}{\pi} \int_{-\pi}^{\pi} e^{2ax} \, dx = \frac{1}{2} a_0 + \sum_{n=1}^{\infty} (a_n^2 + b_n^2)$$

with

$$a_0 = \frac{2 \sinh \pi a}{\pi a}, \quad a_n = \frac{(-1)^n a}{(a^2 + n^2)}, \quad b_n = \frac{(-1)^n n}{(a^2 + n^2)}$$

$$[n = 1, 2, \ldots],$$

which yields

$$\frac{\pi \coth \pi a}{2a} - \frac{1}{2a^2} = \sum_{n=1}^{\infty} \frac{1}{(a^2 + n^2)}.$$

16.11 Fourier Series and Discontinuous Functions

16.11.1 Periodic extensions and convergence of Fourier series

16.11.1.1   Let $f(x)$ be defined on the interval $-\pi \le x \le \pi$. Then since each function in the Fourier series of $f(x)$ is periodic with a period that is a multiple of 2π, the Fourier series itself will be periodic with period 2π. Thus, irrespective of how $f(x)$ is defined outside the **fundamental interval** $-\pi \le x \le 2\pi$, the

Fourier series will replicate the behavior of $f(x)$ in the intervals $(2n-1)\pi \leq x \leq (2n+1)\pi$, for $n = \pm1, \pm2, \ldots$. Each of these intervals is called a **periodic extension** of $f(x)$.

At a point x_0, a Fourier series, which is piecewise continuous with a finite number of jump discontinuities, converges to the number

1. $$\frac{f(x_0 + 0) + f(x_0 - 0)}{2}.$$

Thus, if $f(x)$ is continuous at x_0, the Fourier series converges to the number $f(x_0)$, while if it is discontinuous it follows from 16.11.1.1.1 that it converges to the average of $f(x_0 - 0)$ and $f(x_0 + 0)$. The periodicity of the Fourier series for $f(x)$ implies that these properties that are true in the fundamental interval $-\pi \leq x \leq \pi$ are also true in every periodic extension of $f(x)$.

In particular, if $f(-\pi) \neq f(\pi)$, there will be a jump discontinuity (a saltus) at each end of the fundamental interval (and at the ends of each periodic extension) where the Fourier series will converge to the value

$$\tfrac{1}{2}[f(\pi) + f(-\pi)].$$

These same arguments apply to Fourier series defined on an arbitrary interval $a \leq x \leq b$, and not only to the interval $-\pi \leq x \leq \pi$.

16.11.2 Applications to closed-form summations of numerical series

16.11.2.1   The implications of 16.11.1.1 are best illustrated by means of examples that show how **closed-form summations** may be obtained for certain numerical series.

1. ***Application to $f(x) = x$, $-\pi \leq x \leq \pi$.*** From 16.10.1.1 the Fourier series of $f(x) = x$, $-\pi \leq x \leq \pi$ is known to be

    $$2\sum_{n=1}^{\infty} \frac{(-1)^{n+1}}{n} \sin nx.$$

    A convenient choice of x will reduce this Fourier series to a simple numerical series. Let us choose $x = \pi/2$, at which point $f(x)$ is continuous and the Fourier series simplifies. At $x = \pi/2$ the Fourier series converges to $f(\pi/2) = \pi/2$. Using this result and setting $x = \pi/2$ in the Fourier series gives

    $$\frac{\pi}{2} = 2\sum_{n=1}^{\infty} \frac{(-1)^{n+1} \sin \frac{n\pi}{2}}{n}$$

    or

    $$\frac{\pi}{4} = \sum_{n=1}^{\infty} \frac{(-1)^{n+1}}{(2n-1)} = 1 - \frac{1}{3} + \frac{1}{5} - \cdots.$$

2. **Application to $f(x) = |\sin x|$, $-\pi \le x \le \pi$.** From 16.10.1.4 the Fourier series of $f(x) = |\sin x|$, $-\pi \le x \le \pi$ is known to be

$$\frac{2}{\pi} - \frac{4}{\pi} \sum_{n=1}^{\infty} \frac{\cos 2nx}{4n^2 - 1}.$$

Because $f(x)$ is continuous and $f(-\pi) = f(\pi)$, it follows that the Fourier series converges to $f(x)$ for any choice of x in the interval $-\pi \le x \le \pi$, so let us set $x = 0$ at which point $f(0) = 0$. Then

$$0 = \frac{2}{\pi} - \frac{4}{\pi} \sum_{n=1}^{\infty} \frac{1}{4n^2 - 1}$$

or

$$\frac{1}{2} = \sum_{n=1}^{\infty} \frac{1}{4n^2 - 1} \qquad\qquad \text{(see 1.8.3.3.3).}$$

This result could have been obtained without appeal to Fourier series, because

$$\frac{1}{4n^2 - 1} = \frac{1}{2}\left[\frac{1}{2n - 1} - \frac{1}{2n + 1} \right],$$

so

$$\sum_{n=1}^{\infty} \frac{1}{4n^2 - 1} = \frac{1}{2} \sum_{n=1}^{\infty} \left[\frac{1}{2n - 1} - \frac{1}{2n + 1} \right]$$

$$= \frac{1}{2}\left[\left(1 - \frac{1}{3}\right) + \left(\frac{1}{3} - \frac{1}{5}\right) + \left(\frac{1}{5} - \frac{1}{7}\right) + \cdots \right].$$

After cancellation of terms (**telescoping of the series**) only the first term remains so the sum is seen to be 1/2.

3. **Application to $f(x) = x - x^2$, $0 \le x \le 1$.** From 16.10.1.7 the Fourier series of $f(x) = x - x^2$, $0 \le x \le 1$ is known to be

$$\frac{1}{6} - \frac{1}{\pi^2} \sum_{n=1}^{\infty} \frac{\cos 2n\pi x}{n^2}.$$

Because $f(x)$ is continuous for $0 \le x \le 1$, it follows that at $x = 1/2$ the Fourier series will converge to $f(1/2) = 1/4$. Using this result and setting $x = 1/2$ in the Fourier series gives

$$\frac{1}{4} = \frac{1}{6} - \frac{1}{\pi^2} \sum_{n=1}^{\infty} \frac{\cos n\pi}{n^2}$$

or

$$\frac{1}{4} = \frac{1}{6} - \frac{1}{\pi^2} \sum_{n=1}^{\infty} \frac{(-1)^n}{n^2},$$

and so

$$\frac{\pi^2}{12} = \sum_{n=1}^{\infty} \frac{(-1)^{n+1}}{n^2} = 1 - \frac{1}{4} + \frac{1}{9} - \frac{1}{16} + \cdots.$$

4. ***Application to*** $f(x) = e^{ax}$, $-\pi \leq x \leq \pi$.   From 16.10.1.9 it is known that $f(x) = e^{ax}$, $-\pi \leq x \leq \pi$ has the Fourier series

$$\frac{\sinh \pi a}{\pi a} + \frac{2 \sinh \pi a}{\pi} \sum_{n=1}^{\infty} (-1)^n \left[\left(\frac{a}{a^2 + n^2} \right) \cos nx - \left(\frac{n}{a^2 + n^2} \right) \sin nx \right].$$

The function $f(x)$ is continuous for $-\pi \leq x \leq \pi$, but $f(-\pi) \neq f(\pi)$. Thus, from 16.11.1.1.1, at the ends of the fundamental interval the Fourier series for $f(x)$ will converge to the value

$$\tfrac{1}{2}[f(\pi) + f(-\pi)] = \tfrac{1}{2}[e^{a\pi} + e^{-a\pi}] = \cosh \pi a.$$

Using this result and setting $x = \pi$ in the Fourier series gives

$$\cosh \pi a = \frac{\sinh \pi a}{\pi a} + \frac{2 \sinh \pi a}{\pi} \sum_{n=1}^{\infty} \left(\frac{a}{a^2 + n^2} \right),$$

where use has been made of the result $\cos n\pi = (-1)^n$. Thus

$$\coth \pi a = \frac{1}{\pi} \left[\frac{1}{a} + 2 \sum_{n=1}^{\infty} \frac{a}{a^2 + n^2} \right]$$

or, equivalently,

$$\frac{1}{2} \left[\pi \coth \pi a - \frac{1}{a} \right] = \sum_{n=1}^{\infty} \frac{a}{a^2 + n^2}.$$

Had the value $x = 0$ been chosen, at which point $f(x)$ is continuous, the Fourier series would have converged to $f(0) = 1$, and setting $x = 0$ in the Fourier series would have yielded

$$1 = \frac{\sinh \pi a}{\pi a} + \frac{2 \sinh \pi a}{\pi} \sum_{n=1}^{\infty} \frac{(-1)^n a}{a^2 + n^2},$$

from which it follows that

$$\frac{1}{2a}[1 - \pi a \, \text{cosech} \, \pi a] = \sum_{n=1}^{\infty} \frac{(-1)^{n+1} a}{a^2 + n^2}.$$

17

Bessel Functions

17.1 Bessel's Differential Equation

17.1.1 Different forms of Bessel's equation

17.1.1.1　　In standard form, **Bessel's equation** is either written as

1. $x^2 \dfrac{d^2 y}{dx^2} + x \dfrac{dy}{dx} + (x^2 - v^2)y = 0$

or as

2. $\dfrac{d^2 y}{dx^2} + \dfrac{1}{x}\dfrac{dy}{dx} + \left(1 - \dfrac{v^2}{x^2}\right)y = 0,$

where the real parameter v determines the nature of the two linearly independent solutions of the equation. By convention, v is understood to be any real number that is *not* an integer, and when integral values of this parameter are involved v is replaced by n.

Two linearly independent solutions of 17.1.1.1.1 are the **Bessel functions of the first kind of order v**, written $J_v(x)$ and $J_{-v}(x)$. Another solution of 17.1.1.1.1, to which reference will be made later, is the **Bessel function of the second kind of order** ν, written $Y_\nu(x)$. The general solution of 17.1.1.1.1 is

3. $y = AJ_\nu(x) + BJ_{-\nu}(x)$ [ν *not* an integer].

When the order is an integer ($\nu = n$) the Bessel functions $J_n(x)$ and $J_{-n}(x)$ cease to be linearly independent because

4. $J_{-n}(x) = (-1)^n J_n(x)$.

A second solution of Bessel's equation that is always linearly independent of $J_\nu(x)$ is $Y_\nu(x)$, irrespective of the value of ν. Thus, the general solution of 17.1.1.1.1 may always be written

5. $y = AJ_\nu(x) + BY_\nu(x)$.

The Bessel function of the second kind $Y_\nu(x)$ is defined as

6. $Y_\nu(x) = \dfrac{J_\nu(x)\cos(\nu\pi) - J_{-\nu}(x)}{\sin(\nu\pi)}$,

and when ν is an integer n or zero,

7. $Y_n(x) = \lim_{\nu \to n} Y_\nu(x)$.

A more general form of Bessel's equation that arises in many applications is

8. $x^2 \dfrac{d^2 y}{dx^2} + x\dfrac{dy}{dx} + (\lambda^2 x^2 - \nu^2)y = 0$

or, equivalently,

9. $\dfrac{d^2 y}{dx^2} + \dfrac{1}{x}\dfrac{dy}{dx} + \left(\lambda^2 - \dfrac{\nu^2}{x^2}\right)y = 0$.

These forms may be derived from 17.1.1.1.1 by first making the change of variable $x = \lambda u$, and then replacing u by x. Bessel's equations 17.1.1.1.8–9 always have the general solution

10. $y = AJ_\nu(\lambda x) + BY_\nu(\lambda x)$.

17.2 Series Expansions for $J_\nu(x)$ and $Y_\nu(x)$

17.2.1 Series expansions for $J_n(x)$ and $J_\nu(x)$

17.2.1.1

1. $J_0(x) = \displaystyle\sum_{k=0}^{\infty} \dfrac{(-1)^k x^{2k}}{2^{2k}(k!)^2}$

$= 1 - \dfrac{x^2}{2^2(1!)^2} + \dfrac{x^4}{2^4(2!)^2} - \dfrac{x^6}{2^6(3!)^2} + \cdots$.

2. $J_1(x) = \sum_{k=0}^{\infty} \frac{(-1)^k x^{2k+1}}{2^{2k+1} k!(k+1)!}$

$$= \frac{x}{2} - \frac{x^3}{2^3 1!2!} + \frac{x^5}{2^5 2!3!} - \frac{x^7}{2^7 3!4!} + \cdots.$$

3. $J_2(x) = \frac{x^2}{4} \left[\frac{1}{2!} - \frac{x^2}{2^2 1!3!} + \frac{x^4}{2^4 2!4!} - \frac{x^6}{2^6 3!5!} + \cdots \right].$

4. $J_n(x) = x^n \sum_{k=0}^{\infty} \frac{(-1)^k x^{2k}}{2^{2k+n} k!(n+k)!}.$

5. $J_{-n}(x) = x^n \sum_{k=0}^{\infty} \frac{(-1)^{n+k} x^{2k}}{2^{2k+n} k!(n+k)!}.$

6. $J_v(x) = x^v \sum_{k=0}^{\infty} \frac{(-1)^k x^{2k}}{2^{2k+v} k!\Gamma(v+k+1)}.$

17.2.1.2 Special values.

1. $J_0(0) = 1, \quad J_n(0) = 0 \qquad [n = 1, 2, 3, \ldots]$
2. $J_0'(0) = 0, \quad J_1'(0) = \frac{1}{2}, \quad J_n'(0) = 0 \qquad [n = 2, 3, 4, \ldots]$
3. $\lim_{x \to \infty} J_n(x) = 0, \qquad [n = 0, 1, 2, \ldots]$

17.2.2 Series expansions for $Y_n(x)$ and $Y_v(x)$

17.2.2.1

1. $Y_0(x) = \frac{2}{\pi} \left\{ \left(\ln \frac{x}{2} + \gamma \right) J_0(x) + \sum_{k=1}^{\infty} \frac{(-1)^{(k-1)}}{2^{2k}(k!)^2} \left[1 + \frac{1}{2} + \frac{1}{3} + \cdots + \frac{1}{k} \right] x^{2k} \right\}$

$[\gamma = 0.577215\ldots]$ (see 11.1.1.1.7)

2. $Y_1(x) = \frac{2}{\pi} \left(\ln \frac{x}{2} + \gamma \right) J_1(x) - \frac{2}{\pi x} - \frac{1}{\pi} \sum_{k=1}^{\infty} \frac{(-1)^{(k+1)}}{2^{2k-1}(k!)(k-1)!}$

$$\times \left[\frac{1}{k} + 2 \sum_{m=1}^{k-1} \frac{1}{m} \right] x^{2k-1} \qquad [\gamma = 0.577215\ldots] \qquad \text{(see 11.1.1.1.7)}$$

3. $Y_n(x) = -\frac{\left(\frac{1}{2}x\right)^{-n}}{\pi} \sum_{k=0}^{n-1} \frac{(n-k-1)!}{k!} \left(\frac{1}{4} x^2 \right)^k + \frac{2}{\pi} \ln \frac{x}{2} J_n(x)$

$$- \frac{\left(\frac{1}{2}x\right)^n}{\pi} \sum_{k=0}^{\infty} \{\psi(k+1) + \psi(n+k+1)\} \frac{\left(-\frac{1}{4}x^2\right)^k}{k!(n+k)!}$$

[with $\psi(k)$ given by 11.1.6.1.6]

4. $Y_\nu(x) = \dfrac{1}{\sin(\nu\pi)}\left\{\cos(\nu\pi)\left(\dfrac{x}{2}\right)^\nu \displaystyle\sum_{k=0}^{\infty} \dfrac{(-1)^k x^{2k}}{2^{2k} k!\,\Gamma(\nu+k+1)}\right.$

$$\left. -\left(\dfrac{x}{2}\right)^{-\nu} \sum_{k=0}^{\infty} \dfrac{(-1)^k x^{2k}}{2^{2k} k!\,\Gamma(k-\nu+1)}\right\} \qquad\qquad [\nu\ \textit{not}\ \text{an integer}]$$

17.2.2.2 Special values.

1. $\displaystyle\lim_{x\to 0} Y_n(x) = -\infty \qquad [n = 0, 1, 2, \ldots]$

2. $\displaystyle\lim_{x\to\infty} Y_n(x) = 0 \qquad [n = 0, 1, 2, \ldots]$

17.3 Bessel Functions of Fractional Order

17.3.1 Bessel functions $J_{\pm(n+1/2)}(x)$

17.3.1.1

1. $J_{1/2}(x) = \sqrt{\dfrac{2}{\pi x}}\,\sin x$

2. $J_{-1/2}(x) = \sqrt{\dfrac{2}{\pi x}}\,\cos x$

3. $J_{3/2}(x) = \sqrt{\dfrac{2}{\pi x}}\left(\dfrac{\sin x}{x} - \cos x\right)$

4. $J_{-3/2}(x) = -\sqrt{\dfrac{2}{\pi x}}\left(\dfrac{\cos x}{x} + \sin x\right)$

5. $J_{5/2}(x) = \sqrt{\dfrac{2}{\pi x}}\left[\left(\dfrac{3}{x^2} - 1\right)\sin x - \dfrac{3}{x}\cos x\right]$

6. $J_{-5/2}(x) = \sqrt{\dfrac{2}{\pi x}}\left[\left(\dfrac{3}{x^2} - 1\right)\cos x + \dfrac{3}{x}\sin x\right]$

7. $J_{n+1/2}(x) = \dfrac{2\left(\frac{1}{2}x\right)^{n+1/2}}{\sqrt{\pi}\,n!}\left[\left(1 + \dfrac{d^2}{dx^2}\right)^n \dfrac{\sin x}{x}\right]$

8. $J_{-(n+1/2)}(x) = (-1)^{n+1} Y_{n+1/2}(x)$ (see 17.3.2.1.7)

17.3.2 Bessel functions $Y_{\pm(n+1/2)}(x)$

17.3.2.1

1. $Y_{1/2}(x) = -\sqrt{\dfrac{2}{\pi x}}\,\cos x$

2. $Y_{-1/2}(x) = \sqrt{\dfrac{2}{\pi x}}\,\sin x$

3. $Y_{3/2}(x) = -\sqrt{\dfrac{2}{\pi x}}\left(\dfrac{\cos x}{x} + \sin x\right)$

4. $Y_{-3/2}(x) = \sqrt{\dfrac{2}{\pi x}}\left(-\dfrac{\sin x}{x} + \cos x\right)$

5. $Y_{5/2}(x) = -\sqrt{\dfrac{2}{\pi x}}\left[\left(\dfrac{3}{x^2} - 1\right)\cos x + \dfrac{3}{x}\sin x\right]$

6. $Y_{-5/2}(x) = \sqrt{\dfrac{2}{\pi x}}\left[\left(\dfrac{3}{x^2} - 1\right)\sin x - \dfrac{3}{x}\cos x\right]$

7. $Y_{n+1/2}(x) = -\dfrac{2\left(\frac{1}{2}x\right)^{n+1/2}}{\sqrt{\pi}\,n!}\left\{\left(1 + \dfrac{d^2}{dx^2}\right)^n \dfrac{\cos x}{x}\right\}$

8. $Y_{-(n+1/2)}(x) = (-1)^n J_{n+1/2}(x)$ (see 17.3.1.1.7)

17.4 Asymptotic Representations for Bessel Functions

17.4.1 Asymptotic representations for large arguments

17.4.1.1

1. $J_\nu(x) \sim \sqrt{\dfrac{2}{\pi x}}\left[\cos\left(x - \dfrac{1}{2}\nu\pi - \dfrac{1}{4}\pi\right)\right]$ $[x \gg 0]$

2. $Y_\nu(x) \sim \sqrt{\dfrac{2}{\pi x}}\left[\sin\left(x - \dfrac{1}{2}\nu\pi - \dfrac{1}{4}\pi\right)\right]$ $[x \gg 0]$

17.4.2 Asymptotic representation for large orders

17.4.2.1

1. $J_\nu(x) \sim \dfrac{1}{\sqrt{2\pi\nu}}\left(\dfrac{ex}{2\nu}\right)^\nu$ $[\nu \gg 0]$

2. $Y_\nu(x) \sim -\sqrt{\dfrac{2}{\pi\nu}}\left(\dfrac{ex}{2\nu}\right)^{-\nu}$ $[\nu \gg 0]$

17.5 Zeros of Bessel Functions

17.5.1 Zeros of $J_n(x)$ and $Y_n(x)$

17.5.1.1   Denote the zeros of $J_n(x)$ arranged in order of increasing magnitude by $j_{n,1}, j_{n,2}, j_{n,3}, \ldots$, and those of $Y_n(x)$ when similarly ordered by $y_{n,1}, y_{n,2}, y_{n,3}, \ldots$. Then the zeros of successive orders of Bessel functions of the first kind interlace, as do those of Bessel functions of the second kind, in the sense that

1. $j_{n-1,m} < j_{n,m} < j_{n-1,m+1}$ $[n, m = 1, 2, \ldots]$
2. $y_{n-1,m} < y_{n,m} < y_{n-1,m+1}$ $[n, m = 1, 2, \ldots]$

Table 17.1 lists the zeros $j_{n,m}$ and $y_{n,m}$ for $n = 0, 1, 2$ and $m = 1, 2, \ldots, 10$. Figure 17.1 shows the behavior of $J_n(x)$ and Figure 17.2 the behavior of $Y_n(x)$ for $n = 0, 1, 2$ and $0 \le x \le 10$.

TABLE 17.1   Zeros $j_{n,m}$ and $y_{n,m}$

m	$j_{0,m}$	$j_{1,m}$	$j_{2,m}$	$y_{0,m}$	$y_{1,m}$	$y_{2,m}$
1	2.40483	3.83171	5.13562	0.89358	2.19714	3.38424
2	5.52008	7.01559	8.41724	3.95768	5.42968	6.79381
3	8.65373	10.17347	11.61984	7.08605	8.59601	10.02348
4	11.79153	13.32369	14.79595	10.22235	11.74915	13.20999
5	14.93092	16.47063	17.95982	13.36110	14.89744	16.37897
6	18.07106	19.61586	21.11700	16.50092	18.04340	19.53904
7	21.21164	22.76008	24.27011	19.64131	21.18807	22.69396
8	24.35247	25.90367	22.42057	22.78203	24.33194	25.84561
9	27.49348	29.04683	30.56920	25.92296	27.47529	28.99508
10	30.63461	32.18968	33.71652	29.06403	30.61829	32.14300

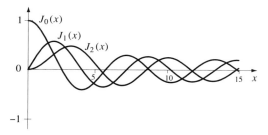

FIGURE 17.1 ∎

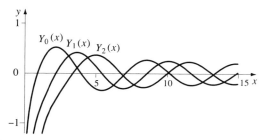

FIGURE 17.2 ∎

17.6 Bessel's Modified Equation

17.6.1 Different forms of Bessel's modified equation

17.6.1.1   In standard form, Bessel's modified equation is either written as

1. $x^2 \dfrac{d^2 y}{dx^2} + x \dfrac{dy}{dx} - (x^2 + v^2)y = 0$

or as

2. $\dfrac{d^2 y}{dx^2} + \dfrac{1}{x}\dfrac{dy}{dx} - \left(1 + \dfrac{v^2}{x^2}\right)y = 0,$

where the real parameter v determines the nature of the two linearly independent solutions of the equation. As in 17.1.1, by convention v is understood to be any real number that is *not* an integer, and when integral values of this parameter are involved v is replaced by n.

Two linearly independent solutions of 17.6.1.1.1 are the **modified Bessel functions of the first kind of order** v, written $I_v(x)$ and $I_{-v}(x)$. Another solution of 17.6.1.1.1, to which reference will be made later, is the **modified Bessel function of the second kind of order** v, written $K_v(x)$.

The general solution of 17.6.1.1.1 is

3. $y = A I_v(x) + B I_{-v}(x)$ [v *not* an integer].

When the order is an integer ($v = n$), the modified Bessel functions $I_n(x)$ and $I_{-n}(x)$ cease to be linearly independent because

4. $I_{-n}(x) = I_n(x).$

A second solution of 17.6.1.1.1 that is always linearly independent of $I_v(x)$ is $K_v(x)$, irrespective of the value of v. Thus, the general solution of 17.6.1.1.1 can always be written

5. $y = A I_v(x) + B K_v(x).$

The modified Bessel function of the second kind $K_v(x)$ is defined as

6. $K_v(x) = \dfrac{\pi}{2}\dfrac{I_{-v}(x) - I_v(x)}{\sin(v\pi)},$

and when v is an integer n or zero,

7. $K_n(x) = \lim\limits_{v \to n} K_v(x).$

A more general form of Bessel's modified equation that arises in many applications is

8. $x^2 \dfrac{d^2 y}{dx^2} + x \dfrac{dy}{dx} - (\lambda^2 x^2 + v^2)y = 0$

or, equivalently,

9. $\dfrac{d^2 y}{dx^2} + \dfrac{1}{x}\dfrac{dy}{dx} - \left(\lambda^2 + \dfrac{v^2}{x^2}\right)y = 0.$

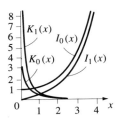

FIGURE 17.3 ∎

These forms may be derived from 17.6.1.1.1 by first making the change of variable $x = \lambda u$, and then replacing u by x. Bessel's modified equations 17.6.1.1.8–9 always have the general solution

10. $y = AI_\nu(\lambda x) + BK_\nu(\lambda x)$.

11.   Figure 17.3 shows the behavior of $I_n(x)$ and $K_n(x)$ for $n = 0, 1$ and $0 \le x \le 4$.

17.7 Series Expansions for $I_\nu(x)$ and $K_\nu(x)$

17.7.1 Series expansions for $I_n(x)$ and $I_\nu(x)$

17.7.1.1

1. $I_0(x) = \displaystyle\sum_{k=0}^{\infty} \frac{2^{2k}}{2^{2k}(k!)^2}$

$$= 1 + \frac{x^2}{2^2(1!)^2} + \frac{x^4}{2^4(2!)^2} + \frac{x^6}{2^6(3!)^2} + \cdots$$

2. $I_1(x) = \displaystyle\sum_{k=0}^{\infty} \frac{2^{2k+1}}{2^{2k+1}k!(k+1)!}$

$$= \frac{x}{2} + \frac{x^3}{2^3 1!2!} + \frac{x^5}{2^5 2!3!} + \frac{x^7}{2^7 3!4!} + \cdots$$

3. $I_n(x) = \displaystyle\sum_{k=0}^{\infty} \frac{x^{n+2k}}{2^{n+2k}k!(n+k)!}$

4. $I_\nu(x) = i^{-\nu} J_\nu(ix) = \displaystyle\sum_{k=0}^{\infty} \frac{x^{\nu+2k}}{2^{\nu+2k}k!\Gamma(\nu+k+1)}$

5. $I_{-\nu}(x) = \displaystyle\sum_{k=0}^{\infty} \frac{x^{2k-\nu}}{2^{2k-\nu}k!\Gamma(k+1-\nu)}$

17.7.1.2

1. $I_0(0) = 1, \quad I_n(0) = 0 \qquad [n = 1, 2, \ldots]$

2. $\displaystyle\lim_{x \to \infty} I_n(x) = 0 \qquad [n = 0, 1, 2, \ldots]$

17.7.2 Series expansions for $K_0(x)$ and $K_n(x)$

17.7.2.1

1. $K_0(x) = -\left[\ln\frac{x}{2} + \gamma\right]I_0(x) + \dfrac{\frac{1}{4}x^2}{1!}$

$$+ \left(1 + \frac{1}{2}\right)\frac{\left(\frac{1}{4}x^2\right)^2}{(2!)^2} + \left(1 + \frac{1}{2} + \frac{1}{3}\right) + \frac{\left(\frac{1}{4}x^2\right)^3}{3!^2} \cdots$$

$[\gamma = 0.577215\ldots]$ (see 11.1.1.1.7)

2. $K_n(x) = \dfrac{1}{2}\left(\dfrac{x}{2}\right)^{-n}\displaystyle\sum_{k=0}^{n-1}\dfrac{(n-k-1)!}{k!}\left(-\dfrac{1}{4}x^2\right)^k$

$$+ (-1)^{n+1}\ln\left(\frac{x}{2}\right)I_n(x) + (-1)^n\frac{1}{2}\left(\frac{x}{2}\right)^n$$

$$\times \sum_{k=0}^{\infty}[\psi(k+1) + \psi(n+k+1)]\frac{\left(-\frac{1}{4}x^2\right)^k}{k!(n+k)!}$$

[with $\psi(k)$ given by 11.1.6.1.6]

3. $K_{-n}(x) = K_n(x)$ $[n = 0, 1, 2, \ldots]$

17.7.2.2 Special cases.

1. $\displaystyle\lim_{x\to 0} K_n \neq (0) = \infty$ $[n = 0, 1, 2, \ldots]$

2. $\displaystyle\lim_{x\to\infty} K_n(x) = 0$ $[n = 0, 1, 2, \ldots]$

17.8 Modified Bessel Functions of Fractional Order

17.8.1 Modified Bessel functions $I_{\pm(n+1/2)}(x)$

17.8.1.1

1. $I_{1/2}(x) = \sqrt{\dfrac{2}{\pi x}}\sinh x$

2. $I_{-1/2}(x) = \sqrt{\dfrac{2}{\pi x}}\cosh x$

3. $I_{3/2}(x) = -\sqrt{\dfrac{2}{\pi x}}\left(\dfrac{\sinh x}{x} - \cosh x\right)$

4. $I_{-3/2}(x) = -\sqrt{\dfrac{2}{\pi x}}\left(\dfrac{\cosh x}{x} - \sinh x\right)$

5. $I_{5/2}(x) = \sqrt{\dfrac{2}{\pi x}} \left[\left(\dfrac{3}{x^2} + 1 \right) \sinh x - \dfrac{3}{x} \cosh x \right]$

6. $I_{-5/2}(x) = \sqrt{\dfrac{2}{\pi x}} \left[\left(\dfrac{3}{x^2} + 1 \right) \cosh x - \dfrac{3}{x} \sinh x \right]$

7. $I_{n+1/2}(x) = \dfrac{1}{\sqrt{2\pi x}} \left[e^x \sum_{k=0}^{n} \dfrac{(-1)^k (n+k)!}{k!(n-k)!(2x)^k} \right.$

$\left. + (-1)^{n+1} e^{-x} \sum_{k=0}^{n} \dfrac{(n+k)!}{k!(n-k)!(2x)^k} \right]$

8. $I_{-(n+1/2)}(x) = \dfrac{1}{\sqrt{2\pi x}} \left[e^x \sum_{k=0}^{n} \dfrac{(-1)^k (n+k)!}{k!(n-k)!(2x)^k} \right.$

$\left. + (-1)^n e^{-x} \sum_{k=0}^{n} \dfrac{(n+k)!}{k!(n-k)!(2x)^k} \right]$

17.8.2 Modified Bessel functions $K_{\pm(n+1/2)}(x)$

17.8.2.1

1. $K_{1/2}(x) = e^{-x} \sqrt{\dfrac{\pi}{2x}}$

2. $K_{3/2}(x) = e^{-x} \sqrt{\dfrac{\pi}{2x}} \left(\dfrac{1}{x} + 1 \right)$

3. $K_{5/2}(x) = e^{-x} \sqrt{\dfrac{\pi}{2x}} \left(\dfrac{3}{x^2} + \dfrac{3}{x} + 1 \right)$

4. $K_{n+1/2}(x) = \sqrt{\dfrac{\pi}{2x}} e^{-x} \sum_{k=0}^{n} \dfrac{(n+k)!}{2^k k!(n-k)! x^k}$

5. $K_{-(n+1/2)}(x) = K_{n+1/2}(x)$

17.9 Asymptotic Representations of Modified Bessel Functions

17.9.1 Asymptotic representations for large arguments

17.9.1.1

1. $I_\nu(x) \sim \dfrac{e^x}{\sqrt{2\pi x}} \left[1 - \dfrac{4\nu^2 - 1}{8x} \right] \qquad [x \gg 0]$

2. $K_\nu(x) \sim \sqrt{\dfrac{\pi}{2x}} e^{-x} \left[1 + \dfrac{4\nu^2 - 1}{8x} \right] \qquad [x \gg 0]$

17.10 Relationships between Bessel Functions

17.10.1 Relationships involving $J_\nu(x)$ and $Y_\nu(x)$

17.10.1.1

1. $J_2(x) = \dfrac{2}{x} J_1(x) - J_0(x)$

2. $Y_2(x) = \dfrac{2}{x} Y_1(x) - Y_0(x)$

3. $\dfrac{d}{dx}[J_0(x)] = -J_1(x)$

4. $\dfrac{d}{dx}[Y_0(x)] = -Y_1(x)$

5. $x J_{\nu-1}(x) + x J_{\nu+1}(x) = 2\nu J_\nu(x)$

6. $x Y_{\nu-1}(x) + x Y_{\nu+1}(x) = 2\nu Y_\nu(x)$

7. $J_{\nu-1}(x) - J_{\nu+1}(x) = 2\dfrac{d}{dx}[J_\nu(x)]$

8. $Y_{\nu-1}(x) - Y_{\nu+1}(x) = 2\dfrac{d}{dx}[Y_\nu(x)]$

9. $x\dfrac{d}{dx}[J_\nu(x)] + \nu J_\nu(x) = x J_{\nu-1}(x)$

10. $x\dfrac{d}{dx}[Y_\nu(x)] + \nu Y_\nu(x) = x Y_{\nu-1}(x)$

11. $x\dfrac{d}{dx}[J_\nu(x)] - \nu J_\nu(x) = -x J_{\nu+1}(x)$

12. $x\dfrac{d}{dx}[Y_\nu(x)] - \nu Y_\nu(x) = -x Y_{\nu+1}(x)$

13. $\left(\dfrac{1}{x}\dfrac{d}{dx}\right)^m [x^\nu J_\nu(x)] = x^{\nu-m} J_{\nu-m}(x)$

14. $\left(\dfrac{1}{x}\dfrac{d}{dx}\right)^m [x^\nu Y_\nu(x)] = x^{\nu-m} Y_{\nu-m}(x)$

15. $\left(\dfrac{1}{x}\dfrac{d}{dx}\right)^m [x^{-\nu} J_\nu(x)] = (-1)^m x^{-\nu-m} J_{\nu+m}(x)$

16. $\left(\dfrac{1}{x}\dfrac{d}{dx}\right)^m [x^{-\nu} Y_\nu(x)] = (-1)^m x^{-\nu-m} Y_{\nu+m}(x)$

17. $J_{-n}(x) = (-1)^n J_n(x)$ 　　　　[n an integer]

18. $Y_{-n}(x) = (-1)^n Y_n(x)$ 　　　　[n an integer]

19. $\dfrac{d}{dx}[x^n J_n(x)] = x^n J_{n-1}(x)$

20. $\dfrac{d}{dx}[x^n Y_n(x)] = x^n Y_{n-1}(x)$

17.10.2 Relationships involving $I_\nu(x)$ and $K_\nu(x)$

17.10.2.1

1. $xI_{\nu-1}(x) - xI_{\nu+1}(x) = 2\nu I_\nu(x)$

2. $I_{\nu-1}(x) + I_{\nu+1}(x) = 2\dfrac{d}{dx}[I_\nu(x)]$

3. $x\dfrac{d}{dx}[I_\nu(x)] + \nu I_\nu(x) = xI_{\nu-1}(x)$

4. $x\dfrac{d}{dx}[I_\nu(x)] - \nu I_\nu(x) = xI_{\nu+1}(x)$

5. $\left(\dfrac{1}{x}\dfrac{d}{dx}\right)^m [x^\nu I_\nu(x)] = x^{\nu-m} I_{\nu-m}(x)$

6. $\left(\dfrac{1}{x}\dfrac{d}{dx}\right)^m [x^{-\nu} I_\nu(x)] = x^{-\nu-m} I_{\nu+m}(x)$

7. $I_{-n}(x) = I_n(x)$ 　　　　[n an integer]

8. $I_2(x) = -\dfrac{2}{x}I_1(x) + I_0(x)$

9. $\dfrac{d}{dx}[I_0(x)] = I_1(x)$

10. $xK_{n-1}(x) - xK_{n+1}(x) = -2nK_n(x)$

11. $K_{n-1}(x) + K_{n+1}(x) = -2\dfrac{d}{dx}[K_n(x)]$

12. $x\dfrac{d}{dx}[K_n(x)] + nK_n(x) = -xK_{n-1}(x)$

13. $x\dfrac{d}{dx}[K_n(x)] - nK_n(x) = -xK_{n+1}(x)$

14. $\left(\dfrac{1}{x}\dfrac{d}{dx}\right)^m [x^n K_n(x)] = (-1)^m x^{n-m} K_{n-m}(x)$

15. $\left(\dfrac{1}{x}\dfrac{d}{dx}\right)^m [x^{-n}K_n(x)] = (-1)^m x^{-n-m}K_{n+m}(x)$

16. $K_{-\nu}(x) = K_\nu(x)$

17. $K_2(x) = \dfrac{2}{x}K_1(x) + K_0(x)$

18. $\dfrac{d}{dx}[K_0(x)] = -K_1(x)$

19. $K_{-n}(x) = K_n(x)$ [n an integer]

17.11 Integral Representations of $J_n(x)$, $I_n(x)$, and $K_n(x)$

17.11.1 Integral representations of $J_n(x)$

17.11.1.1

1. $J_0(x) = \dfrac{2}{\pi}\displaystyle\int_0^{\pi/2} \cos(x\cos\theta)\,d\theta$

2. $J_1(x) = \dfrac{2}{\pi}\displaystyle\int_0^{\pi/2} \sin(x\cos\theta)\cos\theta\,d\theta$

3. $J_n(x) = \dfrac{1}{\pi}\displaystyle\int_0^{\pi} \cos(x\sin\theta - n\theta)\,d\theta$

4. $J_n(x) = \dfrac{1}{\pi}\displaystyle\int_0^{\pi} \cos(x\sin\theta)\cos n\theta\,d\theta$ $[n = 0, 2, 4, \ldots]$

5. $J_n(x) = \dfrac{1}{\pi}\displaystyle\int_0^{\pi} \sin(x\sin\theta)\sin n\theta\,d\theta$ $[n = 1, 3, 5, \ldots]$

6. $J_n(x) = \dfrac{2}{\sqrt{\pi}\Gamma\left(n + \frac{1}{2}\right)}\left(\dfrac{x}{2}\right)^n \displaystyle\int_0^{\pi/2} \cos(x\sin\theta)(\cos\theta)^{2n}\,d\theta$

7. $J_n(x) = \dfrac{2}{\sqrt{\pi}\Gamma\left(n + \frac{1}{2}\right)}\left(\dfrac{x}{2}\right)^n \displaystyle\int_0^{\pi/2} \cos(x\sin\theta)(\sin\theta)^{2n}\,d\theta$

8. $I_n(x) = \dfrac{2}{\sqrt{\pi}\Gamma\left(n + \frac{1}{2}\right)}\left(\dfrac{x}{2}\right)^n \displaystyle\int_0^{\pi/2} \cosh(x\sin\theta)(\cos\theta)^{2n}\,d\theta$

9. $I_n(x) = \dfrac{2}{\sqrt{\pi}\Gamma\left(n + \frac{1}{2}\right)}\left(\dfrac{x}{2}\right)^n \displaystyle\int_0^{\pi/2} \cosh(x\sin\theta)(\sin\theta)^{2n}\,d\theta$

10. $K_n(x) = e^{-x}\sqrt{\dfrac{\pi}{2x}}\dfrac{1}{\Gamma\left(n + \frac{1}{2}\right)}\displaystyle\int_0^{\infty} e^{-u}u^{n-1/2}\left(1 + \dfrac{u}{2x}\right)^{n-1/2}\,du$

17.12 Indefinite Integrals of Bessel Functions

17.12.1 Integrals of $J_n(x)$, $I_n(x)$, and $K_n(x)$

17.12.1.1

1. $\displaystyle \int x J_0(ax)\,dx = \frac{x}{a} J_1(ax)$

2. $\displaystyle \int x^2 J_0(ax)\,dx = \frac{x^2}{a} J_1(ax) + \frac{x}{a^2} J_0(ax) - \frac{1}{a^2} \int J_0(ax)\,dx$

3. $\displaystyle \int \frac{J_0(ax)}{x^2}\,dx = a J_1(ax) - \frac{J_0(ax)}{x} - a^2 \int J_0(ax)\,dx$

4. $\displaystyle \int J_1(ax)\,dx = -\frac{1}{a} J_0(ax)$

5. $\displaystyle \int x J_1(ax)\,dx = -\frac{x}{a} J_0(ax) + \frac{1}{a} \int J_0(ax)\,dx$

6. $\displaystyle \int \frac{J_1(ax)}{x}\,dx = -J_1(ax) + a \int J_0(ax)\,dx$

7. $\displaystyle \int x J_n(ax) J_n(bx)\,dx = \frac{x\left[a J_n(bx) J_n'(ax) - b J_n(ax) J_n'(bx)\right]}{b^2 - a^2}$ $[a \neq b]$

8. $\displaystyle \int x J_n^2(ax)\,dx = \frac{x^2}{2}\left[J_n'(ax)\right]^2 + \frac{x^2}{2}\left(1 - \frac{n^2}{a^2 x^2}\right)[J_n(ax)]^2$

9. $\displaystyle \int x^n J_{n-1}(ax)\,dx = \frac{x^n}{a} J_n(ax)$

10. $\displaystyle \int x^{-n} J_{n+1}(ax)\,dx = -\frac{x^{-n}}{a} J_n(ax)$

11. $\displaystyle \int x^n I_{n-1}(ax)\,dx = \frac{x^n}{a} I_n(ax)$

12. $\displaystyle \int x^{-n} I_{n+1}(ax)\,dx = \frac{x^{-n}}{a} I_n(ax)$

13. $\displaystyle \int x^n K_{n-1}(ax)\,dx = -\frac{x^n}{a} K_n(ax)$

14. $\displaystyle \int x^{-n} K_{n+1}(ax)\,dx = -\frac{x^{-n}}{a} K_n(ax)$

The integrals involving $J_n(x)$ are also true when $J_n(x)$ is replaced by $Y_n(x)$.

17.13 Definite Integrals Involving Bessel Functions

17.13.1 Definite integrals involving $J_n(x)$ and elementary functions

17.13.1.1

1. $\displaystyle\int_0^a J_1(x)\,dx = 1 - J_0(a) \qquad [a > 0]$

2. $\displaystyle\int_a^\infty J_1(x)\,dx = J_0(a) \qquad [a > 0]$

3. $\displaystyle\int_0^\infty J_n(ax)\,dx = \frac{1}{a} \qquad [n > -1, a > 0]$

4. $\displaystyle\int_0^\infty \frac{J_n(ax)}{x}\,dx = \frac{1}{n} \qquad [n = 1, 2, \ldots]$

5. $\displaystyle\int_0^1 x J_n(ax) J_n(bx)\,dx = \begin{cases} 0, & a \neq b, \\ \frac{1}{2}[J_{n+1}(a)]^2, & a = b,\ J_n(a) = J_n(b) = 0 \end{cases}$

$[n > -1]$

6. $\displaystyle\int_0^\infty e^{-ax} J_0(bx)\,dx = \frac{1}{\sqrt{a^2 + b^2}}$

7. $\displaystyle\int_0^\infty e^{-ax} J_n(bx)\,dx = \frac{1}{\sqrt{a^2 + b^2}}\left[\frac{\sqrt{a^2 + b^2} - a}{b}\right]^n \qquad [a > 0, n = 0, 1, 2, \ldots]$

8. $\displaystyle\int_0^\infty J_n(ax) \sin bx\,dx = \begin{cases} \dfrac{\sin[n \arcsin(b/a)]}{\sqrt{a^2 - b^2}}, & 0 < b < a \\[4mm] \dfrac{a^n \cos(n\pi/2)}{\sqrt{b^2 - a^2}(b + \sqrt{b^2 - a^2})^n}, & 0 < a < b \end{cases}$

$[n > -2]$

9. $\displaystyle\int_0^\infty J_n(ax) \cos bx\,dx = \begin{cases} \dfrac{\cos[n \arccos(b/a)]}{\sqrt{a^2 - b^2}}, & 0 < b < a \\[4mm] \dfrac{-a^n \sin(n\pi/2)}{\sqrt{b^2 - a^2}(b + \sqrt{b^2 - a^2})^n}, & 0 < a < b \end{cases}$

$[n > -2]$

10. $\displaystyle\int_0^\infty \frac{J_m(x) J_n(x)}{x}\,dx = \begin{cases} \dfrac{2}{\pi(m^2 - n^2)} \sin \dfrac{(m - n)\pi}{2}, & m \neq n \\[4mm] 1/2m, & m = n \end{cases} \qquad [m+n > 0]$

11. $\displaystyle\int_0^\infty J_0(ax)J_1(bx)\,dx = \begin{cases} 1/b, & b^2 > a^2 \\ 0, & b^2 < a^2 \end{cases}$

12. $\displaystyle\int_0^\infty J_0(ax)J_1(ax)\,dx = \dfrac{1}{2a} \qquad [a > 0]$

CHAPTER 18

Orthogonal Polynomials

18.1 Introduction

18.1.1 Definition of a system of orthogonal polynomials

18.1.1.1   Let $\{\Phi_n(x)\}$ be a system of polynomials defined for $a \leq x \leq b$ such that $\Phi_n(x)$ is of degree n, and let $w(x) > 0$ be a function defined for $a \leq x \leq b$. Define the positive numbers $\|\Phi_n\|^2$ as

1. $\|\Phi_n\|^2 = \displaystyle\int_a^b [\Phi_n(x)]^2 w(x)\, dx.$

Then the system of polynomials $\{\Phi_n(x)\}$ is said to be **orthogonal over $a \leq x \leq b$ with respect to the weight function $w(x)$** if

2. $\displaystyle\int_a^b \Phi_m(x)\Phi_n(x)w(x)\, dx = \begin{cases} 0, & m \neq n \\ \|\Phi_n\|^2, & m = n \end{cases} \qquad [m, n = 0, 1, 2, \ldots].$

The **normalized** system of polynomials $[\phi_n(x)]$, where $\phi_n(x) = \Phi_n(x)/\|\Phi_n\|$, is said to be **orthonormal** over $a \leq x \leq b$ with respect to the **weight function** $w(x)$, where $\|\Phi_n\|$ is called the **norm** of $\Phi_n(x)$ and it follows from 18.1.1.1.1–2 that

3. $\displaystyle\int_a^b \phi_m(x)\phi_n(x)w(x)\,dx = \begin{cases} 0, & m \neq n \\ 1, & m = n \end{cases}$ $[m, n = 0, 1, 2, \ldots].$

Orthogonal polynomials are special solutions of linear variable coefficient second-order differential equations defined on the interval $a \leq x \leq b$, in which n appears as a parameter. These polynomials can be generated by differentiation of a suitable sequence of functions (**Rodrigue's formula**), and they satisfy recurrence relations.

18.2 Legendre Polynomials $P_n(x)$

18.2.1 Differential equation satisfied by $P_n(x)$

18.2.1.1   The **Legendre polynomials** $P_n(x)$ satisfy the differential equation

1. $\displaystyle (1 - x^2)\frac{d^2 y}{dx^2} - 2x\frac{dy}{dx} + n(n + 1)y = 0$

defined on the interval $-1 \leq x \leq 1$, with $n = 0, 1, 2, \ldots.$

18.2.2 Rodrigue's formula for $P_n(x)$

18.2.2.1   The Legendre polynomial $P_n(x)$ of degree n is given by **Rodrigue's formula:**

1. $\displaystyle P_n(x) = \frac{1}{2^n n!}\frac{d^n}{dx^n}[(x^2 - 1)^n].$

18.2.3 Orthogonality relation for $P_n(x)$

18.2.3.1   The **weight function** for Legendre polynomials is $w(x) = 1$, and the **orthogonality relation** is

$\displaystyle\int_{-1}^1 P_m(x)P_n(x)\,dx = \begin{cases} 0, & \text{for } m \neq n \\ \frac{2}{2n+1}, & \text{for } m = n \end{cases}$ $[n = 0, 1, 2, \ldots].$

18.2.4 Explicit expressions for $P_n(x)$

18.2.4.1

1. $\displaystyle P_n(x) = \frac{1}{2^n}\sum_{k=0}^{[n/2]} \frac{(-1)^k (2n - 2k)!}{k!(n - k)!(n - 2k)!}x^{n-2k}$

$\displaystyle = \frac{(2n)!}{2^n (n!)^2}\left[x^n - \frac{n(n - 1)}{2(2n - 1)}x^{n-2} + \frac{n(n - 1)(n - 2)(n - 3)}{2 \cdot 4(2n - 1)(2n - 3)}x^{n-4} - \cdots\right]$

$\displaystyle \left[\left[\frac{n}{2}\right] \text{ signifies the integral part of } \frac{n}{2}\right]$

2. $P_{2n}(x) = (-1)^n \dfrac{(2n-1)!!}{2^n n!} \left[1 - \dfrac{2n(2n+1)}{2!} x^2 \right.$

   $\left. + \dfrac{2n(2n-2)(2n+1)(2n+3)}{4!} x^4 - \cdots \right]$

   $[(2n-1)!! = 1 \cdot 3 \cdot 5 \cdots (2n-1)]$

3. $P_{2n+1}(x) = (-1)^n \dfrac{(2n+1)!!}{2^n n!} \left[x - \dfrac{2n(2n+3)}{3!} x^3 \right.$

   $\left. + \dfrac{2n(2n-2)(2n+3)(2n+5)}{5!} x^5 - \cdots \right]$

   $[(2n+1)!! = 1 \cdot 3 \cdot 5 \cdots (2n+1)]$

18.2.4.2 Special cases and graphs of $P_n(x)$.

Notation: $x = \cos\theta$.

1. $P_0(x) = 1$
2. $P_1(x) = x = \cos\theta$
3. $P_2(x) = \frac{1}{2}(3x^2 - 1) = \frac{1}{4}(3\cos 2\theta + 1)$
4. $P_3(x) = \frac{1}{2}(5x^3 - 3x) = \frac{1}{8}(5\cos 3\theta + 3\cos\theta)$
5. $P_4(x) = \frac{1}{8}(35x^4 - 30x^2 + 3) = \frac{1}{64}(35\cos 4\theta + 20\cos 2\theta + 9)$
6. $P_5(x) = \frac{1}{8}(63x^5 - 70x^3 + 15x) = \frac{1}{128}(63\cos 5\theta + 35\cos 3\theta + 30\cos\theta)$
7. $P_6(x) = \frac{1}{16}(231x^6 - 315x^4 + 105x^2 - 5)$

   $= \frac{1}{512}(231\cos 6\theta + 126\cos 4\theta + 105\cos 2\theta + 50)$

Graphs of these polynomials are shown in Figure 18.1.

Normalization

For $x = 1$, $P_\ell = 1$

$x = -1$, $P_\ell = (\pm)1$ for $\ell \begin{smallmatrix}even\\odd\end{smallmatrix}$

18.2.5 Recurrence relations satisfied by $P_n(x)$

18.2.5.1

1. $(n+1)P_{n+1}(x) = (2n+1)x P_n(x) - n P_{n-1}(x)$

2. $(x^2 - 1)\dfrac{d}{dx}[P_n(x)] = nx P_n(x) - n P_{n-1}(x)$

   $= \dfrac{n(n+1)}{2n+1}[P_{n+1}(x) - P_{n-1}(x)]$

3. $\dfrac{d}{dx}[P_{n+1}(x)] - x\dfrac{d}{dx}[P_n(x)] = (n+1)P_n(x)$

4. $x\dfrac{d}{dx}[P_n(x)] - \dfrac{d}{dx}[P_{n-1}(x)] = n P_n(x)$

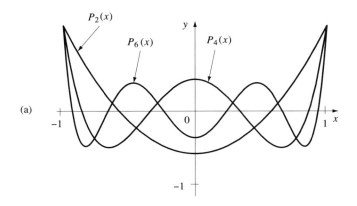

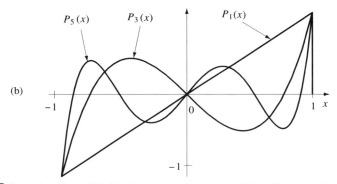

FIGURE 18.1 ■ Legendre polynomials $P_n(x)$: (a) even polynomials and (b) odd polynomials.

5. $\dfrac{d}{dx}[P_{n+1}(x) - P_{n-1}(x)] = (2n + 1)P_n(x)$

18.2.6 Generating function for $P_n(x)$

18.2.6.1   The Legendre polynomial $P_n(x)$ occurs as the multiplier of t^n in the expansion of the **generating function:**

1. $\dfrac{1}{\sqrt{1 - 2tx + t^2}} = \displaystyle\sum_{k=0}^{\infty} P_k(x)t^k \qquad \left[|t| < \min\left|x \pm \sqrt{x^2 - 1}\right|\right]$

18.2.7 Legendre functions of the second kind $Q_n(x)$

18.2.7.1   There is a nonpolynomial solution to the Legendre differential equation 18.2.1.1.1 that is linearly independent of the Legendre polynomial $P_n(x)$ and that has singularities at $x = \pm 1$. This solution, denoted by $Q_n(x)$, is called the **Legendre function of the second kind** of order n, to distinguish it from $P_n(x)$, which is the corresponding function of the first kind.

The general solution of the Legendre differential equation 18.2.1.1.1 on the interval $-1 < x < 1$ with $n = 0, 1, 2, \ldots$, is

1. $\quad y(x) = AP_n(x) + BQ_n(x),$

where A and B are arbitrary constants.

The functions $Q_n(x)$, the first six of which are listed below, satisfy the same recurrence relations as those for $P_n(x)$ given in 18.2.5.1.

2. $\quad Q_0(x) = \dfrac{1}{2}\ln\left(\dfrac{1+x}{1-x}\right)$

3. $\quad Q_1(x) = \dfrac{x}{2}\ln\left(\dfrac{1+x}{1-x}\right) - 1$

4. $\quad Q_2(x) = \dfrac{1}{4}(3x^2 - 1)\ln\left(\dfrac{1+x}{1-x}\right) - \dfrac{3}{2}x$

5. $\quad Q_3(x) = \dfrac{1}{4}(5x^3 - 3x)\ln\left(\dfrac{1+x}{1-x}\right) - \dfrac{5}{2}x^2 + \dfrac{2}{3}$

6. $\quad Q_4(x) = \dfrac{1}{16}(35x^4 - 30x^2 + 3)\ln\left(\dfrac{1+x}{1-x}\right) - \dfrac{35}{8}x^3 + \dfrac{55}{24}x$

7. $\quad Q_5(x) = \dfrac{1}{16}(63x^5 - 70x^3 + 15x)\ln\left(\dfrac{1+x}{1-x}\right) - \dfrac{63}{8}x^4 + \dfrac{49}{8}x^2 - \dfrac{8}{15}$

Graphs of $Q_n(x)$ for $-0.95 \le x \le 0.95$ and $n = 0, 1, 2, 3$ are shown in Figure 18.2.

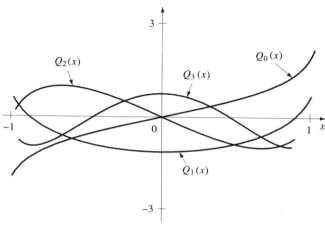

FIGURE 18.2 Legendre functions $Q_n(x)$.

18.3 Chebyshev Polynomials $T_n(x)$ and $U_n(x)$

18.3.1 Differential equation satisfied by $T_n(x)$ and $U_n(x)$

18.3.1.1   The **Chebyshev polynomials** $T_n(x)$ and $U_n(x)$ satisfy the differential equation

1. $(1 - x^2)\dfrac{d^2 y}{dx^2} - x\dfrac{dy}{dx} + n^2 y = 0,$

defined on the interval $-1 \leq x \leq 1$, with $n = 0, 1, 2, \ldots$. The functions $T_n(x)$ and $\sqrt{1 - x^2}U_n(x)$ are two linearly independent solutions of 18.3.1.1.1

18.3.2 Rodrigue's formulas for $T_n(x)$ and $U_n(x)$

18.3.2.1   The Chebyshev polynomials $T_n(x)$ and $U_n(x)$ of degree n are given by **Rodrigue's formulas:**

1. $T_n(x) = \dfrac{(-1)^n \sqrt{\pi}(1 - x^2)^{1/2}}{2^n \Gamma\left(n + \frac{1}{2}\right)}\dfrac{d^n}{dx^n}[(1 - x^2)^{n-1/2}]$

2. $U_n(x) = \dfrac{(-1)^n \sqrt{\pi}(n + 1)(1 - x^2)^{-1/2}}{2^{n+1} \Gamma\left(n + \frac{3}{2}\right)}\dfrac{d^n}{dx^n}[(1 - x^2)^{n+1/2}]$

18.3.3 Orthogonality relations for $T_n(x)$ and $U_n(x)$

18.3.3.1   The **weight function** $w(x)$ for the Chebyshev polynomials $T_n(x)$ and $U_n(x)$ is $w(x) = (1 - x^2)^{-1/2}$, and the **orthogonality relations** are

1. $\displaystyle\int_{-1}^{1} T_m(x)T_n(x)(1 - x^2)^{-1/2}\, dx = \begin{cases} 0, & m \neq n \\ \pi/2, & m = n \neq 0 \\ \pi, & m = n = 0 \end{cases}$

2. $\displaystyle\int_{-1}^{1} U_m(x)U_n(x)(1 - x^2)^{-1/2}\, dx = \begin{cases} 0, & m \neq n \\ \pi/8, & m = n \end{cases}$

18.3.4 Explicit expressions for $T_n(x)$ and $U_n(x)$

18.3.4.1

1. $T_n(x) = \cos(n \arccos x) = \frac{1}{2}\left[\left(x + i\sqrt{1 - x^2}\right)^n + \left(x - i\sqrt{1 - x^2}\right)^n\right]$

        $= x^n - \dbinom{n}{2}x^{n-2}(1 - x^2) + \dbinom{n}{4}x^{n-4}(1 - x^2)^2$

           $- \dbinom{n}{6}x^{n-6}(1 - x^2)^3 + \cdots.$

2. $U_n(x) = \dfrac{\sin[(n+1)\arccos x]}{\sin[\arccos x]}$

$$= \frac{1}{2i\sqrt{1-x^2}}\left[(x+i\sqrt{1-x^2})^{n+1} - (x-i\sqrt{1-x^2})^{n+1}\right]$$

$$= \binom{n+1}{1}x^n - \binom{n+1}{3}x^{n-2}(1-x^2) + \binom{n+1}{5}x^{n-4}(1-x^2)^2 - \cdots.$$

18.3.4.2 Special cases and graphs of $T_n(x)$ and $U_n(x)$.

1. $T_0(x) = 1$
2. $T_1(x) = x$
3. $T_2(x) = 2x^2 - 1$
4. $T_3(x) = 4x^3 - 3x$
5. $T_4(x) = 8x^4 - 8x^2 + 1$
6. $T_5(x) = 16x^5 - 20x^3 + 5x$
7. $T_6(x) = 32x^6 - 48x^4 + 18x^2 - 1$
8. $T_7(x) = 64x^7 - 112x^5 + 56x^3 - 7x$
9. $T_8(x) = 128x^8 - 256x^6 + 160x^4 - 32x^2 + 1$
10. $U_0(x) = 1$
11. $U_1(x) = 2x$
12. $U_2(x) = 4x^2 - 1$
13. $U_3(x) = 8x^3 - 4x$
14. $U_4(x) = 16x^4 - 12x^2 + 1$
15. $U_5(x) = 32x^5 - 32x^3 + 6x$
16. $U_6(x) = 64x^6 - 80x^4 + 24x^2 - 1$
17. $U_7(x) = 128x^7 - 192x^5 + 80x^3 - 8x$
18. $U_8(x) = 256x^8 - 448x^6 + 240x^4 - 40x^2 + 1$

See Figures 18.3 and 18.4.

18.3.4.3 Particular values.

1. $T_n(1) = 1$
2. $T_n(-1) = (-1)^n$
3. $T_{2n}(0) = (-1)^n$
4. $T_{2n+1}(0) = 0$
5. $U_{2n+1}(0) = 0$
6. $U_{2n}(0) = (-1)^n$

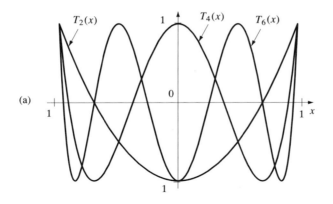

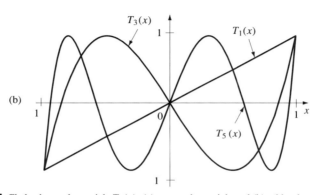

FIGURE 18.3 ▌ Chebyshev polynomials $T_n(x)$: (a) even polynomials and (b) odd polynomials.

18.3.5 Recurrence relations satisfied by $T_n(x)$ and $U_n(x)$

18.3.5.1

1. $T_{n+1}(x) - 2xT_n(x) + T_{n-1}(x) = 0$
2. $U_{n+1}(x) - 2xU_n(x) + U_{n-1}(x) = 0$
3. $T_n(x) = U_n(x) - xU_{n-1}(x)$
4. $(1 - x^2)U_{n-1}(x) = xT_n(x) - T_{n+1}(x)$

18.3.6 Generating functions for $T_n(x)$ and $U_n(x)$

18.3.6.1   The Chebyshev polynomials $T_n(x)$ and $U_n(x)$ occur as the multipliers of t^n in the expansions of the respective **generating functions**:

1. $\dfrac{1 - t^2}{1 - 2tx + t^2} = T_0(x) + 2\displaystyle\sum_{k=1}^{\infty} T_k(x)t^k$

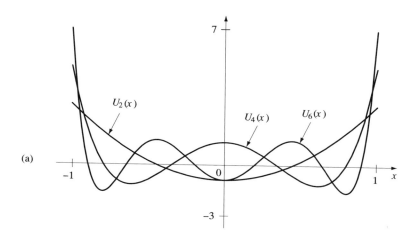

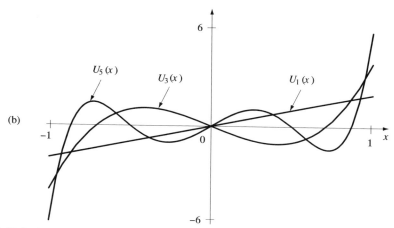

FIGURE 18.4 ▌ Chebyshev polynomials $U_n(x)$: (a) even polynomials and (b) odd polynomials.

$$2. \quad \frac{1}{1 - 2tx + t^2} = \sum_{k=0}^{\infty} U_k(x)t^k.$$

18.4 Laguerre Polynomials $L_n(x)$

18.4.1 Differential equation satisfied by $L_n(x)$

18.4.1.1   The **Laguerre polynomials** $L_n(x)$ satisfy the differential equation

$$1. \quad x\frac{d^2y}{dx^2} + (1 - x)\frac{dy}{dx} + ny = 0$$

defined on the interval $0 \leq x < \infty$, with $n = 0, 1, 2, \ldots$.

18.4.2 Rodrigue's formula for $L_n(x)$

18.4.2.1   The Laguerre Polynomial $L_n(x)$ of degree n is given by the following **Rodrigue's formula:**

1. $L_n(x) = \dfrac{e^x}{n!} \dfrac{d^n}{dx^n}[e^{-x}x^n]$

A definition also in use in place of 18.4.2.1.1 is

2. $L_n(x) = e^x \dfrac{d^n}{dx^n}[e^{-x}x^n].$

This leads to the omission of the scale factor $1/n!$ in 18.4.4.1 and to a modification of the recurrence relations in 18.4.5.1.

18.4.3 Orthogonality relation for $L_n(x)$

18.4.3.1   The **weight function** for Laguerre polynomials is $w(x) = e^{-x}$, and the **orthogonality relation** is

1. $\displaystyle\int_0^\infty e^{-x} L_m(x)L_n(x)\,dx = \begin{cases} 0, & m \neq n \\ 1, & m = n. \end{cases}$

18.4.4 Explicit expressions for $L_n(x)$

18.4.4.1
1. $L_0(x) = 1$
2. $L_1(x) = 1 - x$
3. $L_2(x) = \frac{1}{2!}(2 - 4x + x^2)$
4. $L_3(x) = \frac{1}{3!}(6 - 18x + 9x^2 - x^3)$
5. $L_4(x) = \frac{1}{4!}(24 - 96x + 72x^2 - 16x^3 + x^4)$
6. $L_5(x) = \frac{1}{5!}(120 - 600x + 600x^2 - 200x^3 + 25x^4 - x^5)$
7. $L_6(x) = \frac{1}{6!}(720 - 4320x + 5400x^2 - 2400x^3 + 450x^4 - 36x^5 + x^6)$
8. $L_7(x) = \frac{1}{7!}(5040 - 35280x + 52920x^2 - 29400x^3 + 7350x^4 - 882x^5 + 49x^6 - x^7)$

18.4.5 Recurrence relations satisfied by $L_n(x)$

18.4.5.1
1. $(n + 1)L_{n+1}(x) = (2n + 1 - x)L_n(x) - nL_{n-1}(x)$

2. $x\dfrac{d}{dx}[L_n(x)] = nL_n(x) - nL_{n-1}(x)$

18.4.6 Generating function for $L_n(x)$

18.4.6.1 The Laguerre polynomial $L_n(x)$ occurs as the multiplier of t^n in the expansion of the **generating function:**

1. $\dfrac{1}{(1-t)} \exp\left(\dfrac{xt}{t-1}\right) = \displaystyle\sum_{k=0}^{\infty} L_k(x) t^k.$

18.5 Hermite Polynomials $H_n(x)$

18.5.1 Differential equation satisfied by $H_n(x)$

18.5.1.1 The **Hermite polynomials** $H_n(x)$ satisfy the differential equation

1. $\dfrac{d^2 y}{dx^2} - 2x \dfrac{dy}{dx} + 2ny = 0$

defined on the interval $-\infty < x < \infty$, with $n = 0, 1, 2, \ldots$.

18.5.2 Rodrigue's formula for $H_n(x)$

18.5.2.1 The Hermite polynomial $H_n(x)$ of degree n is given by the following Rodrigue's formula:

1. $H_n(x) = (-1)^n e^{x^2} \dfrac{d^n}{dx^n} \left[e^{-x^2} \right].$

18.5.3 Orthogonality relation for $H_n(x)$

18.5.3.1 The **weight function** $w(x)$ for the Hermite polynomials $H_n(x)$ is $w(x) = e^{-x^2}$, and the **orthogonality relation** is

1. $\displaystyle\int_{-\infty}^{\infty} e^{-x^2} H_m(x) H_n(x)\, dx = \begin{cases} 0, & m \neq n \\ \sqrt{\pi}\, 2^n n!, & m = n. \end{cases}$

18.5.4 Explicit expressions for $H_n(x)$

18.5.4.1

1. $H_0(x) = 1$
2. $H_1(x) = 2x$
3. $H_2(x) = 4x^2 - 2$
4. $H_3(x) = 8x^3 - 12x$
5. $H_4(x) = 16x^4 - 48x^2 + 12$
6. $H_5(x) = 32x^5 - 160x^3 + 120x$
7. $H_6(x) = 64x^6 - 480x^4 + 720x^2 - 120$
8. $H_7(x) = 128x^7 - 1344x^5 + 3360x^3 - 1680x$
9. $H_8(x) = 256x^8 - 3584x^6 + 13440x^4 - 13440x^2 + 1680$

18.5.5 Recurrence relations satisfied by $H_n(x)$

18.5.5.1

1. $H_{n+1}(x) = 2x H_n(x) - 2n H_{n-1}(x)$

2. $\dfrac{d}{dx}[H_n(x)] = 2n H_{n-1}(x)$

3. $n H_n(x) = -n \dfrac{d}{dx}[H_{n-1}(x)] + x \dfrac{d}{dx}[H_n(x)]$

4. $H_n(x) = 2x H_{n-1}(x) - \dfrac{d}{dx}[H_{n-1}(x)]$

18.5.6 Generating function for $H_n(x)$

18.5.6.1   The Hermite polynomial $H_n(x)$ occurs as the multiplier of $t^n/n!$ in the expansion of the **generating function:**

1. $\exp(-t^2 + 2tx) = \displaystyle\sum_{k=0}^{\infty} H_k(x) \dfrac{t^k}{k!}.$

19

Laplace Transformation

19.1 Introduction

19.1.1 Definition of the Laplace transform

19.1.1.1   The **Laplace transform** of the function $f(x)$, denoted by $F(s)$, is defined as the improper integral

1. $F(s) = \displaystyle\int_0^\infty f(x)e^{-sx}\,dx \qquad [\mathrm{Re}\{s\} > 0].$

The functions $f(x)$ and $F(s)$ are called a **Laplace transform pair,** and knowledge of either one enables the other to be recovered.

    If f can be integrated over all finite intervals, and there is a constant c for which

2. $\displaystyle\int_0^\infty f(x)e^{-c|x|}\,dx,$

is finite, then the Laplace transform exists when $s = \sigma + i\tau$ is such that $a \geq c$.

    Setting

3. $F(s) = \mathcal{L}[f(x); s],$

to emphasize the nature of the transform, we have the symbolic inverse result

4. $f(x) = \mathcal{L}^{-1}[F(s); x]$.

The inversion of the Laplace transform is accomplished for analytic functions $F(s)$ that behave asymptotically like s^{-k} (or order $O(s^{-k})$) by means of the **inversion integral**

5. $f(x) = \dfrac{1}{2\pi i} \displaystyle\int_{\gamma - i\infty}^{\gamma + i\infty} F(s)e^{sx}\, dx,$

where γ is a real constant that exceeds the real part of all the singularities of $F(s)$.

19.1.2 Basic properties of the Laplace transform

19.1.2.1

1.   For a and b arbitrary constants,

$$\mathcal{L}[af(x) + bg(x); s] = aF(s) + bG(s) \qquad \text{(linearity).}$$

2.   If $n > 0$ is an integer and $\lim_{x \to \infty} f(x)e^{-sx} = 0$, then for $x > 0$,

$$\mathcal{L}[f^{(n)}(x); s] = s^n F(s) - s^{n-1} f(0) - s^{n-2} f^{(1)}(0) - \cdots - f^{(n-1)}(0)$$

$$\text{(transform of a derivative).}$$

3.   If $\lim_{x \to \infty}[e^{-sx} \int_0^x f(\xi)\, d\xi] = 0$, then

$$\mathcal{L}\left[\int_0^x f(\xi)\, d\xi; s\right] = \frac{1}{s} F(s) \qquad \text{(transform of an integral).}$$

4. $\mathcal{L}[e^{-ax} f(x); s] = F(s + a)$ \qquad (first shift theorem).

5.   Let $\mathcal{L}[f(x); s] = F(s)$ for $s > s_0$, and take $a > 0$ to be an arbitrary nonnegative number.
     Then, if

$$H(x - a) = \begin{cases} 0, & x < a \\ 1, & x \geq a \end{cases}$$

is the **Heaviside step function,**

$$\mathcal{L}[H(x - a) f(x - a); s] = e^{-as} F(s) \qquad [s > s_0]$$

$$\text{(second shift theorem).}$$

6.   Let $\mathcal{L}[f(x); s] = F(s)$ for $s > s_0$. Then,

$$\mathcal{L}[(-x)^n f(x); s] = \frac{d^n}{ds^n}[F(s)] \quad [s > s_0] \quad \text{(differentiation of a transform).}$$

7. Let $f(x)$ be a piecewise-continuous function for $x \geq 0$ and periodic with period X. Then

$$\mathcal{L}[f(x); s] = \frac{1}{1 - e^{-sX}} \int_0^X f(x)e^{-sx}\,dx$$

(transform of a periodic function).

8. The **Laplace convolution** $f * g$ of two functions $f(x)$ and $g(x)$ is defined as the integral

$$f * g(x) = \int_0^x f(x - \xi)g(\xi)\,d\xi,$$

and it has the property that $f * g = g * f$ and $f * (g * h) = (f * g) * h$. In terms of the convolution operation

$$\mathcal{L}[f * g(x); s] = F(s)G(s) \qquad \text{[convolution (Faltung) theorem].}$$

19.1.3 The Dirac delta function $\delta(x)$

19.1.3.1　　The **Dirac delta function** $\delta(x)$, which is particularly useful when working with the Laplace transform, has the following properties:

1. $\delta(x - a) = 0 \qquad [x \neq a]$

2. $\displaystyle\int_{-\infty}^{\infty} \delta(x - a)\,dx = 1$

3. $\displaystyle\int_{-\infty}^{x} \delta(\zeta - a)\,d\xi = H(x - a),$

   where $H(x - a)$ is the Heaviside step function defined in 19.1.2.1.5.

4. $\displaystyle\int_{-\infty}^{\infty} f(x)\delta(x - a)\,dx = f(a)$

The delta function, which can be regarded as an impulse function, is not a function in the usual sense, but a generalized function or distribution.

19.1.4 Laplace transform pairs

Table 19.1 lists Laplace transform pairs, and it can either be used to find the Laplace transform $F(s)$ of a function $f(x)$ shown in the left-hand column or, conversely, to find the inverse Laplace transform $f(x)$ of a given Laplace transform $F(s)$ shown in the right-hand column. To assist in the task of determining inverse Laplace transforms, many commonly occurring Laplace transforms $F(s)$ of an algebraic nature have been listed together with their more complicated inverse Laplace transforms $f(x)$.

The list of both Laplace transforms and inverse transforms may be extended by appeal to the theorems listed in 19.1.2.

TABLE 19.1   Table of Laplace transform pairs.

$f(x)$	$F(s)$
1. 1	$1/s$
2. x^n, $n = 0, 1, 2, \ldots$	$n!/s^{n+1}$, $\mathrm{Re}\{s\} > 0$
3. x^ν, $\nu > -1$	$\Gamma(\nu + 1)/s^{\nu+1}$, $\mathrm{Re}\{s\} > 0$
4. $x^{n-\frac{1}{2}}$	$\left(\dfrac{\sqrt{\pi}}{2}\right)\left(\dfrac{3}{2}\right)\left(\dfrac{5}{2}\right)\cdots\left(\dfrac{n-1}{2}\right)\Big/ s^{n+1/2}$, $\mathrm{Re}\{s\} > 0$
5. e^{-ax}	$1/(s + a)$, $\mathrm{Re}\{s\} > -\mathrm{Re}\{a\}$
6. xe^{-ax}	$1/(s + a)^2$, $\mathrm{Re}\{s\} > -\mathrm{Re}\{a\}$
7. $(e^{-ax} - e^{-bx})/(b - a)$	$(s + a)^{-1}(s + b)^{-1}$, $\mathrm{Re}\{s\} > \{-\mathrm{Re}\{a\}, -\mathrm{Re}\{b\}\}$
8. $(ae^{-ax} - be^{-bx})/(b - a)$	$s(s + a)^{-1}(s + b)^{-1}$, $-\mathrm{Re}\{s\} > \{-\mathrm{Re}\{a\}, -\mathrm{Re}\{b\}\}$
9. $(e^{ax} - 1)/a$	$s^{-1}(s - a)^{-1}$, $\mathrm{Re}\{s\} > \mathrm{Re}\{a\}$
10. $(e^{ax} - ax - 1)/a^2$	$s^{-2}(s - a)^{-1}$, $\mathrm{Re}\{s\} > \mathrm{Re}\{a\}$
11. $\left(e^{ax} - \frac{1}{2}a^2x^2 - ax - 1\right)\Big/a^3$	$s^{-3}(s - a)^{-1}$, $\mathrm{Re}\{s\} > \mathrm{Re}\{a\}$
12. $(1 + ax)e^{ax}$	$s(s - a)^{-2}$, $\mathrm{Re}\{s\} > \mathrm{Re}\{a\}$
13. $[1 + (ax - 1)e^{ax}]/a^2$	$s^{-1}(s - a)^{-2}$, $\mathrm{Re}\{s\} > \mathrm{Re}\{a\}$
14. $[2 + ax + (ax - 2)e^{ax}]/a^3$	$s^{-2}(s - a)^{-2}$, $\mathrm{Re}\{s\} > \mathrm{Re}\{a\}$
15. $x^n e^{ax}$, $n = 0, 1, 2, \ldots$	$n!(s - a)^{-(n+1)}$, $\mathrm{Re}\{s\} > \mathrm{Re}\{a\}$
16. $\left(x + \frac{1}{2}ax^2\right)e^{ax}$	$s(s - a)^{-3}$, $\mathrm{Re}\{s\} > \mathrm{Re}\{a\}$
17. $\left(1 + 2ax + \frac{1}{2}a^2x^2\right)e^{ax}$	$s^2(s - a)^{-3}$, $\mathrm{Re}\{s\} > \mathrm{Re}\{a\}$
18. $\frac{1}{6}x^3 e^{ax}$	$(s - a)^{-4}$, $\mathrm{Re}\{s\} > \mathrm{Re}\{a\}$
19. $\left(\frac{1}{2}x^2 + \frac{1}{6}ax^3\right)e^{ax}$	$s(s - a)^{-4}$, $\mathrm{Re}\{s\} > \mathrm{Re}\{a\}$
20. $\left(x + ax^2 + \frac{1}{6}a^2x^3\right)e^{ax}$	$s^2(s - a)^{-4}$, $\mathrm{Re}\{s\} > \mathrm{Re}\{a\}$
21. $\left(1 + 3ax + \frac{3}{2}a^2x^2 + \frac{1}{6}a^3x^3\right)e^{ax}$	$s^3(s - a)^{-4}$, $\mathrm{Re}\{s\} > \mathrm{Re}\{a\}$
22. $(ae^{ax} - be^{bx})/(a - b)$	$s(s - a)^{-1}(s - b)^{-1}$, $\mathrm{Re}\{s\} > \{\mathrm{Re}\{a\}, \mathrm{Re}\{b\}\}$
23. $\left(\dfrac{1}{a}e^{ax} - \dfrac{1}{b}e^{bx} + \dfrac{1}{b} - \dfrac{1}{a}\right)\Big/(a - b)$	$s^{-1}(s - a)^{-1}(s - b)^{-1}$, $\mathrm{Re}\{s\} > \{\mathrm{Re}\{a\}, \mathrm{Re}\{b\}\}$
24. $x^{\nu-1}e^{-ax}$, $\mathrm{Re}\,\nu > 0$	$\Gamma(\nu)(s + a)^{-\nu}$, $\mathrm{Re}\{s\} > -\mathrm{Re}\{a\}$
25. $xe^{-x^2/(4a)}$, $\mathrm{Re}\{a\} > 0$	$2a - 2\pi^{\frac{1}{2}}a^{\frac{3}{2}}se^{as^2}\,\mathrm{erfc}(sa^{\frac{1}{2}})$
26. $\sin(ax)$	$a(s^2 + a^2)^{-1}$, $\mathrm{Re}\{s\} > |\mathrm{Im}\{a\}|$
27. $\cos(ax)$	$s(s^2 + a^2)^{-1}$, $\mathrm{Re}\{s\} > |\mathrm{Im}\{a\}|$
28. $|\sin(ax)|$, $a > 0$	$a(s^2 + a^2)^{-1}\coth\left(\dfrac{\pi s}{2a}\right)$, $\mathrm{Re}\{s\} > 0$

(Continues)

TABLE 19.1 (*Continued*)

$f(x)$	$F(s)$
29. $\lvert\cos(ax)\rvert, a > 0$	$(s^2 + a^2)^{-1}\left[s + a\,\mathrm{csch}\left(\dfrac{\pi}{2a}\right)\right], \mathrm{Re}\{s\} > 0$
30. $[1 - \cos(ax)]/a^2$	$s^{-1}(s^2 + a^2)^{-1}, \mathrm{Re}\{s\} > \lvert\mathrm{Im}\{a\}\rvert$
31. $[ax - \sin(ax)]/a^3$	$s^{-2}(s^2 + a^2)^{-1}, \mathrm{Re}\{s\} > \lvert\mathrm{Im}\{a\}\rvert$
32. $[\sin(ax) - ax\cos(ax)]/(2a^3)$	$(s^2 + a^2)^{-2}, \mathrm{Re}\{s\} > \lvert\mathrm{Im}\{a\}\rvert$
33. $[x\sin(ax)]/(2a)$	$s(s^2 + a^2)^{-2}, \mathrm{Re}\{s\} > \lvert\mathrm{Im}\{a\}\rvert$
34. $[\sin(ax) + ax\cos(ax)]/(2a)$	$s^2(s^2 + a^2)^{-2}, \mathrm{Re}\{s\} > \lvert\mathrm{Im}\{a\}\rvert$
35. $x\cos(ax)$	$(s^2 - a^2)(s^2 + a^2)^{-2}, \mathrm{Re}\{s\} > \lvert\mathrm{Im}\{a\}\rvert$
36. $[\cos(ax) - \cos(bx)]/(b^2 - a^2)$	$s(s^2 + a^2)^{-1}(s^2 + b^2)^{-1}, \mathrm{Re}\{s\} > \{\lvert\mathrm{Im}\{a\}\rvert, \lvert\mathrm{Im}\{b\}\rvert\}$
37. $\left[\frac{1}{2}a^2x^2 - 1 + \cos(ax)\right]/a^4$	$s^{-3}(s^2 + a^2)^{-1}, \mathrm{Re}\{s\} > \lvert\mathrm{Im}\{a\}\rvert$
38. $\left[1 - \cos(ax) - \frac{1}{2}ax\sin(ax)\right]/a^4$	$s^{-1}(s^2 + a^2)^{-2}, \mathrm{Re}\{s\} > \lvert\mathrm{Im}\{a\}\rvert$
39. $\left[\dfrac{1}{b}\sin(bx) - \dfrac{1}{a}\sin(ax)\right]/(a^2 - b^2)$	$(s^2 + a^2)^{-1}(s^2 + b^2)^{-1}, \mathrm{Re}\{s\} > \{\lvert\mathrm{Im}\{a\}\rvert, \lvert\mathrm{Im}\{b\}\rvert\}$
40. $\left[1 - \cos(ax) + \frac{1}{2}ax\sin(ax)\right]/a^2$	$s^{-1}(s^2 + a^2)^{-2}(2s^2 + a^2), \mathrm{Re}\{s\} > \lvert\mathrm{Im}\{a\}\rvert$
41. $[a\sin(ax) - b\sin(bx)]/(a^2 - b^2)$	$s^2(s^2 + a^2)^{-1}(s^2 + b^2)^{-1}, \mathrm{Re}\{s\} > \{\lvert\mathrm{Im}\{a\}\rvert, \lvert\mathrm{Im}\{b\}\rvert\}$
42. $\sin(a + bx)$	$(s\sin a + b\cos a)(s^2 + b^2)^{-1}, \mathrm{Re}\{s\} > \lvert\mathrm{Im}\{b\}\rvert$
43. $\cos(a + bx)$	$(s\cos a - b\sin a)(s^2 + b^2)^{-1}, \mathrm{Re}\{s\} > \lvert\mathrm{Im}\{b\}\rvert$
44. $\left[\dfrac{1}{a}\sinh(ax) - \dfrac{1}{b}\sin(bx)\right]/(a^2 + b^2)$	$(s^2 - a^2)^{-1}(s^2 + b^2)^{-1}, \mathrm{Re}\{s\} > \{\lvert\mathrm{Re}\{a\}\rvert, \lvert\mathrm{Im}\{b\}\rvert\}$
45. $[\cosh(ax) - \cos(bx)]/(a^2 + b^2)$	$s(s^2 - a^2)^{-1}(s^2 + b^2)^{-1}, \mathrm{Re}\{s\} > \{\lvert\mathrm{Re}\{a\}\rvert, \lvert\mathrm{Im}\{b\}\rvert\}$
46. $[a\sinh(ax) + b\sin(bx)]/(a^2 + b^2)$	$s^2(s^2 - a^2)^{-1}(s^2 + b^2)^{-1}, \mathrm{Re}\{s\} > \{\lvert\mathrm{Re}\{a\}\rvert, \lvert\mathrm{Im}\{b\}\rvert\}$
47. $\sin(ax)\sin(bx)$	$2abs[s^2 + (a - b)^2]^{-1}[s^2 + (a + b)^2]^{-1}, \mathrm{Re}\{s\} > \{\lvert\mathrm{Im}\{a\}\rvert, \lvert\mathrm{Im}\{b\}\rvert\}$
48. $\cos(ax)\cos(bx)$	$s(s^2 + a^2 + b^2)[s^2 + (a - b)^2]^{-1}[s^2 + (a+b)^2]^{-1}, \mathrm{Re}\{s\} > \{\lvert\mathrm{Im}\{a\}\rvert, \lvert\mathrm{Im}\{b\}\rvert\}$
49. $\sin(ax)\cos(bx)$	$a(s^2 + a^2 - b^2)[s^2 + (a - b)^2]^{-1}[s^2 + (a + b)^2]^{-1}\mathrm{Re}\{s\} > \{\lvert\mathrm{Im}\{a\}\rvert, \lvert\mathrm{Im}\{b\}\rvert\}$
50. $\sin^2(ax)$	$2a^2s^{-1}(s^2 + 4a^2)^{-1}, \mathrm{Re}\{s\} > \lvert\mathrm{Im}\{a\}\rvert$
51. $\cos^2(ax)$	$(s^2 + 2a^2)s^{-1}(s^2 + 4a^2)^{-1}, \mathrm{Re}\{s\} > \lvert\mathrm{Im}\{a\}\rvert$
52. $\sin(ax)\cos(ax)$	$a(s^2 + 4a^2)^{-1}, \mathrm{Re}\{s\} > \lvert\mathrm{Im}\{a\}\rvert$
53. $e^{-ax}\sin(bx)$	$b[(s + a)^2 + b^2]^{-1}, \mathrm{Re}\{s\} > \{-\mathrm{Re}\{a\}, \lvert\mathrm{Im}\{b\}\rvert\}$
54. $e^{-ax}\cos(bx)$	$(s + a)[(s + a)^2 + b^2]^{-1}, \mathrm{Re}\{s\} > \{-\mathrm{Re}\{a\}, \lvert\mathrm{Im}\{b\}\rvert\}$
55. $\sinh(ax)$	$a(s^2 - a^2)^{-1}, \mathrm{Re}\{s\} > \lvert\mathrm{Re}\{a\}\rvert$
56. $\cosh(ax)$	$s(s^2 - a^2)^{-1}, \mathrm{Re}\{s\} > \lvert\mathrm{Re}\{a\}\rvert$
57. $x^{\nu-1}\cosh(ax), \mathrm{Re}\{\nu\} > 0$	$\frac{1}{2}\Gamma(\nu)[(s - a)^{-\nu} + (s + a)^{-\nu}], \mathrm{Re}\{s\} > \lvert\mathrm{Re}\{a\}\rvert$
58. $x\sinh(ax)$	$2as(s^2 - a^2)^{-2}, \mathrm{Re}\{s\} > \lvert\mathrm{Re}\{a\}\rvert$
59. $x\cosh(ax)$	$(s^2 + a^2)(s^2 - a^2)^{-2}, \mathrm{Re}\{s\} > \lvert\mathrm{Re}\{a\}\rvert$
60. $\sinh(ax) - \sin(ax)$	$2a^3(s^4 - a^4)^{-1}, \mathrm{Re}\{s\} > \{\lvert\mathrm{Re}\{a\}\rvert, \lvert\mathrm{Im}\{a\}\rvert\}$

(*Continues*)

TABLE 19.1   (*Continued*)

$f(x)$	$F(s)$
61. $\cosh(ax) - \cos(ax)$	$2a^2 s(s^4 - a^4)^{-1}, \operatorname{Re}\{s\} > \{|\operatorname{Re}\{a\}|, |\operatorname{Im}\{a\}|\}$
62. $\sinh(ax) + ax\cosh(ax)$	$2as^2(a^2 - s^2)^{-2}, \operatorname{Re}\{s\} > |\operatorname{Re}\{a\}|$
63. $ax\cosh(ax) - \sinh(ax)$	$2a^3(a^2 - s^2)^{-2}, \operatorname{Re}\{s\} > |\operatorname{Re}\{a\}|$
64. $x\sinh(ax) - \cosh(ax)$	$s(a^2 + 2a - s^2)(a^2 - s^2)^{-2}, \operatorname{Re}\{s\} > |\operatorname{Re}\{a\}|$
65. $\left[\dfrac{1}{a}\sinh(ax) - \dfrac{1}{b}\sinh(bx)\right]\!\Big/(a^2 - b^2)$	$(a^2 - s^2)^{-1}(b^2 - s^2)^{-1}, \operatorname{Re}\{s\} > \{|\operatorname{Re}\{a\}|, |\operatorname{Re}\{b\}|\}$
66. $[\cosh(ax) - \cosh(bx)]/(a^2 - b^2)$	$s(s^2 - a^2)^{-1}(s^2 - b^2)^{-1}, \operatorname{Re}\{s\} > \{|\operatorname{Re}\{a\}|, |\operatorname{Re}\{b\}|\}$
67. $[a\sinh(ax) - b\sinh(bx)]/(a^2 - b^2)$	$s^2(s^2 - a^2)^{-1}(s^2 - b^2)^{-1}, \operatorname{Re}\{s\} > \{|\operatorname{Re}\{a\}|, |\operatorname{Re}\{b\}|\}$
68. $\sinh(a + bx)$	$(b\cosh a + s\sinh a)(s^2 - b^2)^{-1}, \operatorname{Re}\{s\} > |\operatorname{Re}\{b\}|$
69. $\cosh(a + bx)$	$(s\cosh a + b\sinh a)(s^2 - b^2)^{-1}, \operatorname{Re}\{s\} > |\operatorname{Re}\{b\}|$
70. $\sinh(ax)\sinh(bx)$	$2abs[s^2 - (a + b)^2]^{-1}[s^2 - (a - b)^2]^{-1}, \operatorname{Re}\{s\} > \{|\operatorname{Re}\{a\}|, |\operatorname{Re}\{b\}|\}$
71. $\cosh(ax)\cosh(bx)$	$s(s^2 - a^2 - b^2)[s^2 - (a + b)^2]^{-1}[s^2 - (a - b)^2]^{-1}, \operatorname{Re}\{s\} > \{|\operatorname{Re}\{a\}|, |\operatorname{Re}\{b\}|\}$
72. $\sinh(ax)\cosh(bx)$	$a(s^2 - a^2 + b^2)[s^2 - (a + b)^2]^{-1}[s^2 - (a-b)^2]^{-1}, \operatorname{Re}\{s\} > \{|\operatorname{Re}\{a\}|, |\operatorname{Re}\{b\}|$
73. $\sinh^2(ax)$	$2a^2 s^{-1}(s^2 - 4a^2)^{-1}, \operatorname{Re}\{s\} > |\operatorname{Re}\{a\}|$
74. $\cosh^2(ax)$	$(s^2 - 2a^2)^{-1}(s^2 - 4a^2)^{-1}, \operatorname{Re}\{s\} > |\operatorname{Re}\{a\}|$
75. $\sinh(ax)\cosh(ax)$	$a(s^2 - 4a^2)^{-1}, \operatorname{Re}\{s\} > |\operatorname{Re}\{a\}|$
76. $[\cosh(ax) - 1]/a^2$	$s^{-1}(s^2 - a^2)^{-1}, \operatorname{Re}\{s\} > |\operatorname{Re}\{a\}|$
77. $[\sinh(ax) - ax]/a^3$	$s^{-2}(s^2 - a^2)^{-1}, \operatorname{Re}\{s\} > |\operatorname{Re}\{a\}|$
78. $\left[\cosh(ax) - \tfrac{1}{2}a^2 x^2 - 1\right]\!\Big/a^4$	$s^{-3}(s^2 - a^2)^{-1}, \operatorname{Re}\{s\} > |\operatorname{Re}\{a\}|$
79. $\left[1 - \cosh(ax) + \tfrac{1}{2}ax\sinh(ax)\right]\!\Big/a^4$	$s^{-1}(s^2 - a^2)^{-2}, \operatorname{Re}\{s\} > |\operatorname{Re}\{a\}|$
80. $H(x - a) = \begin{cases} 0, & x < a \\ 1, & x > a \end{cases}$ (Heaviside step function)	$s^{-1}e^{-as}, a \geq 0$
81. $\delta(x)$ (Dirac delta function)	1
82. $\delta(x - a)$	$e^{-as}, a \geq 0$
83. $\delta'(x - a)$	$se^{-as}, a \geq 0$
84. $\operatorname{erf}\left(\dfrac{x}{2a}\right) = \dfrac{2}{\sqrt{\pi}} \int_0^{x/(2a)} e^{-t^2}\, dt$	$s^{-1}e^{a^2 s^2}\operatorname{erfc}(as), \operatorname{Re}\{s\} > 0, |\arg a| < \pi/4$
85. $\operatorname{erf}(a\sqrt{x})$	$as^{-1}(s + a^2)^{-1/2}, \operatorname{Re}\{s\} > \{0, -\operatorname{Re}\{a^2\}\}$
86. $\operatorname{erfc}(a\sqrt{x})$	$1 - a(s + a^2)^{-1/2}, \operatorname{Re}\{s\} > 0$
87. $\operatorname{erfc}(a/\sqrt{x})$	$s^{-1}e^{-2\sqrt{s}}, \operatorname{Re}\{s\} > 0, \operatorname{Re}\{a\} > 0$
88. $J_\nu(ax)$	$a^\nu[s + (s^2 + a^2)^{1/2}]^{-\nu}(s^2 + a^2)^{-1/2}, \operatorname{Re}\{s\} > |\operatorname{Im}\{a\}|, \operatorname{Re}\{\nu\} > -1$
89. $xJ_\nu(ax)$	$a^\nu[s + \nu(s^2 + a^2)^{1/2}][s + (s^2 + a^2)^{1/2}]^{-\nu}, (s^2 + a^2)^{-3/2},$ $\operatorname{Re}\{s\} > |\operatorname{Im}\{a\}|, \operatorname{Re}\{\nu\} > -2$

(*Continues*)

TABLE 19.1   (*Continued*)

$f(x)$	$F(s)$
90. $x^{-1}J_\nu(ax)$	$a^\nu \nu^{-1}[s + (s^2 + a^2)^{1/2}]^{-\nu}$, $\mathrm{Re}\{s\} \geq |\mathrm{Im}\{a\}|$
91. $x^n J_n(ax)$	$1 \cdot 3 \cdot 5 \cdots (2n - 1)a^n(s^2 + a^2)^{-(n+1/2)}$, $\mathrm{Re}\{s\} \geq |\mathrm{Im}\{a\}|$
92. $x^\nu J_\nu(ax)$	$2^\nu \pi^{-1/2}\Gamma\left(\nu + \frac{1}{2}\right)a^\nu(S^2 + a^2)^{-(\nu+1/2)}\mathrm{Re}\{s\} > |\mathrm{Im}\{a\}|$, $\mathrm{Re}\{\nu\} > -\frac{1}{2}$
93. $x^{\nu+1}J_\nu(ax)$	$2^{\nu+1}\pi^{-1/2}\Gamma\left(\nu + \frac{3}{2}\right)a^\nu s(s^2 + a^2)^{-(\nu+3/2)}$, $\mathrm{Re}\{s\} > |\mathrm{Im}\{a\}|$, $\mathrm{Re}\{\nu\} > -1$
94. $I_\nu(ax)$	$a^\nu(s^2 + a^2)^{-1/2}[s + (s^2 - a^2)^{1/2}]^{-\nu}$, $\mathrm{Re}\{s\} > |\mathrm{Re}\{a\}|$, $\mathrm{Re}\{\nu\} > -1$
95. $x^\nu I_\nu(ax)$	$2^\nu \pi^{-1/2}\Gamma\left(\nu + \frac{1}{2}\right)a^\nu(s^2 - a^2)^{-(\nu+1/2)}$, $\mathrm{Re}\{s\} > |\mathrm{Re}\{a\}|$, $\mathrm{Re}\{\nu\} > -\frac{1}{2}$
96. $x^{-(\nu+1)}I_\nu(ax)$	$2^{\nu+1}\pi^{-1/2}\Gamma\left(\nu + \frac{3}{2}\right)a^\nu s(s^2 - a^2)^{-(\nu+3/2)}$, $\mathrm{Re}\{s\} > |\mathrm{Re}\{a\}|$, $\mathrm{Re}\{\nu\} > -1$
97. $x^{-1}I_\nu(ax)$	$\nu^{-1}a^\nu[s + (s^2 - a^2)^{1/2}]^{-\nu}\mathrm{Re}\{s\} > |\mathrm{Re}\{a\}|$, $\mathrm{Re}\{\nu\} > 0$
98. $J_0(ax) - axJ_1(ax)$	$s^2(s^2 + a^2)^{-3/2}$, $\mathrm{Re}\{s\} > |\mathrm{Im}\{a\}|$
99. $I_0(ax) + axI_1(ax)$	$s^2(s^2 - a^2)^{-3/2}$, $\mathrm{Re}\{s\} > |\mathrm{Im}\{a\}|$

20

Fourier Transforms

20.1 Introduction

20.1.1 Fourier exponential transform

20.1.1.1 Let $f(x)$ be a bounded function such that in any interval (a, b) it has only a finite number of maxima and minima and a finite number of discontinuities (it satisfies the **Dirichlet conditions**). Then if $f(x)$ is absolutely integrable on $(-\infty, \infty)$, so that

$$\int_{-\infty}^{\infty} |f(x)|\, dx < \infty,$$

the **Fourier transform** of $f(x)$, also called the **exponential Fourier transform,** is the $F(\omega)$ defined as

1. $F(\omega) = \dfrac{1}{\sqrt{2\pi}} \displaystyle\int_{-\infty}^{\infty} f(x)e^{i\omega x}\, dx.$

The functions $f(x)$ and $F(\omega)$ are called a **Fourier transform pair,** and knowledge of either one enables the other to be recovered.

Setting

2. $F(\omega) = \mathcal{F}[f(x); \omega]$,

where $\mathcal{F}$ is used to denote the operation of finding the Fourier transform, we have the symbolic inverse result

3. $f(x) = \mathcal{F}^{-1}[f(\omega); x]$.

The inversion of the Fourier transform is accomplished by means of the **inversion integral**

4. $\dfrac{1}{2}[f(x+) + f(x-)] = \dfrac{1}{\sqrt{2\pi}} \displaystyle\int_{-\infty}^{\infty} F(\omega)e^{-i\omega x}\, d\omega,$

where $f(a+)$ and $f(a-)$ signify the values of $f(x)$ to the immediate right and left, respectively, of $x = a$.

At points of continuity of $f(x)$ the above result reduces to

5. $f(x) = \dfrac{1}{\sqrt{2\pi}} \displaystyle\int_{-\infty}^{\infty} F(\omega)e^{-i\omega x}\, d\omega.$

20.1.2 Basic properties of the Fourier transforms

20.1.2.1

1.   For a, b arbitrary constants and $f(x), g(x)$ functions with the respective Fourier transforms $F(\omega), G(\omega)$,

$$\mathcal{F}[af(x) + bg(x); \omega] = aF(\omega) + bG(\omega) \qquad \text{(linearity).}$$

2.   If $a \neq 0$ is an arbitrary constant, then

$$\mathcal{F}[f(ax); \omega] = \frac{1}{|a|}F\left(\frac{\omega}{a}\right) \qquad \text{(scaling).}$$

3.   For any real constant a,

$$\mathcal{F}[f(x - a); \omega] = e^{i\omega a} F(\omega) \qquad \text{(spatial shift).}$$

4.   For any real constant a,

$$\mathcal{F}[e^{iax} f(x); \omega] = F(\omega + a) \qquad \text{(frequency shift).}$$

5.   If $n > 0$ is an integer, $f^{(n)}(x)$ is piecewise-continuously differentiable, and each of the derivatives $f^{(r)}(x)$ with $r = 0, 1, \ldots, n$ is absolutely integrable for $-\infty < x < \infty$, then

$$\mathcal{F}[f^{(n)}(x); \omega] = (-i\omega)^n F(\omega) \qquad \text{(differentiation).}$$

6. $\displaystyle\int_{-\infty}^{\infty} |F(\omega)|^2\, d\omega = \int_{-\infty}^{\infty} |f(x)|^2\, dx.$ \qquad (Parseval's relation).

7.   The **Fourier convolution** of two integrable functions $f(x)$ and $g(x)$ is defined as

$$f * (g(x)) = \frac{1}{\sqrt{2\pi}} \int_{-\infty}^{\infty} f(x - u)g(u)\, du,$$

and it has the following properties

(i) $f * (kg(x)) = (kf) * (g(x)) = kf * (g(x))$ (scaling; k = const.)

(ii) $f * (g(x) + h(x)) = f * (g(x)) + f * (h(x))$ (linearity)

(iii) $f * (g(x)) = g * (f(x))$ (commutability)

If $f(x)$, $g(x)$ have the respective Fourier transforms $F(\omega)$, $G(\omega)$,

$$\mathcal{F}[f * (g(x)); \omega] = F(\omega)G(\omega)$$ [convolution (Faltung) theorem].

20.1.3 Fourier transform pairs

20.1.3.1   Table 20.1 lists some elementary Fourier transform pairs, and it may either be used to find the Fourier transform $F(\omega)$ of a function $f(x)$ shown in the left-hand column or, conversely, to find the inverse Fourier transform $f(x)$ of a given Fourier transform $F(\omega)$ shown in the right-hand column. The list may be extended by appeal to the properties given in 20.1.2.

Care is necessary when using general tables of Fourier transform pairs because of the different choices made for the numerical normalization factors multiplying the integrals, and the signs of the exponents in the integrands. The product of the numerical factors multiplying the transform and its inverse need only equal $1/(2\pi)$, so the Fourier transform and its inverse may be defined as

1. $F(\omega) = \dfrac{\alpha}{2\pi} \displaystyle\int_{-\infty}^{\infty} f(x)e^{i\omega x}\, dx$

and

2. $\dfrac{1}{2}[f(x+) + f(x-)] = \dfrac{1}{\alpha} \displaystyle\int_{-\infty}^{\infty} F(\omega)e^{-i\omega x}\, d\omega,$

where α is an arbitrary real number. Throughout this section we have set $\alpha = \sqrt{2\pi}$, but other reference works set $\alpha = 2\pi$ or $\alpha = 1$. Another difference in the notation used elsewhere involves the choice of the sign prefixing i in 20.1.3.1.1 and 20.1.3.1.2, which is sometimes reversed.

In many physical applications of the Fourier integral it is convenient to write $\omega = 2\pi n$ and $\alpha = 2\pi$, when 20.1.3.1.1 and 20.1.3.1.2 become

3. $F(n) = \displaystyle\int_{-\infty}^{\infty} f(x)e^{2\pi i n x}\, dx$

and

4. $\dfrac{1}{2}[f(x+) + f(x-)] = \displaystyle\int_{-\infty}^{\infty} F(n)e^{-2\pi i n x}\, dn.$

20.1.4 Fourier cosine and sine transforms

20.1.4.1   When a function $f(x)$ satisfies the Dirichlet conditions of 20.1.1, and it is either an even function or it is only defined for positive values of x, its Fourier cosine transform is defined as

TABLE 20.1   Table of Fourier Transform Pairs

$$f(x) = \frac{1}{\sqrt{2\pi}} \int_{-\infty}^{\infty} F(\omega)e^{-i\omega x}\, d\omega \qquad F(\omega) = \frac{1}{\sqrt{2\pi}} \int_{-\infty}^{\infty} f(x)e^{i\omega x}\, dx$$

	$f(x)$		$F(\omega)$
1.	$\dfrac{1}{a^2 + x^2}$	$[a > 0]$	$\sqrt{\dfrac{\pi}{2}}\,\dfrac{e^{-a|\omega|}}{a}$
2.	$\dfrac{x}{a^2 + x^2}$	$[a > 0]$	$-\dfrac{i}{2}\sqrt{\dfrac{\pi}{2}}\,\dfrac{\omega e^{-a|\omega|}}{a}$
3.	$\dfrac{1}{x(a^2 + x^2)}$		$\sqrt{\dfrac{\pi}{2}}\,\dfrac{i(1 - e^{-\omega|a|})}{a^2}$
4.	$f(x) = \begin{cases} 1, & |x| < a \\ 0, & |x| > a \end{cases}$		$\sqrt{\dfrac{2}{\pi}}\,\dfrac{\sin a\omega}{\omega}$
5.	$\dfrac{1}{x}$		$i\sqrt{\dfrac{\pi}{2}}\,\mathrm{sgn}(\omega)$
6.	$\dfrac{1}{|x|}$		$\dfrac{1}{|\omega|}$
7.	$\dfrac{1}{x^n}$	$[n \text{ a positive integer }]$	$i\sqrt{\dfrac{\pi}{2}}\,\mathrm{sgn}(\omega)\,\dfrac{(i\omega)^{n-1}}{(n-1)!}$
8.	$x^n\,\mathrm{sgn}(x)$	$[n \text{ a positive integer}]$	$\sqrt{\dfrac{2}{\pi}}\,\dfrac{n!}{(-i\omega)^{n+1}}$
9.	$|x|^a$	$[a < 1 \text{ not a negative integer}]$	$\sqrt{\dfrac{2}{\pi}}\,\dfrac{\Gamma(a+1)}{|\omega|^{a+1}}\cos\left[\dfrac{\pi}{2}(a+1)\right]$
10.	$x^a H(x)$	$[a < 1 \text{ not a negative integer}]$	$\dfrac{\Gamma(a+1)}{\sqrt{2\pi}}\,\dfrac{\exp[-\pi i(a+1)\mathrm{sgn}(\omega)]}{|\omega|^{a+1}}$
11.	$e^{-a|x|}$	$[a > 0]$	$\sqrt{\dfrac{2}{\pi}}\,\dfrac{a}{a^2 + \omega^2}$
12.	$xe^{-a|x|}$	$[a > 0]$	$-\sqrt{\dfrac{2}{\pi}}\,\dfrac{2ia\omega}{(a^2 + \omega^2)^2}$
13.	$|x|e^{-a|x|}$	$[a > 0]$	$\sqrt{\dfrac{2}{\pi}}\,\dfrac{a^2 - \omega^2}{(a^2 + \omega^2)^2}$
14.	$e^{-ax} H(x)$	$[a > 0]$	$\dfrac{1}{\sqrt{2\pi}}\,\dfrac{1}{a - i\omega}$
15.	$e^{ax} H(-x)$	$[a > 0]$	$\dfrac{1}{\sqrt{2\pi}}\,\dfrac{1}{a + i\omega}$
16.	$xe^{-ax} H(x)$	$[a > 0]$	$\dfrac{1}{\sqrt{2\pi}}\,\dfrac{1}{(a - i\omega)^2}$
17.	$-xe^{ax} H(-x)$	$[a > 0]$	$\dfrac{1}{\sqrt{2\pi}}\,\dfrac{1}{(a + i\omega)^2}$
18.	$e^{-a^2 x^2}$	$[a > 0]$	$\dfrac{1}{a\sqrt{2}}\,e^{-\omega^2/(4a^2)}$
19.	$\delta(x - a)$		$\dfrac{1}{\sqrt{2\pi}}\,e^{-i\omega a}$
20.	$\delta(px + q)$	$[p \neq 0]$	$\dfrac{1}{|p|}\,\dfrac{1}{\sqrt{2\pi}}\,e^{-iq\omega/p}$

(Continues)

TABLE 20.1   (*Continued*)

$f(x) = \dfrac{1}{\sqrt{2\pi}} \displaystyle\int_{-\infty}^{\infty} F(\omega)e^{-i\omega x}\, d\omega$	$F(\omega) = \dfrac{1}{\sqrt{2\pi}} \displaystyle\int_{-\infty}^{\infty} f(x)e^{i\omega x}\, dx$
21. $\cos(ax^2)$	$\dfrac{1}{\sqrt{2a}}\cos\left(\dfrac{\omega^2}{4a} - \dfrac{\pi}{4}\right)$
22. $\sin(ax^2)$	$\dfrac{1}{\sqrt{2a}}\sin\left(\dfrac{\omega^2}{4a} + \dfrac{\pi}{4}\right)$
23. $x\operatorname{csch} x$	$\sqrt{(2\pi^3)}\,\dfrac{e^{\pi\omega}}{(1+e^{\pi\omega})^2}$
24. $\operatorname{sech}(ax)$ $[a > 0]$	$\dfrac{1}{a}\sqrt{\dfrac{\pi}{2}}\operatorname{sech}\left(\dfrac{\pi\omega}{2a}\right)$
25. $\tanh(ax)$ $[a > 0]$	$\dfrac{i}{a}\sqrt{\dfrac{\pi}{2}}\operatorname{csch}\left(\dfrac{\pi\omega}{2a}\right)$
26. $\sqrt{x}\,J_{-1/4}\left(\tfrac{1}{2}x^2\right)$	$\sqrt{\omega}\,J_{-1/4}\left(\tfrac{1}{2}\omega^2\right)$

1. $F_c(\omega) = \sqrt{\dfrac{2}{\pi}} \displaystyle\int_0^{\infty} f(x)\cos(\omega x)\, dx,$

and we write

2. $F_c(\omega) = \mathcal{F}_c[f(x); \omega].$

The functions $f(x)$ and $F_c(\omega)$ are called a **Fourier cosine transform pair,** and knowledge of either one enables the other to be recovered.

The **inversion integral** for the Fourier cosine transform is

3. $f(x) = \sqrt{\dfrac{2}{\pi}} \displaystyle\int_0^{\infty} F_c(\omega)\cos(\omega x)\, d\omega$ $[x > 0].$

Similarly, if $f(x)$ is either an odd function or it is only defined for positive values of x, its **Fourier sine transform** is defined as

4. $F_s(\omega) = \sqrt{\dfrac{2}{\pi}} \displaystyle\int_0^{\infty} f(x)\sin(\omega x)\, dx,$

and we write

5. $F_s(\omega) = \mathcal{F}_s[f(x); \omega].$

The functions $f(x)$ and $F_s(\omega)$ are called a **Fourier sine transform** pair and knowledge of either one enables the other to be recovered. The **inversion integral** for the Fourier sine transform is

6. $f(x) = \sqrt{\dfrac{2}{\pi}} \displaystyle\int_0^{\infty} F_s(\omega)\sin(\omega x)\, d\omega$ $[x > 0].$

20.1.5 Basic properties of the Fourier cosine and sine transforms

20.1.5.1

1. For a, b arbitrary constants and $f(x)$, $g(x)$ functions with the respective Fourier cosine and sine transforms $F_c(\omega)$, $G_c(\omega)$, $F_s(\omega)$, $G_s(\omega)$,

$$\mathcal{F}_c[af(x) + bg(x)] = aF_c(\omega) + bG_c(\omega)$$
$$\mathcal{F}_s[af(x) + bg(x)] = aF_s(\omega) + bG_s(\omega)$$

(linearity).

2. If $a > 0$, then

$$\mathcal{F}_c[f(ax); \omega] = \frac{1}{a}F_c\left(\frac{\omega}{a}\right)$$

$$\mathcal{F}_c[f(ax); \omega] = \frac{1}{a}F_s\left(\frac{\omega}{a}\right)$$

(scaling).

3. $\mathcal{F}_s[\cos(ax)f(x); \omega] = \frac{1}{2}[F_s(\omega + a) + F_s(\omega - a)]$

$\mathcal{F}_s[\sin(ax)f(x); \omega] = \frac{1}{2}[F_c(\omega - a) - F_c(\omega + a)]$

$\mathcal{F}_c[\cos(ax)f(x); \omega] = \frac{1}{2}[F_c(\omega + a) + F_c(\omega - a)]$

$\mathcal{F}_c[\sin(ax)f(x); \omega] = \frac{1}{2}[F_s(a + \omega) + F_s(a - \omega)]$

(frequency shift).

4. $\mathcal{F}_c[f'(x); \omega] = \omega F_s(\omega) - \sqrt{\dfrac{2}{\pi}}f(0)$

$\mathcal{F}_s[f'(x); \omega] = -\omega F_c(\omega)$

(differentiation).

5. $\mathcal{F}_c^{-1} = \mathcal{F}_c$

$\mathcal{F}_s^{-1} = \mathcal{F}_s$

(symmetry).

6. $\displaystyle\int_0^\infty |F_c(\omega)|^2\, d\omega = \int_0^\infty |f(x)|^2\, dx$

$\displaystyle\int_0^\infty |F_c(\omega)|^2\, d\omega = \int_0^\infty |f(x)|^2\, dx$

(Parseval's relation).

20.1.6 Fourier cosine and sine transform pairs

20.1.6.1 Tables 20.2 and 20.3 list some elementary Fourier cosine and sine transform pairs, and they may either be used to find the required transform of a function $f(x)$ in the left-hand column or, conversely, given either $F_c(\omega)$ or $F_s(\omega)$, to find the corresponding inverse transform $f(x)$.

Tables 20.2 and 20.3 may be extended by use of the properties listed in 20.1.5. As with the exponential Fourier transform, care must be taken when using other reference works because of the use of different choices for the normalizing factors multiplying the Fourier cosine and sine transforms and their inverses. The product of the normalizing factors multiplying either the Fourier cosine or sine transform and the corresponding inverse transform is $2/\pi$.

TABLE 20.2   Table of Fourier Cosine Pairs

$$f(x) = \sqrt{\frac{2}{\pi}} \int_0^\infty F_c(\omega) \cos(\omega x)\, d\omega \qquad F_c(\omega) = \sqrt{\frac{2}{\pi}} \int_0^\infty f(x) \cos(\omega x)\, dx$$

	$f(x)$		$F_c(\omega)$
1.	$1/\sqrt{x}$		$1/\sqrt{\omega}$
2.	x^{a-1}	$[0 < a < 1]$	$\sqrt{\dfrac{2}{\pi}} \dfrac{\Gamma(a) \cos(a\pi/2)}{\omega^a}$
3.	$f(x) = \begin{cases} 1, & 0 < x < a \\ 0, & x > a \end{cases}$		$\sqrt{\dfrac{2}{\pi}} \dfrac{\sin(a\omega)}{\omega}$
4.	$\dfrac{1}{a^2 + x^2}$	$[a > 0]$	$\sqrt{\dfrac{\pi}{2}} \dfrac{e^{-a\omega}}{a}$
5.	$\dfrac{1}{(a^2 + x^2)^2}$	$[a > 0]$	$\dfrac{1}{2}\sqrt{\dfrac{\pi}{2}} \dfrac{e^{-a\omega}(1 + a\omega)}{a^3}$
6.	e^{-ax}	$[a > 0]$	$\sqrt{\dfrac{2}{\pi}} \dfrac{a}{a^2 + \omega^2}$
7.	xe^{-ax}	$[a > 0]$	$\sqrt{\dfrac{2}{\pi}} \dfrac{a^2 - \omega^2}{(a^2 + \omega^2)^2}$
8.	$x^{n-1}e^{-ax}$	$[a > 0,\ n > 0]$	$\sqrt{\dfrac{2}{\pi}} \dfrac{\Gamma(n) \cos[n\arctan(\omega/a)]}{(a^2 + \omega^2)^{n/2}}$
9.	$e^{-a^2 x^2}$		$\dfrac{1}{|a|\sqrt{2}} e^{-\omega^2/(4a^2)}$
10.	$\cos(ax)^2$	$[a > 0]$	$\dfrac{1}{2}\dfrac{1}{\sqrt{a}}\left[\cos\dfrac{\omega^2}{4a} + \sin\dfrac{\omega^2}{4a}\right]$
11.	$\sin(ax)^2$	$[a > 0]$	$\dfrac{1}{2}\dfrac{1}{\sqrt{a}}\left[\cos\dfrac{\omega^2}{4a} - \sin\dfrac{\omega^2}{4a}\right]$
12.	$\operatorname{sech}(ax)$	$[a > 0]$	$\sqrt{\dfrac{\pi}{2}} \dfrac{\operatorname{sech}(\pi\omega/2a)}{a}$
13.	$x\operatorname{csch}x$		$\sqrt{2\pi^3} \dfrac{e^{\pi\omega}}{(1 + e^{\pi\omega})^2}$
14.	$\dfrac{e^{-bx} - e^{-ax}}{x}$	$[a > b]$	$\dfrac{1}{\sqrt{2\pi}} \ln\left(\dfrac{a^2 + \omega^2}{b^2 + \omega^2}\right)$
15.	$\dfrac{\sin(ax)}{x}$	$[a > 0]$	$\begin{cases} \sqrt{\dfrac{\pi}{2}}, & \omega < a \\ \dfrac{1}{2}\sqrt{\dfrac{\pi}{2}}, & \omega = a \\ 0, & \omega > a \end{cases}$
16.	$\dfrac{\sinh(ax)}{\sinh(bx)}$	$[0 < a < b]$	$\dfrac{1}{b}\sqrt{\dfrac{\pi}{2}} \dfrac{\sin(\pi a/b)}{\cosh(\pi\omega/b) + \cos(\pi a/b)}$
17.	$\dfrac{\cosh(ax)}{\cosh(bx)}$	$[0 < a < b]$	$\dfrac{\sqrt{2\pi}}{b} \dfrac{\cos(\pi a/2b)\cosh(\pi\omega/2b)}{\cosh(\pi\omega/b) + \cos(\pi\omega/b)}$
18.	$\dfrac{J_0(ax)}{b^2 + x^2}$	$[a, b > 0]$	$\sqrt{\dfrac{\pi}{2}} \dfrac{e^{-b\omega} I_0(ab)}{b} \quad (\omega > a)$

TABLE 20.3 Table of Fourier Sine Pairs

$$f(x) = \sqrt{\frac{2}{\pi}} \int_0^\infty F_s(\omega) \sin \omega x \, d\omega \qquad F_s(\omega) = \sqrt{\frac{2}{\pi}} \int_0^\infty f(x) \sin \omega x \, dx$$

	$f(x)$		$F_s(\omega)$
1.	$1/\sqrt{x}$		$1/\sqrt{\omega}$
2.	x^{a-1}	$[0 < a < 1]$	$\sqrt{\frac{2}{\pi}} \dfrac{\Gamma(a) \sin(a\pi/2)}{\omega^a}$
3.	$f(x) = \begin{cases} 1, & 0 < x < a \\ 0, & x > a \end{cases}$		$\sqrt{\frac{2}{\pi}} \dfrac{1 - \cos(a\omega)}{\omega}$
4.	$1/x$		$\sqrt{\dfrac{\pi}{2}} \operatorname{sgn}(\omega)$
5.	$\dfrac{x}{a^2 + x^2}$	$[a > 0]$	$\sqrt{\dfrac{\pi}{2}} e^{-a\omega} \qquad [\omega > 0]$
6.	$\dfrac{x}{(a^2 + x^2)^2}$	$[a > 0]$	$\sqrt{\dfrac{\pi}{8}} \dfrac{\omega e^{-a\omega}}{a} \qquad [\omega > 0]$
7.	$\dfrac{1}{x(a^2 + x^2)}$		$\sqrt{\dfrac{\pi}{2}} \dfrac{(1 - e^{-\omega|a|})}{a^2} \operatorname{sgn}(\omega)$
8.	$\dfrac{x^2 - a^2}{x(x^2 + a^2)}$		$\sqrt{2\pi} \left[e^{-|a\omega|} - \frac{1}{2} \right] \operatorname{sgn}(\omega)$
9.	e^{-ax}	$[a > 0]$	$\sqrt{\dfrac{2}{\pi}} \dfrac{\omega}{a^2 + \omega^2}$
10.	xe^{-ax}	$[a > 0]$	$\sqrt{\dfrac{2}{\pi}} \dfrac{2a\omega}{(a^2 + \omega^2)^2}$
11.	$xe^{-x^2/2}$		$\omega e^{-\omega^2/2}$
12.	$\dfrac{e^{-ax}}{x}$	$[a > 0]$	$\dfrac{2}{\pi} \arctan(\omega/a)$
13.	$x^{n-1} e^{-ax}$	$[a > 0, n > 0]$	$\sqrt{\dfrac{2}{\pi}} \dfrac{\Gamma(n) \sin[n \arctan(\omega/a)]}{(a^2 + \omega^2)^{n/2}}$
14.	$\dfrac{e^{-bx} - e^{-ax}}{x}$	$[a > b]$	$\sqrt{\dfrac{2}{\pi}} \arctan\left[\dfrac{(a - b)\omega}{ab + \omega^2} \right]$
15.	$\operatorname{csch} x$		$\sqrt{\dfrac{\pi}{2}} \tanh(\pi\omega/2)$
16.	$x \operatorname{csch} x$		$\sqrt{2\pi^3} \dfrac{e^{\pi\omega}}{(1 + e^{\pi\omega})^2}$
17.	$\dfrac{\sin(ax)}{x}$	$[a > 0]$	$\dfrac{1}{\sqrt{2\pi}} \ln\left(\dfrac{\omega + a}{\omega - a} \right)$
18.	$\dfrac{\sin(ax)}{x^2}$	$[a > 0]$	$\begin{cases} \sqrt{\dfrac{\pi}{2}} \omega, & \omega < a \\ \sqrt{\dfrac{\pi}{2}} a, & \omega > a \end{cases}$
19.	$\dfrac{\sinh(ax)}{\cosh(bx)}$	$[0 < a < b]$	$\dfrac{\sqrt{2\pi}}{b} \dfrac{\sin(\pi a/2b) \sinh(\pi\omega/b)}{\cos(\pi a/b) + \cosh(\pi\omega/b)}$
20.	$\dfrac{\cosh(ax)}{\sinh(bx)}$	$[0 < a < b]$	$\dfrac{1}{b} \sqrt{\dfrac{\pi}{2}} \dfrac{\sinh(\pi\omega/b)}{\cosh(\pi\omega/b) + \cos(\pi a/b)}$
21.	$\dfrac{x J_0(ax)}{(b^2 + x^2)}$	$[a, b > 0]$	$\sqrt{\dfrac{\pi}{2}} e^{-b\omega} I_0(a\omega) \qquad [\omega > a]$

21

Numerical Integration

21.1 Classical Methods

21.1.1 Open- and closed-type formulas

The **numerical integration (quadrature) formulas** that follow are of the form

1. $$I = \int_a^b f(x)\,dx = \sum_{k=0}^{n} w_k^{(n)} f(x_k) + R_n,$$

with $a \leq x_0 < x_1 < \cdots < x_n \leq b$. The weight coefficients $w_k^{(n)}$ and the abscissas x_k, with $k = 0, 1, \ldots, n$ are known numbers independent of $f(x)$ that are determined by the numerical integration method to be used and the number of points at which $f(x)$ is to be evaluated. The **remainder term,** R_n, when added to the summation, makes the result exact. Although, in general, the precise value of R_n is unknown, its analytical form is usually known and can be used to determine an upper bound to $|R_n|$ in terms of n.

An integration formula is said to be of the **closed type** if it requires the function values to be determined at the end points of the interval of integration, so that $x_0 = a$

and $x_n = b$. An integration formula is said to be of **open type** if $x_0 > a$ and $x_n < b$, so that in this case the function values are not required at the end points of the interval of integration.

Many fundamental numerical integration formulas are based on the assumption that an interval of integration contains a specified number of abscissas at which points the function must be evaluated. Thus, in the basic Simpson's rule, the interval $a \le x \le b$ is divided into two subintervals of equal length at the ends of which the function has to be evaluated, so that $f(x_0)$, $f(x_1)$, and $f(x_2)$ are required, with $x_0 = a$, $x_1 = \frac{1}{2}(a + b)$, and $x_2 = b$. To control the error, the interval $a \le x \le b$ is normally subdivided into a number of smaller intervals, to each of which the basic integration formula is then applied and the results summed to yield the numerical estimate of the integral. When this approach is organized so that it yields a single integration formula, it is usual to refer to the result as a **composite** integration formula.

21.1.2 Composite midpoint rule (open type)

1. $$\int_a^b f(x)\,dx = 2h \sum_{k=0}^{n/2} f(x_{2k}) + R_n,$$

   with n an even integer, $1 + (n/2)$ subintervals of length $h = (b - a)/(n + 2)$, $x_k = a + (k + 1)h$ for $k = -1, 0, 1, \ldots, n + 1$, and the remainder term

2. $$R_n = \frac{(b - a)h^2}{6} f^{(2)}(\xi) = \frac{(n + 2)h^3}{6} f^{(2)}(\xi),$$

   for some ξ such that $a < \xi < b$.

21.1.3 Composite trapezoidal rule (closed type)

1. $$\int_a^b f(x)\,dx = \frac{h}{2}\left[f(a) + f(b) + 2 \sum_{k=1}^{n-1} f(x_k) \right] + R_n,$$

   with n an integer (even or odd), $h = (b - a)/n$, $x_k = a + kh$ for $k = 0, 1, 2, \ldots, n$, and the remainder term

2. $$R_n = -\frac{(b - a)h^2}{12} f^{(2)}(\xi) = -\frac{nh^3}{12} f^{(2)}(\xi),$$

   for some ξ such that $a < \xi < b$.

21.1.4 Composite Simpson's rule (closed type)

1. $$\int_a^b f(x)\,dx = \frac{h}{3}\left[f(a) + f(b) + 2 \sum_{k=1}^{(n/2)-1} f(x_{2k}) + 4 \sum_{k=1}^{n/2} f(x_{2k-1}) \right] + R_n,$$

   with n an even integer, $h = (b - a)/n$, $x_k = a + kh$ for $k = 0, 1, 2, \ldots, n$, and the remainder term

2. $$R_n = -\frac{(b - a)h^4}{180} f^{(4)}(\xi) = -\frac{nh^5}{180} f^{(4)}(\xi),$$

   for some ξ such that $a < \xi < b$.

21.1.5 Newton–Cotes formulas

Closed types.

1. $\int_{x_0}^{x_3} f(x)\,dx = \frac{3h}{8}(f(x_0) + 3f(x_1) + 3f(x_2) + f(x_3)) + R,$

   $h = (x_3 - x_0)/3$, $x_k = x_0 + kh$ for $k = 0, 1, 2, 3$, and the remainder term

2. $R = -\frac{3h^5}{80} f^{(4)}(\xi),$

   for some ξ such that $x_0 < \xi < x_3$.   (**Simpson's $\frac{3}{8}$ rule**)

3. $\int_{x_0}^{x_4} f(x)\,dx = \frac{2h}{45}(7f(x_0) + 32f(x_1) + 12f(x_2) + 32f(x_3) + 7f(x_4)) + R,$

   $h = (x_4 - x_0)/4$, $x_k = x_0 + kh$ for $k = 0, 1, 2, 3, 4$, and the remainder term

4. $R = -\frac{8h^7}{945} f^{(6)}(\xi),$

   for some ξ such that $x_0 < \xi < x_4$.   (**Bode's rule**)

5. $\int_{x_0}^{x_5} f(x)\,dx = \frac{5h}{288}(19f(x_0) + 75f(x_1) + 50f(x_2) + 50f(x_3)$

   $\qquad\qquad\qquad + 75f(x_4) + 19f(x_5)) + R,$

   $h = (x_5 - x_0)/5$, $x_k = x_0 + kh$ for $k = 0, 1, 2, 3, 4, 5$, and the remainder term

6. $R = -\frac{275h^7}{12096} f^{(6)}(\xi),$

   for some ξ such that $x_0 < \xi < x_5$.

Open types.

7. $\int_{x_0}^{x_3} f(x)\,dx = \frac{3h}{2}(f(x_1) + f(x_2)) + R,$

   $h = (x_3 - x_0)/3$, $x_k = x_0 + kh$, $k = 1, 2$, and the remainder term

8. $R = \frac{h^3}{4} f^{(2)}(\xi),$

   for some ξ such that $x_0 < \xi < x_3$.

9. $\int_{x_0}^{x_4} f(x)\,dx = \frac{4h}{3}(2f(x_1) - f(x_2) + 2f(x_3)) + R,$

   $h = (x_4 - x_0)/4$, $x_k = x_0 + kh$ for $k = 1, 2, 3$, and the remainder term

10. $R = \frac{28h^5}{90} f^{(4)}(\xi),$

    for some ξ such that $x_0 < \xi < x_4$.

11. $\int_{x_0}^{x_5} f(x)\,dx = \frac{5h}{24}(11f(x_1) + f(x_2) + f(x_3) + 11f(x_4)) + R,$

$$h = (x_5 - x_0)/5, \ x_k = x_0 + kh \text{ for } k = 1, 2, 3, 4, \text{ and the remainder term}$$

12. $R = \dfrac{95h^5}{144} f^{(4)}(\xi),$

for some ξ such that $x_0 < \xi < x_5$.

21.1.6 Gaussian quadrature (open-type)

The fundamental Gaussian quadrature formula applies to an integral over the interval $[-1, 1]$. It is a highly accurate method, but unlike the other integration formulas given here it involves the use of abscissas that are unevenly spaced throughout the interval of integration.

1. $\displaystyle\int_{-1}^{1} f(x) \, dx = \sum_{k=1}^{n} w_k^{(n)} f\left(x_k^{(n)}\right) + R_n,$

where the abscissa $x_k^{(n)}$ is the k'th zero of the Legendre polynomial $P_n(x)$, and the weight $w_k^{(n)}$ is given by

2. $w_k^{(n)} = 2\left[P_n'\left(x_k^{(n)}\right)\right]^2 / \left(1 - \left(x_k^{(n)}\right)^2\right).$

The remainder term is

3. $R_n = \dfrac{2^{2n+1}(n!)^4}{(2n+1)[(2n)!]^3} f^{(2n)}(\xi),$

for some ξ such that $-1 < \xi < 1$.

To apply this result to an integral over the interval $[a, b]$ the substitution

4. $y_k^{(n)} = \frac{1}{2}(b-a)x_k^{(n)} + \frac{1}{2}(b+a)$

is made, yielding the result

5. $\displaystyle\int_{a}^{b} f(y) \, dy = \frac{1}{2}(b-a) \sum_{k=1}^{n} w_k^{(n)} f[y^{(n)}] + R_n,$

where the remainder term is now

$$R_n = \dfrac{(b-a)^{2n+1}(n!)^4}{(2n+1)[(2n)!]^3} f^{(2n)}(\xi),$$

for some $a < \xi < b$.

The abscissas and weights for $n = 2, 3, 4, 5, 6$ are given in Table 21.1.

21.1.7 Romberg integration (closed-type)

A robust and efficient method for the evaluation of the integral

1. $I = \displaystyle\int_{a}^{b} f(x) \, dx$

is provided by the process of **Romberg integration.** The method proceeds in stages, with an increase in accuracy of the numerical estimate of I occurring at the end of each successive stage. The process may be continued until the result is accurate to

TABLE 21.1 Gaussian Abscissas and Weights

n	k	$x_k^{(n)}$	$w_k^{(n)}$
2	1	0.5773502692	1.0000000000
	2	−0.5773502692	1.0000000000
3	1	0.7745966692	0.5555555555
	2	0.0000000000	0.8888888888
	3	−0.7745966692	0.5555555555
4	1	0.8611363116	0.3478548451
	2	0.3399810436	0.6521451549
	3	−0.3399810436	0.6521451549
	4	−0.8611363116	0.3478548451
5	1	0.9061798459	0.2369268850
	2	0.5384693101	0.4786286205
	3	0.0000000000	0.5688888889
	4	−0.5384693101	0.4786286205
	5	−0.9061798459	0.2369268850
6	1	0.9324695142	0.1713244924
	2	0.6612093865	0.3607615730
	3	0.2386191861	0.4679139346
	4	0.2386191861	0.4679139346
	5	−0.6612093865	0.3607615730
	6	−0.9324695142	0.1713244924

the required number of decimal places, provided that at the nth stage the derivative $f^{(n)}(x)$ is nowhere singular in the interval of integration.

Romberg integration is based on the composite trapezoidal rule and an extrapolation process (Richardson extrapolation), and is well suited to implementation on a computer. This is because of the efficient use it makes of the function evaluations that are necessary at each successive stage of the computation, and its speed of convergence, which enables the numerical estimate of the integral to be obtained to the required degree of precision relatively quickly.

The Romberg method. At the mth stage of the calculation, the interval of integration $a \le x \le b$ is divided into 2^m intervals of length $(b - a)/2^m$. The corresponding composite trapezoidal estimate for I is then given by

1. $I_{0,0} = \dfrac{(b - a)}{2}[f(a) + f(b)]$ and

$$I_{0,m} = \frac{(b - a)}{2^m}\left[\frac{1}{2}f(a) + \frac{1}{2}f(b) + \sum_{r=1}^{2^m - 1} f(x_r)\right],$$

where $x_r = a + r[(b - a)/2^m]$ for $r = 1, 2, \ldots, 2^m - 1$. Here the initial suffix represents the computational step reached at the m'th stage of the calculation, with the value zero indicating the initial trapezoidal estimate. The second suffix starts with the number of subintervals on which the initial trapezoidal estimate is based and the steps down to zero at the end of the mth stage.

Define

2. $$I_{k,m} = \frac{4^k I_{k-1,m+1} - I_{k-1,m}}{4^k - 1},$$

for $k = 1, 2, \ldots$ and $m = 1, 2, \ldots$.

For some preassigned error $\varepsilon > 0$ and some integer N, $I_{N,0}$ is the required estimate of the integral to within an error ε if

3. $|I_{N,0} - I_{N-1,1}| < \varepsilon$.

The pattern of the calculation proceeds as shown here. Each successive entry in the r'th row provides an increasingly accurate estimate of the integral, with the final estimate at the end of the r'th stage of the calculation being provided by $I_{r,0}$:

$$I_{0,0}$$
$$I_{0,1} \searrow I_{1,0}$$
$$I_{0,2} \searrow I_{1,1} \searrow I_{2,0}$$
$$I_{0,3} \searrow I_{1,2} \searrow I_{2,1} \searrow I_{3,0}$$
$$I_{0,4} \searrow I_{1,3} \searrow I_{2,2} \searrow I_{3,1} \searrow I_{4,0}$$

$$\cdots\cdots\cdots\cdots\cdots\cdots\cdots\cdots$$

To illustrate the method consider the integral

$$I = \int_1^5 \frac{\ln(1 + \sqrt{x})}{1 + x^2}\, dx.$$

Four stages of Romberg integration lead to the following results, inspection of which shows that the approximation $I_{4,0} = 0.519256$ has converged to five decimal places. Notice that the accuracy achieved in $I_{4,0}$ was obtained as a result of only 17 function evaluations at the end of the 16 subintervals involved, coupled with the use of relation 21.1.7.2. To obtain comparable accuracy using only the trapezoidal rule would involve the use of 512 subintervals.

$I_{0,m}$	$I_{1,m}$	$I_{2,m}$	$I_{3,m}$	$I_{4,m}$
0.783483				
0.592752	0.529175			
0.537275	0.518783	0.518090		
0.523633	0.519086	0.519106	0.519122	
0.520341	0.519243	0.519254	0.519255	0.519256

22

Solutions of Standard Ordinary Differential Equations

22.1 Introduction

22.1.1 Basic definitions

22.1.1.1 An ***n*th-order ordinary differential equation** (ODE) for the function $y(x)$ is an equation defined on some interval I that relates $y^{(n)}(x)$ and some or all of $y^{(n-1)}(x)$, $y^{(n-2)}(x)$, ..., $y^{(1)}(x)$, $y(x)$ and x, where $y^{(r)}(x) = d^r y/dx^r$. In its most general form, such an equation can be written

1. $F\left(x, y(x), y^{(1)}(x), \ldots, y^{(n)}(x)\right) = 0,$

where $y^{(n)}(x) \not\equiv 0$ and F is an arbitrary function of its arguments.

The **general solution** of 22.1.1.1.1 is an n times differentiable function $Y(x)$, defined on I, that contains n arbitrary constants, with the property that when $y = Y(x)$ is substituted into the differential equation it reduces it to an identity in x. A **particular solution** is a special case of the general solution in which the arbitrary constants have been assigned specific numerical values.

22.1.2 Linear dependence and independence

22.1.2.1   A set of functions $y_1(x), y_2(x), \ldots, y_n(x)$ defined on some interval I is said to be **linearly dependent** on the interval I if there are constants $c_1, c_2, \ldots, c_n$, not all zero, such that

1. $c_1 y_1(x) + c_2 y_2(x) + \cdots + c_n y_n(x) = 0.$

The set of functions is said to be **linearly independent** on I if 22.1.2.1.1 implies that $c_1 = c_2 = \cdots = c_n = 0$ for all x in I.

The following **Wronskian test** provides a test for linear dependence.

Wronskian test.   Let the n functions $y_1(x), y_2(x), \ldots, y_n(x)$ defined on an interval I be continuous together with their derivatives of every order up to and including those of order n. Then the functions are linearly dependent on I if the **Wronskian determinant** $W[y_1, y_2, \ldots, y_n] = 0$ for all x in I, where

2. $$W[y_1, y_2, \ldots, y_n] = \begin{vmatrix} y_1 & y_2 & \cdots & y_n \\ y_1^{(1)} & y_2^{(1)} & \cdots & y_n^{(1)} \\ \vdots & \vdots & \vdots & \vdots \\ y_1^{(n-1)} & y_2^{(n-1)} & \cdots & y_n^{(n-1)} \end{vmatrix}$$

EXAMPLE 1.   The functions $1, \sin^2 x$, and $\cos^2 x$ are linearly dependent for all x, because setting $y_1 = 1, y_2 = \sin^2 x, y_3 = \cos^2 x$ gives us the following Wronskian:

$$W[y_1, y_2, y_3] = \begin{vmatrix} 1 & \sin^2 x & \cos^2 x \\ 0 & 2 \sin x \cos x & -2 \sin x \cos x \\ 0 & 2(\cos^2 x - \sin^2 x) & -2(\cos^2 x - \sin^2 x) \end{vmatrix} = 0.$$

The linear dependence of these functions is obvious from the fact that

$$1 = \sin^2 x + \cos^2 x,$$

because when this is written in the form

$$y_1 - y_2 - y_3 = 0$$

and compared with 22.1.2.1.1 we see that $c_1 = 1, c_2 = -1$, and $c_3 = -1$.

EXAMPLE 2.   The functions $1, \sin x$, and $\cos x$ are linearly independent for all x, because setting $y_1 = 1, y_2 = \sin x, y_3 = \cos x$, gives us the following Wronskian:

$$W[y_1, y_2, y_3] = \begin{vmatrix} 1 & \sin x & \cos x \\ 0 & \cos x & -\sin x \\ 0 & -\sin x & -\cos x \end{vmatrix} = -1.$$

EXAMPLE 3.   Let us compute the Wronskian of the functions

$$y_1 = \begin{cases} x, & x < 0 \\ 0, & x \geq 0 \end{cases} \quad \text{and} \quad y_2 = \begin{cases} 0, & x < 0 \\ x, & x \geq 0, \end{cases}$$

which are defined for all x. For $x < 0$ we have

$$W[y_1, y_2] = \begin{vmatrix} x & 0 \\ 1 & 0 \end{vmatrix} = 0,$$

whereas for $x \geq 0$

$$W[y_1, y_2] = \begin{vmatrix} 0 & x \\ 0 & 1 \end{vmatrix} = 0,$$

so for all x

$$W[y_1, y_2] = 0.$$

However, despite the vanishing of the Wronskian for all x, the functions y_1 and y_2 are *not* linearly dependent, as can be seen from 22.1.2.1.1, because there exist no nonzero constants c_1 and c_2 such that

$$c_1 y_1 + c_2 y_2 = 0$$

for all x. This is not a failure of the Wronskian test, because although y_1 and y_2 are continuous, their first derivatives are not continuous as required by the Wronskian test.

22.2 Separation of Variables

22.2.1

A differential equation is said to have **separable variables** if it can be written in the form

1. $F(x)G(y)\, dx + f(x)g(y)\, dy = 0.$

The general solution obtained by direct integration is

2. $\displaystyle\int \frac{F(x)}{f(x)}\, dx + \int \frac{g(y)}{G(y)}\, dy = \text{const.}$

22.3 Linear First-Order Equations

22.3.1

The **general linear first-order equation** is of the form

1. $\displaystyle\frac{dy}{dx} + P(x)y = Q(x).$

This equation has an **integrating factor**

2. $\mu(x) = e^{\int P(x)dx},$

and in terms of $\mu(x)$ the general solution becomes

3. $y(x) = \dfrac{c}{\mu(x)} + \dfrac{1}{\mu(x)} \displaystyle\int Q(x)\mu(x)\,dx,$

where c is an arbitrary constant. The first term on the right-hand side is the complementary function $y_c(x)$, and the second term is the particular integral of $y_p(x)$ (see
22.7.1).

EXAMPLE.   Find the general solution of

$$\frac{dy}{dx} + \frac{2}{x} y = \frac{1}{x^2(1+x^2)}.$$

In this case $P(x) = 2/x$ and $Q(x) = x^{-2}(1+x^2)^{-1}$, so

$$\mu(x) = \exp\left[\int \frac{2}{x}\,dx\right] = x^2$$

and

$$y(x) = \frac{c}{x^2} + \frac{1}{x^2} \int \frac{dx}{1+x^2}$$

so that

$$y(x) = \frac{c}{x^2} + \frac{1}{x^2} \arctan x.$$

Notice that the arbitrary additive integration constant involved in the determination
of $\mu(x)$ has been set equal to zero. This is justified by the fact that, had this not
been done, the constant factor so introduced could have been incorporated into the
arbitrary constant c.

22.4 Bernoulli's Equation

22.4.1

Bernoulli's equation is a *nonlinear* first-order equation of the form

1. $\dfrac{dy}{dx} + p(x)y = q(x)y^{\alpha},$

with $\alpha \neq 0$ and $\alpha \neq 1$.
      Division of 22.4.1.1 by y^{α} followed by the change of variable

2. $z = y^{1-\alpha}$

converts Bernoulli's equation to the linear first-order equation

3. $\dfrac{dz}{dx} + (1-\alpha)p(x)z = (1-\alpha)q(x)$

which may be solved for $z(x)$ by the method of 22.3.1. The solution of the original
equation is given by

4. $y(x) = z^{1/(1-\alpha)}.$

22.5 Exact Equations

22.5.1

An **exact equation** is of the form

1. $P(x, y)\, dx + Q(x, y)\, dy = 0,$

where $P(x, y)$ and $Q(x, y)$ satisfy the condition

2. $\dfrac{\partial P}{\partial y} = \dfrac{\partial Q}{\partial x}.$

Thus the left-hand side of 22.5.1.1 is the total differential of some function $F(x, y) = $ constant, with $P(x, y) = \partial F/\partial x$ and $Q(x, y) = \partial F/\partial y$.

The general solution is

3. $\displaystyle\int P(x, y)\, dx + \int \left[Q(x, y) - \frac{\partial}{\partial y} \int P(x, y)\, dx \right] dy = $ const.,

where integration with respect to x implies y is to be regarded as a constant, while integration with respect to y implies x is to be regarded as a constant.

EXAMPLE. Find the general solution of

$$3x^2 y\, dx + (x^3 - y^2)\, dy = 0.$$

Setting $P(x, y) = 3x^2 y$ and $Q(x, y) = x^3 - y^2$ we see the equation is exact because $\partial P/\partial y = \partial Q/\partial x = 3x^2$. It follows that

$$\int P(x, y)\, dx = \int 3x^2 y\, dx = 3y \int x^2\, dx = x^3 y + c_1,$$

where c_1 is an arbitrary constant, whereas

$$\int \left[Q(x, y) - \frac{\partial}{\partial y} \int P(x, y)\, dx \right] dy = \int [x^3 - y^2 - x^3]\, dy = -\frac{1}{3} y^3 + c_2,$$

where c_2 is an arbitrary constant. Thus, the required solution is

$$x^3 y - \tfrac{1}{3} y^3 = c,$$

where $c = -(c_1 + c_2)$ is an arbitrary constant.

22.6 Homogeneous Equations

22.6.1

In its simplest form, a **homogeneous equation** may be written

1. $\dfrac{dy}{dx} = F\left(\dfrac{y}{x}\right),$

where F is a function of the single variable

2. $u = y/x.$

More generally, a homogeneous equation is of the form

3. $P(x, y)\, dx + Q(x, y)\, dy = 0,$

where P and Q are both algebraically homogeneous functions of the same degree. Here, by requiring P and Q to be **algebraically homogeneous of degree n,** we mean that

4. $P(kx, ky) = k^n P(x, y) \quad \text{and} \quad Q(kx, ky) = k^n Q(x, y),$

with k a constant.

The solution of 22.6.1.1 is given in implicit form by

5. $\ln x = \displaystyle\int \frac{du}{F(u) - u}.$

The solution of 22.6.1.3 also follows from this same result by setting $F = -P/Q$.

22.7 Linear Differential Equations

22.7.1

An **nth-order variable coefficient differential equation** is *linear* if it can be written in the form

1. $\tilde{a}_0(x)y^{(n)} + \tilde{a}_1(x)y^{(n-1)} + \cdots + \tilde{a}_n(x)y = \tilde{f}(x),$

where $\tilde{a}_0, \tilde{a}_1, \ldots, \tilde{a}_n$ and $\tilde{f}$ are real-valued functions defined on some interval I. Provided $\tilde{a}_0 \neq 0$ this equation can be rewritten as

2. $y^{(n)} + a_1(x)y^{(n-1)} + \cdots + a_n(x)y = f(x),$

where $a_r = \tilde{a}_r/\tilde{a}_0,\, r = 1, 2, \ldots, n$, and $f = \tilde{f}/\tilde{a}_0$.

It is convenient to write 22.7.1.2 in the form

3. $L[y(x)] = f(x),$

where

4. $L[y(x)] = y^{(n)} + a_1(x)y^{(n-1)} + \cdots + a_n(x)y.$

The nth-order linear differential equation 22.7.1.2 is called **inhomogeneous (non-homogeneous)** when $f(x) \neq 0$, and **homogeneous** when $f(x) \equiv 0$.

The homogeneous equation corresponding to 22.7.1.2 is

5. $L[y(x)] = 0.$

This equation has n linearly independent solutions $y_1(x), y_2(x), \ldots, y_n(x)$ and its general solution, also called the **complementary function,** is

6. $y_c(x) = c_1 y_1(x) + c_2 y_2(x) + \cdots + c_n y_n(x),$

where $c_1, c_2, \ldots, c_n$ are arbitrary constants.

The **general solution** of the inhomogeneous equation 22.7.1.2 is

7. $y(x) = y_c(x) + y_p(x),$

where the form of the **particular integral** $y_p(x)$ is determined by the inhomogeneous term $f(x)$ and contains no arbitrary constants.

An **initial value problem** for 22.7.1.3 involves the determination of a solution satisfying the **initial conditions** at $x = x_0$,

8. $y(x_0) = y_0, \quad y^{(1)}(x_0) = y_1, \ldots, y^{(n-1)}(x_0) = y_{n-1}$

for some specified values of $y_0, y_1, \ldots, y_{n-1}$.

A **two-point boundary value problem** for a differential equation involves the determination of a solution satisfying suitable values of y and certain of its derivatives at the two distinct points $x = a$ and $x = b$.

22.8 Constant Coefficient Linear Differential Equations— Homogeneous Case

22.8.1

A special case of 22.7.1.3 arises when the coefficients $a_1, a_2, \ldots, a_n$ are real-valued absolute constants. Such equations are called **linear constant coefficient differential equations**. The determination of the n linearly independent solutions $y_1, y_2, \ldots, y_n$ that enter into the general solution

1. $y_c = c_1 y_1(x) + c_2 y_2(x) + \cdots + c_n y_n(x)$

of the associated constant coefficient homogeneous equation $L[y(x)] = 0$ may be obtained as follows:

(i)   Form the n'th degree **characteristic polynomial** $P_n(\lambda)$ defined as

(ii) $P_n(\lambda) = \lambda^n + a_1\lambda^{n-1} + \cdots + a_n,$

where $a_1, a_2, \ldots, a_n$ are the constant coefficients occurring in $L[y(x)] = 0$.

(iii) Factor $P_n(\lambda)$ into a product of the form

$$P_n(\lambda) = (\lambda - \lambda_1)^p (\lambda - \lambda_2)^q \cdots (\lambda - \lambda_m)^r,$$

where

$$p + q + \cdots + r = n,$$

and $\lambda_1, \lambda_2, \ldots, \lambda_m$ are either real roots of $P_n(\lambda) = 0$ or, if they are complex, they occur in complex conjugate pairs [because the coefficients of $P_n(\lambda)$ are all real]. Roots $\lambda_1, \lambda_2, \ldots, \lambda_m$ are said to have **multiplicities** $p, q, \ldots, r$, respectively.

(iv) To every real root $\lambda = \mu$ of $P_n(\lambda) = 0$ with multiplicity M there corresponds the M linearly independent solutions of $L[y(x)] = 0$:

$$y_1(x) = e^{\mu x}, \quad y_2(x) = xe^{\mu x}, \quad y_3(x) = x^2 e^{\mu x}, \quad \ldots, \quad y_M(x) = x^{M-1}e^{\mu x}.$$

(v)   To every pair of complex conjugate roots of $P_n(\lambda) = 0$, $\lambda = \alpha + i\beta$, and $\lambda = \alpha - i\beta$, each with multiplicity N, there corresponds the $2N$ linearly independent solutions of $L[y(x)] = 0$

$$y_1(x) = e^{\alpha x} \cos \beta x \qquad\qquad \overline{y}_1(x) = e^{\alpha x} \sin \beta x$$
$$y_2(x) = xe^{\alpha x} \cos \beta x \qquad\qquad \overline{y}_2(x) = xe^{\alpha x} \sin \beta x$$

$$\vdots \qquad\qquad\qquad\qquad \vdots$$

$$y_N(x) = x^{N-1}e^{\alpha x} \cos \beta x \qquad \overline{y}_N(x) = x^{N-1}e^{\alpha x} \sin \beta x$$

(vi)   The general solution of the homogeneous equation $L[y(x)] = 0$ is then the sum of each of the linearly independent solutions found in (v) and (vi) multiplied by a real arbitrary constant. If a solution of an initial value problem is required for the homogeneous equation $L[y(x)] = 0$, the arbitrary constants in $y_c(x)$ must be chosen so that $y_c(x)$ satisfies the initial conditions.

EXAMPLE 1.   Find the general solution of the homogeneous equation

$$y^{(3)} - 2y^{(2)} - 5y^{(1)} + 6y = 0,$$

and the solution of the initial value problem in which $y(0) = 1$, $y^{(1)}(0) = 0$, and $y^{(2)}(0) = 2$.

The characteristic polynomial

$$P_3(\lambda) = \lambda^3 - 2\lambda^2 - 5\lambda + 6$$
$$= (\lambda - 1)(\lambda + 2)(\lambda - 3),$$

so the roots of $P_3(\lambda) = 0$ are $\lambda_1 = 1$, $\lambda_2 = -2$, and $\lambda_3 = 3$, which are all real with multiplicity 1. Thus, the general solution is

$$y(x) = c_1 e^x + c_2 e^{-2x} + c_3 e^{3x}.$$

To solve the initial value problem the constants c_1, c_2, and c_3 must be chosen such that when $x = 0$, $y = 1$, $dy/dx = 0$, and $d^2y/dx^2 = 2$.

$$y(0) = 1 \qquad \text{implies} \quad 1 = c_1 + c_2 + c_3$$
$$y^{(1)}(0) = 0 \qquad \text{implies} \quad 0 = c_1 - 2c_2 + 3c_3$$
$$y^{(2)}(0) = 0 \qquad \text{implies} \quad 2 = c_1 + 4c_2 + 9c_3$$

so $c_1 = 2/3$, $c_2 = 1/3$, and $c_3 = 0$ leading to the solution

$$y(x) = \tfrac{2}{3}e^x + \tfrac{1}{3}e^{-2x}.$$

EXAMPLE 2.   Find the general solution of the homogeneous equation

$$y^{(4)} + 5y^{(3)} + 8y^{(2)} + 4y^{(1)} = 0.$$

The characteristic polynomial

$$P_4(\lambda) = \lambda^4 + 5\lambda^3 + 8\lambda^2 + 4\lambda$$
$$= \lambda(\lambda + 1)(\lambda + 2)^2,$$

so the roots of $P_4(\lambda) = 0$ are $\lambda_1 = 0$, $\lambda_2 = -1$, and $\lambda_2 = -2$ (twice), which are all real, but λ_2 has multiplicity 2. Thus, the general solution is

$$y(x) = c_1 + c_2 e^{-x} + c_3 e^{-2x} + c_4 x e^{-2x}.$$

EXAMPLE 3.   Find the general solution of the homogeneous equation

$$y^{(3)} + 3y^{(2)} + 9y^{(1)} - 13y = 0.$$

The characteristic polynomial

$$P_3(\lambda) = \lambda^3 + 3\lambda^2 + 9\lambda - 13$$
$$= (\lambda - 1)(\lambda^2 + 4\lambda + 13),$$

so the roots of $P_3(\lambda) = 0$ are $\lambda_1 = 1$ and the pair of complex conjugate roots of the real quadratic factor, which are $\lambda_2 = -2 - 3i$ and $\lambda_3 = -2 + 3i$. All the roots have multiplicity 1. Thus, the general solution is

$$y(x) = c_1 e^x + c_2 e^{-2x} \cos 3x + c_3 e^{-2x} \sin 3x.$$

EXAMPLE 4.   Find the general solution of the homogeneous equation

$$y^{(5)} - 4y^{(4)} + 14y^{(3)} - 20y^{(2)} + 25y^{(1)} = 0.$$

The characteristic polynomial

$$P_5(\lambda) = \lambda^5 - 4\lambda^4 + 14\lambda^3 - 20\lambda^2 + 25\lambda$$
$$= \lambda(\lambda^2 - 2\lambda + 5)^2,$$

so the roots of $P_5(\lambda) = 0$ are the single real root $\lambda_1 = 0$ and the complex conjugate roots of the real quadratic factor $\lambda_2 = 1 - 2i$ and $\lambda_3 = 1 + 2i$, each with multiplicity 2. Thus, the general solution is

$$y(x) = c_1 + c_2 e^x \cos 2x + c_3 x e^x \cos 2x + c_4 e^x \sin 2x + c_5 x e^x \sin 2x.$$

EXAMPLE 5.   Use the Laplace transform to solve the initial value problem in Example 1.

Taking the Laplace transform of the differential equation

$$\frac{d^3 y}{dx^3} - 2\frac{d^2 y}{dx^2} - 5\frac{dy}{dx} + 6y = 0$$

gives, after using 19.1.21.2,

$$\underbrace{s^3 Y(s) - y^{(2)}(0) - sy^{(1)}(0) - s^2 y(0)}_{\mathcal{L}[d^3 y/dx^3]} - \underbrace{2(s^2 Y(s) - y^{(1)}(0) - sy(0))}_{2\mathcal{L}[d^2 y/dx^3]}$$

$$- \underbrace{5(sY(s) - y(0))}_{5\mathcal{L}[dy/dx]} + \underbrace{6Y(s)}_{6\mathcal{L}[y]} = 0.$$

After substituting the initial values $y(0) = 1$, $y^{(1)}(0) = 0$, and $y^{(2)}(0) = 2$ this reduces to

$$(s^3 - 2s^2 - 5s + 6)Y(s) = s^2 - 2s - 3,$$

so

$$Y(s) = \frac{s^2 - 2s - 3}{s^3 - 2s^2 - 5s + 6} = \frac{s + 1}{(s - 1)(s + 2)}.$$

When expressed in terms of partial fractions (see 1.7.2) this becomes

$$Y(s) = \frac{2}{3}\frac{1}{s-1} + \frac{1}{3}\frac{1}{s+2}.$$

Inverting this result by means of entry 5 in Table 19.1 leads to

$$y(x) = \tfrac{2}{3}e^x + \tfrac{1}{3}e^{-2x},$$

which is, of course, the solution obtained in Example 1.

22.9 Linear Homogeneous Second-Order Equation

22.9.1

An important elementary equation that arises in many applications is the linear homogeneous second-order equation

1. $\dfrac{d^2 y}{dx^2} + a\dfrac{dy}{dx} + by = 0,$

where a, b are real-valued constants.

The characteristic polynomial is

2. $P_2(\lambda) = \lambda^2 + a\lambda + b,$

and the nature of the roots of $P_2(\lambda) = 0$ depends on the **discriminant** $a^2 - 4b$, which may be positive, zero, or negative.

Case 1.   If $a^2 - 4b > 0$, the characteristic equation $P_2(\lambda) = 0$ has the two distinct real roots

3. $m_1 = \tfrac{1}{2}[-a - \sqrt{a^2 - 4b}]$ and $m_2 = \tfrac{1}{2}[-a + \sqrt{a^2 - 4b}],$

and the solution is

4. $y(x) = c_1 e^{m_1 x} + c_2 e^{m_2 x}.$

This solution decays to zero as $x \to +\infty$ if $m_1, m_2 < 0$, and becomes infinite if a root is positive.

Case 2.   If $a^2 - 4b = 0$, the characteristic equation $P_2(\lambda) = 0$ has the two identical real roots (multiplicity 2)

5. $m_1 = \tfrac{1}{2}a,$

and the solution is

6. $y(x) = c_1 e^{m_1 x} + c_2 x e^{m_1 x}.$

This solution is similar to that of Case 1 in that it decays to zero as $x \to +\infty$ if $m_1 < 0$ and becomes infinite if $m_1 > 0$.

Case 3.   If $a^2 - 4b < 0$, the characteristic equation $P_2(\lambda) = 0$ has the complex conjugate roots

7. $m_1 = \alpha + i\beta$ and $m_2 = \alpha - i\beta,$

where

8. $\alpha = -\frac{1}{2}a$ and $\beta = \frac{1}{2}\sqrt{4b - a^2}$,

and the solution is

9. $y(x) = e^{\alpha x}(c_1 \cos \beta x + c_2 \sin \beta x)$.

This solution is oscillatory and decays to zero as $x \to +\infty$ if $\alpha < 0$, and becomes infinite if $\alpha > 0$.

22.10 Constant Coefficient Linear Differential Equations— Inhomogeneous Case

22.10.1

The general solution of the inhomogeneous constant coefficient n'th-order linear differential equation

1. $\cdot y^{(n)} + a_1 y^{(n-1)} + a_2 y^{(n-2)} + \cdots + a_n y = f(x)$

is of the form

2. $y(x) = y_c(x) + y_p(x)$,

where the $y_c(x)$ is the general solution of the homogeneous form of the equation $L[y(x)] = 0$ and $y_p(x)$ is a **particular integral** whose form is determined by the inhomogeneous term $f(x)$. The solution $y_c(x)$ may be found by the method given in 22.8.1. The particular integral $y_p(x)$ may be obtained by using the method of **variation of parameters.** The result is

3. $y_p(x) = \sum_{r=1}^{n} y_r(x) \int \dfrac{D_r(x) f(x)}{W[y_1, y_2, \ldots, y_n]} \, dx$,

where $y_1, y_2, \ldots, y_n$ are n linearly independent solutions of $L[y(x)] = 0$, $W[y_1, y_2, \ldots, y_n]$ is the Wronskian of these solutions (see 20.1.2.1.2), and $D_r(x)$ is the determinant obtained by replacing the r'th column of the Wronskian by $(0, 0, \ldots, 1)$.

If an **initial value problem** is to be solved, the constants in the general solution $y(x) = y_c(x) + y_p(x)$ must be chosen so that $y(x)$ satisfies the initial conditions.

EXAMPLE. Find the particular integral and general solution of

$$\frac{d^3 y}{dx^3} + 4\frac{dy}{dx} = \tan 2x, \qquad [-\pi/4 < x < \pi/4],$$

and solve the associated initial value problem in which $y(0) = y^{(1)}(0) = y^{(2)}(0) = 0$. In this case

$$L[y(x)] = \frac{d^3 y}{dx^3} + 4\frac{dy}{dx},$$

so the characteristic polynomial

$$P_3(\lambda) = \lambda^3 + 4\lambda = \lambda(\lambda^2 + 4).$$

The roots of $P_3(\lambda) = 0$ are $\lambda_1 = 0$, $\lambda_2 = 2i$, and $\lambda_3 = -2i$, so it follows that the three linearly independent solutions of $L[y(x)] = 0$ are

$$y_1(x) = 1, \quad y_2(x) = \cos 2x, \quad \text{and} \quad y_3(x) = \sin 2x,$$

and hence

$$y_c(x) = c_1 + c_2 \cos 2x + c_3 \sin 2x.$$

The Wronskian

$$W[y_1, y_2, y_3] = \begin{vmatrix} 1 & \cos 2x & \sin 2x \\ 0 & -2\sin 2x & 2\cos 2x \\ 0 & -4\cos 2x & -4\sin 2x \end{vmatrix} = 8,$$

while

$$D_1 = \begin{vmatrix} 0 & \cos 2x & \sin 2x \\ 0 & -2\sin 2x & 2\cos 2x \\ 1 & -4\cos 2x & -4\sin 2x \end{vmatrix} = 2,$$

$$D_2 = \begin{vmatrix} 1 & 0 & \sin 2x \\ 0 & 0 & 2\cos 2x \\ 0 & 1 & -4\sin 2x \end{vmatrix} = -2\cos 2x,$$

$$D_3 = \begin{vmatrix} 1 & \cos 2x & 0 \\ 0 & -2\sin 2x & 0 \\ 0 & -4\cos 2x & 1 \end{vmatrix} = -2\sin 2x,$$

From 22.10.1.3 we have

$$y_p(x) = \frac{1}{4} \int \tan 2x \, dx - \frac{1}{4} \cos 2x \int \cos 2x \tan 2x \, dx$$

$$- \frac{1}{4} \sin 2x \int \sin 2x \tan 2x \, dx,$$

so

$$y_p(x) = 1 - \tfrac{1}{8} \ln[\cos 2x] - \tfrac{1}{8} \sin 2x \ln[\sec 2x + \tan 2x].$$

The general solution is

$$y(x) = c_1 + c_2 \cos 2x + c_3 \sin 2x - \tfrac{1}{8} \ln[\cos 2x] - \tfrac{1}{8} \sin 2x \ln[\sec 2x + \tan 2x]$$

for $-\pi/4 < x < \pi/4$, where the constant 1 in $y_p(x)$ has been incorporated into the arbitrary constant c_1.

To solve the initial value problem, the constants c_1, c_2, and c_3 in the general solution $y(x)$ must be chosen such that $y(0) = y^{(1)}(0) = y^{(2)}(0) = 0$. The equations for c_1, c_2, and c_3 that result when these conditions are used are:

$$\begin{aligned}
(y(0) = 0) && c_1 + c_2 = 0 \\
(y^{(1)}(0) = 0) && 2c_3 = 0 \\
(y^{(2)}(0) = 0) && -4c_2 - \tfrac{1}{2} = 0,
\end{aligned}$$

so $c_1 = \frac{1}{8}$, $c_2 = -\frac{1}{8}$, and $c_3 = 0$, and the solution of the initial value problem is

$$y(x) = \frac{1}{8} - \frac{1}{8}\cos 2x - \frac{1}{8}\ln[\cos 2x] - \frac{1}{8}\sin 2x[\sec 2x + \tan 2x].$$

22.11 Linear Inhomogeneous Second-Order Equation

22.11.1

When the method of 22.10.1 is applied to the solution of the inhomogeneous second-order constant coefficient equation

1. $\dfrac{d^2y}{dx^2} + a\dfrac{dy}{dx} + by = f(x),$

the general solution assumes a simpler form depending on the discriminant $a^2 - 4b$.

Case 1. If $a^2 - 4b > 0$ the general solution is

2. $y(x) = c_1 e^{m_1 x} + c_2 e^{m_2 x} + \dfrac{e^{m_1 x}}{m_1 - m_2} \displaystyle\int e^{-m_1 x} f(x)\, dx$

$$- \frac{e^{m_2 x}}{m_1 - m_2} \int e^{-m_2 x} f(x)\, dx,$$

where

3. $m_1 = \frac{1}{2}[-a - \sqrt{a^2 - 4b}]$ and $m_2 = \frac{1}{2}[-a + \sqrt{a^2 - 4b}].$

Case 2. If $a^2 - 4b = 0$ the general solution is

4. $y(x) = c_1 e^{m_1 x} + c_2 x e^{m_1 x} - e^{m_1 x} \displaystyle\int x e^{-m_1 x} f(x)\, dx$

$$+ e^{m_1 x} \int e^{-m_1 x} f(x)\, dx,$$

where

5. $m_1 = -\frac{1}{2}a.$

Case 3. If $a^2 - 4b < 0$ the general solution is

6. $y(x) = e^{\alpha x}(c_1 \cos \beta x + c_2 \sin \beta x) + \dfrac{e^{\alpha x}\sin \beta x}{\beta} \displaystyle\int e^{-\alpha x} f(x)\cos \beta x\, dx$

$$- \frac{e^{\alpha x}\cos \beta x}{\beta} \int e^{\alpha x} f(x)\sin \beta x\, dx,$$

where

7. $\alpha = -\frac{1}{2}a$ and $\beta = \frac{1}{2}\sqrt{4b - a^2}.$

22.12 Determination of Particular Integrals by the Method of Undetermined Coefficients

22.12.1

An alternative method may be used to find a particular integral when the inhomogeneous term $f(x)$ is simple in form. Consider the linear inhomogeneous n'th-order constant coefficient differential equation

1. $y^{(n)} + a_1 y^{(n-1)} + a_2 y^{(n-2)} + \cdots + a_n y = f(x)$,

in which the inhomogeneous term $f(x)$ is a polynomial, an exponential, a product of a power of x, and a trigonometric function of the form $x^s \cos qx$ or $x^s \sin qx$ or a sum of any such terms. Then the particular integral may be obtained by the **method of undetermined coefficients (constants)**.

To arrive at the general form of the particular integral it is necessary to proceed as follows:

(i)   If $f(x) =$ constant, include in $y_p(x)$ the undetermined constant term C.

(ii)   If $f(x)$ is a polynomial of degree r then:

(a)   if $L[y(x)]$ contains an undifferentiated term y, include in $y_p(x)$ terms of the form

$$A_0 x^r + A_1 x^{r-1} + \cdots + A_r,$$

where $A_0, A_1, \ldots, A_r$ are undetermined constants;

(b)   if $L[y(x)]$ does not contain an undifferentiated y, and its lowest order derivative is $y^{(s)}$, include in $y_p(x)$ terms of the form

$$A_0 x^{r+s} + A_1 x^{r+s-1} + \cdots + A_r x^s,$$

where $A_0, A_1, \ldots, A_r$ are undetermined constants.

(iii)   If $f(x) = e^{\alpha x}$ then:

(a)   if e^{ax} is not contained in the complementary function (it is not a solution of the homogeneous form of the equation), include in $y_p(x)$ the term Be^{ax}, where B is an undetermined constant;

(b)   if the complementary function contains the terms $e^{ax}, xe^{ax}, \ldots, x^m e^{ax}$, include in $y_p(x)$ the term $Bx^{m+1}e^{ax}$, where B is an undetermined constant.

(iv)   If $f(x) = \cos qx$ and/or $\sin qx$ then:

(a)   if $\cos qx$ and/or $\sin qx$ are not contained in the complementary function (they are not solutions of the homogeneous form of the equation), include in $y_p(x)$ terms of the form

$$C \cos qx \quad \text{and} \quad \sin qx,$$

where C and D are undetermined constants;

(b)   if the complementary function contains terms $x^s \cos qx$ and/or $x^s \sin qs$ with $s = 0, 1, 2, \ldots, m$, include in $y_p(x)$ terms of the form

$$x^{m+1}(C \cos qx + D \sin qx),$$

where C and D are undetermined constants.

(v)   The general form of the particular integral is then the sum of all the terms generated in (i) to (iv).

(vi)   The unknown constant coefficients occurring in $y_p(x)$ are found by substituting $y_p(x)$ into 22.12.1.1, and then choosing them so that the result becomes an identity in x.

EXAMPLE.   Find the complementary function and particular integral of

$$\frac{d^2y}{dx^2} + \frac{dy}{dx} - 2y = x + \cos 2x + 3e^x,$$

and solve the associated initial value problem in which $y(0) = 0$ and $y^{(1)}(0) = 1$.
The characteristic polynomial

$$P_2(\lambda) = \lambda^2 + \lambda - 2$$
$$= (\lambda - 1)(\lambda + 2),$$

so the linearly independent solutions of the homogeneous form of the equation are

$$y_1(x) = e^x \quad \text{and} \quad y_2(x) = e^{-2x},$$

and the complementary function is

$$y_c(x) = c_1 e^x + c_2 e^{-2x}.$$

Inspection of the inhomogeneous term shows that only the exponential e^x is contained in the complementary function. It then follows from (ii) that, corresponding to the term x, we must include in $y_p(x)$ terms of the form

$$A + Bx.$$

Similarly, from (iv)(a), corresponding to the term $\cos 2x$, we must include in $y_p(x)$ terms of the form

$$C \cos 2x + D \sin 2x.$$

Finally, from (iii)(b), it follows that, corresponding to the term e^x, we must include in $y_p(x)$ a term of the form

$$Exe^x.$$

Then from (v) the general form of the particular integral is

$$y_p(x) = A + Bx + C \cos 2x + D \sin 2x + Exe^x.$$

To determine the unknown coefficients A, B, C, D, and E we now substitute $y_p(x)$ into the original equation to obtain

$$(-6C + 2D) \cos 2x - (2C + 6D) \sin 2x + 3Ee^x - 2Bx + B - 2A$$
$$= x + \cos 2x + 3e^x.$$

For this to become an identity, the coefficients of corresponding terms on either side of this expression must be identical:

(Coefficients of $\cos 2x$)	$-6C + 2D = 1$
(Coefficients of $\sin 2x$)	$-2C - 6D = 0$
(Coefficients of x)	$-2B = 1$
(Coefficients of e^x)	$3E = 3$
(Coefficients of constant terms)	$B - 2A = 0$

Thus $A = -\frac{1}{4}$, $B = -\frac{1}{2}$, $C = -\frac{3}{20}$, $D = \frac{1}{20}$, $E = 1$, and hence the required particular integral is

$$y_p(x) = -\frac{1}{4} - \frac{1}{2}x - \frac{3}{20}\cos 2x + \frac{1}{20}\sin 2x + xe^x,$$

while the general solution is

$$y(x) = y_c(x) + y_p(x).$$

The solution of the initial value problem is obtained by selecting the constants c_1 and c_2 in $y(x)$ so that $y(0) = 0$ and $y^{(1)}(0) = 1$. The equations for c_1 and c_2 that result when these conditions are used are:

$$
\begin{aligned}
(y(0) = 0) \qquad & c_1 + c_2 - \tfrac{2}{5} = 0 \\
(y'(0) = 1) \qquad & c_1 - 2c_2 + \tfrac{3}{5} = 1,
\end{aligned}
$$

so $c_1 = 2/5$ and $c_2 = 0$ and the solution of the initial value problem is

$$y(x) = e^x\left[x + \tfrac{2}{5}\right] - \tfrac{3}{20}\cos 2x + \tfrac{1}{20}\sin 2x - \tfrac{1}{2}x - \tfrac{1}{4}.$$

22.13 The Cauchy–Euler Equation

22.13.1

The **Cauchy–Euler equation** of order n is of the form

1. $x^n y^{(n)} + a_1 x^{n-1} y^{n-1} + \cdots + a_n y = f(x),$

where $a_1, a_2, \ldots, a_n$ are constants.

The change of variable

2. $x = e^t$

reduces the equation to a linear constant coefficient equation with the inhomogeneous term $f(e^t)$, which may be solved by the method described in 22.10.1.

In the special case of the Cauchy–Euler equation of order 2, which may be written

3. $x^2 \dfrac{d^2 y}{dx^2} + a_1 x \dfrac{dy}{dx} + a_2 y = f(x),$

the change of variable 22.13.1.2. reduces it to

$$\frac{d^2 y}{dt^2} + (a_1 - 1)\frac{dy}{dt} + a_2 y = f(e^t).$$

The following solution of the homogeneous Cauchy–Euler equation of order 2 is often useful: If

4. $x^2 \dfrac{d^2 y}{dx^2} + a_1 x \dfrac{dy}{dx} + a_2 y = 0,$

and λ_1, λ_2 are the roots of the polynomial equation

5. $\lambda^2 + (a_1 - 1)\lambda + a_2 = 0,$

then, provided $x \neq 0$, the general solution of 22.13.1.4 is

6. $\begin{aligned} y_c(x) &= c_1 |x|^{\lambda_1} + c_2 |x|^{\lambda_2}, & \text{if } \lambda_1 \neq \lambda_2 \text{ are both real,} \\ y_c(x) &= (c_1 + c_2 \ln|x|)|x|^{\lambda_1}, & \text{if } \lambda_1 \text{ and } \lambda_2 \text{ are real and } \lambda_1 = \lambda_2, \\ y_c(x) &= x^\alpha [c_1 \cos(\beta \ln|x|) \\ & \quad + c_2 \sin(\beta \ln|x|)], \\ & & \text{if } \lambda_1 \text{ and } \lambda_2 \text{ are complex conjugates} \end{aligned}$

with $\lambda_1 = \alpha + i\beta$ and $\lambda_2 = \alpha - i\beta$.

22.14 Legendre's Equation

22.14.1

Legendre's equation is

1. $(1 - x^2)\dfrac{d^2 y}{dx^2} - 2x \dfrac{dy}{dx} + n(n+1)y = 0,$

where $n = 0, 1, 2, \ldots$.
The general solution is

2. $y(x) = c_1 P_n(x) + c_2 Q_n(x),$

where $P_n(x)$ is the Legendre polynomial of degree n and $Q_n(x)$ is Legendre function of the second kind (see 18.2.4.2 and 18.2.7.1).

22.15 Bessel's Equations

22.15.1

Bessel's equation is

1. $x^2 \dfrac{d^2 y}{dx^2} + x \dfrac{dy}{dx} + (\lambda^2 x^2 - \nu^2)y = 0.$

The general solution is

2. $y(x) = c_1 J_\nu(\lambda x) + c_2 Y_\nu(\lambda x),$

where J_ν is the Bessel function of the first kind of order ν and Y_ν is the Bessel function of the second kind of order ν (see 17.2.1 and 17.2.2). **Bessel's modified equation is**

3. $x^2 \dfrac{d^2 y}{dx^2} + x \dfrac{dy}{dx} - (\lambda^2 x^2 + \nu^2) y = 0.$

The general solution is

4. $y(x) = c_1 I_\nu(\lambda x) + c_2 K_\nu(\lambda x),$

where I_ν is the modified Bessel function of the first kind of order ν and K_ν is the modified Bessel function of the second kind of order ν (see 17.7.1 and 17.7.2).

     EXAMPLE 1.   Solve the two-point boundary value problem

$$x^2 \dfrac{d^2 y}{dx^2} + x \dfrac{dy}{dx} + \lambda^2 x^2 = 0,$$

subject to the boundary conditions $y(0) = 2$ and $y^{(1)}(a) = 0$.

     The equation is Bessel's equation of order 0, so the general solution is

$$y(x) = c_1 J_0(\lambda x) + c_2 Y_0(\lambda x).$$

The boundary condition $y(0) = 2$ requires the solution to be finite at the origin, but $Y_0(0)$ is *infinite* (see 17.2.2.2.1) so we must set $c_2 = 0$ and require that $2 = c_1 J_0(0)$, so $c_1 = 2$ because $J_0(0) = 1$ (see 17.2.1.2.1). Boundary condition $y(a) = 0$ then requires that

$$0 = 2 J_0(\lambda a),$$

but the zeros of $J_0(x)$ are $j_{0,1} \; j_{0,2}, \ldots, j_{0,m}, \ldots$ (see Table 17.1), so $\lambda a = j_{0,m}$, or $\lambda_m = j_{0,m}/a$, for $m = 1, 2, \ldots$. Consequently, the required solutions are

$$y_m(x) = 2 J_0 \left(\dfrac{j_{0,m} x}{a} \right) \qquad [m = 1, 2, \ldots].$$

The numbers $\lambda_1, \lambda_2, \ldots$ are the **eigenvalues** of the problem and the functions $y_1, y_2, \ldots$, are the corresponding **eigenfunctions.** Because any constant multiple of $y_m^{(x)}$ is also a solution, it is usual to omit the constant multiplier 2 in these eigenfunctions.

     EXAMPLE 2.   Solve the two-point boundary value problem

$$x^2 \dfrac{d^2 y}{dx^2} + x \dfrac{dy}{dx} - (x^2 + 4) y = 0,$$

subject to the boundary conditions $y(a) = 0$ and $y(b) = 0$ with $0 < a < b < \infty$.

     The equation is Bessel's modified equation of order 2, so the general solution is

$$y(x) = c_1 I_2(\lambda x) + c_2 K_2(\lambda x).$$

Since $0 < a < b < \infty$, I_2, and K_2 are *finite* in the interval $a \le x \le b$, so both must be retained in the general solution (see 17.7.1.2 and 17.7.2.2).

     Application of the boundary conditions leads to the conditions:

$$(y(a) = 0) \qquad c_1 I_2(\lambda a) + c_2 K_2(\lambda a) = 0$$
$$(y(b) = 0) \qquad c_1 I_2(\lambda b) + c_2 K_2(\lambda b) = 0,$$

and this homogeneous system only has a nontrivial solution (c_1, c_2 not both zero) if

$$\begin{vmatrix} I_2(\lambda a) & K_2(\lambda a) \\ I_2(\lambda b) & K_2(\lambda b) \end{vmatrix} = 0.$$

Thus, the required values $\lambda_1, \lambda_2, \ldots$ of λ must be the zeros of the transcendental equation

$$I_2(\lambda a) K_2(\lambda b) - I_2(\lambda b) K_2(\lambda a) = 0.$$

For a given a, b it is necessary to determine $\lambda_1, \lambda_2, \ldots$, numerically. Here also the numbers $\lambda_1, \lambda_2, \ldots$, are the **eigenvalues** of the problem, and the functions

$$y_m(x) = c_1 \left[I_2(\lambda_m x) - \frac{I_2(\lambda_m a)}{K_2(\lambda_m a)} K_2(\lambda_m x) \right]$$

or, equivalently,

$$y_m(x) = c_1 \left[I_2(\lambda_m x) - \frac{I_2(\lambda_m b)}{K_2(\lambda_m b)} K_2(\lambda_m x) \right],$$

are the corresponding **eigenfunctions.** As in Example 1, because any multiple of $y_m(x)$ is also a solution, it is usual to set $c_1 = 1$ in the eigenfunctions.

22.16 Power Series and Frobenius Methods

22.16.1

To appreciate the need for the Frobenius method when finding solutions of linear variable coefficient differential equations, it is first necessary to understand the power series method and the reason for its failure in certain circumstance.

To illustrate the **power series method,** consider seeking a solution of

$$\frac{d^2 y}{dx^2} + x \frac{dy}{dx} + y = 0$$

in the form of a power series about the origin

$$y(x) = \sum_{n=0}^{\infty} a_n x^n.$$

Differentiation of this expression to find dy/dx and d^2y/dx^2, followed by substitution into the differential equation and the grouping of terms leads to the result

$$(a_0 + 2a_2) + \sum_{n=1}^{\infty} [(n+1)(n+2)a_{n+2} + (n+1)a_n]x^n = 0.$$

For $y(x)$ to be a solution, this expression must be an identity for all x, which is only possible if $a_0 + 2a_2 = 0$ and the coefficient of every power of x vanishes, so that

$$(n+1)(n+2)a_{n+2} + (n+1)a_n = 0 \qquad [n = 1, 2, \ldots].$$

Thus, $a_2 = -\frac{1}{2}a_0$, and in general the coefficients a_n are given recursively by

$$a_{n+2} = -\left(\frac{1}{n+2}\right)a_n,$$

from which it follows that $a_2, a_4, a_6, \ldots$ are all expressible in terms of an arbitrary nonzero constant a_0, while $a_1, a_3, a_5, \ldots$ are all expressible in terms of an arbitrary nonzero constant a_1. Routine calculations then show that

$$a_{2m} = \frac{(-1)^m}{2^m m!} a_0 \qquad [m = 1, 2, \ldots],$$

while

$$a_{2m+1} = \frac{(-1)^m}{1.3.5 \cdots (2m + 1)} a_1 \qquad [m = 1, 2, \ldots].$$

After substituting for the a_n in the power series for $y(x)$ the solution can be written

$$y(x) = a_0 \sum_{m=0}^{\infty} \frac{(-1)^m}{2^m m!} x^{2m} + a_1 \sum_{m=0}^{\infty} \frac{(-1)^m}{1.3.5 \cdots (2m + 1)} x^{2m+1}$$

$$= a_0 \left(1 - \tfrac{1}{2}x + \tfrac{1}{8}x^2 - \cdots\right) + a_1 \left(x - \tfrac{1}{3}x^3 + \tfrac{1}{15}x^5 \cdots\right).$$

Setting

$$y_1(x) = 1 - \tfrac{1}{2}x + \tfrac{1}{8}x^2 - \cdots \quad \text{and} \quad y_2(x) = x - \tfrac{1}{3}x^3 + \tfrac{1}{15}x^5 - \cdots,$$

it follows that y_1 and y_2 are linearly independent, because y_1 is an even function and y_2 is an odd function. Thus, because a_0, a_1 are arbitrary constants, the general solution of the differential equation can be written

$$y(x) = a_0 y_1(x) + a_2 y_2(x).$$

The power series method was successful in this case because the coefficients of d^2y/dx^2, dy/dx and y in the differential equation were all capable of being expressed as Maclaurin series, and so could be combined with the power series expression for $y(x)$ and its derivatives that led to $y_1(x)$ and $y_2(x)$. (In this example, the coefficients 1, x, and 1 in the differential equation are their own Maclaurin series.) The method fails if the variable coefficients in a differential equation cannot be expressed as Maclaurin series (they are not analytic at the origin).

The **Frobenius method** overcomes the difficulty just outlined for a wide class of variable coefficient differential equations by generalizing the type of solution that is sought. To proceed further, it is first necessary to define regular and singular points of a differential equation.

The second-order linear differential equation with variable coefficients

1. $$\frac{d^2y}{dx^2} + p(x)\frac{dy}{dx} + q(x)y = 0$$

is said to have a **regular point** at the origin if $p(x)$ and $q(x)$ are analytic at the origin. If the origin is *not* a regular point it is called a **singular point.** A singular point at the origin is called a **regular singular point** if

2. $\lim_{x \to 0} \{xp(x)\}$ is finite

and

3. $\lim_{x \to 0} \{x^2 q(x)\}$ is finite.

Singular points that are not regular are said to be **irregular.** The behavior of a solution in a neighborhood of an irregular singular point is difficult to determine and very erratic, so this topic is not discussed further.

There is no loss of generality involved in only considering an equation with a regular singular point located at the origin, because if such a point is located at x_0 the transformation $X = x - x_0$ will shift it to the origin.

If the behavior of a solution is required for large x (an asymptotic solution), making the transformation $x = 1/z$ in the differential equation enables the behavior for large x to be determined by considering the case of small z. If $z = 0$ is a regular point of the transformed equation, the original equation is said to have **regular point at infinity.** If, however, $z = 0$ is a regular singular point of the transformed equation, the original equation is said to have a **regular singular point at infinity.**

The **Frobenius method** provides the solution of the differential equation

4. $\dfrac{d^2 y}{dx^2} + p(x)\dfrac{dy}{dx} + q(x)y = 0$

in a neighborhood of a regular singular point located at the origin. Remember that if the regular singular point is located at x_0, the transformation $X = x - x_0$ reduces the equation to the preceding case.

A solution is sought in the form

5. $y(x) = x^\lambda \displaystyle\sum_{n=0}^{\infty} a_n x^n \qquad [a_0 \neq 0],$

where the number λ may be either real or complex, and is such that $a_0 \neq 0$.

Expressing $p(x)$ and $q(x)$ as their Maclaurin series

6. $p(x) = p_0 + p_1(x) + p_2 x^2 + \cdots \quad$ and $\quad q(x) = q_0 + q_1(x) + q_2 x^2 + \cdots,$

and substituting for $y(x)$, $p(x)$ and $q(x)$ in 22.16.1.4, as in the power series method, leads to an identity involving powers of x, in which the coefficient of the lowest power x^λ is given by

$[\lambda(\lambda - 1) + p_0\lambda + q_0]a_0 = 0.$

Since, by hypothesis, $a_0 \neq 0$, this leads to the **indicial equation**

7. $\lambda(\lambda - 1) + p_0\lambda + q_0 = 0,$

from which the permissible values of the exponent λ may be found.

The form of the two linearly independent solutions $y_1(x)$ and $y_2(x)$ of 22.16.1.4 is determined as follows:

Case 1. If the indicial equation has distinct roots λ_1 and λ_2 that do not differ by an integer, then

8. $y_1(x) = x^{\lambda_1} \displaystyle\sum_{n=0}^{\infty} a_n x^n$

and

9. $y_2(x) = x^{\lambda_2} \displaystyle\sum_{n=0}^{\infty} b_n x^n,$

where the coefficients a_n and b_n are found recursively in terms of a_0 and b_0, as in the power series method, by equating to zero the coefficient of the general term in the identity in powers of x that led to the indicial equation, and first setting $\lambda = \lambda_1$ and then $\lambda = \lambda_2$. Here the coefficients in the solution for $y_2(x)$ have been denoted by b_n, instead of a_n, to avoid confusion with the coefficients in $y_1(x)$. The same convention is used in Cases 2 and 3, which follow.

   Case 2.   If the indicial equation has a double root $\lambda_1 = \lambda_2 = \lambda$, then

10. $y_1(x) = x^\lambda \displaystyle\sum_{n=0}^{\infty} a_n x^n$

and

11. $y_2(x) = y_1(x) \ln |x| + x^\lambda \displaystyle\sum_{n=1}^{\infty} b_n x^n.$

   Case 3.   If the indicial equation has roots λ_1 and λ_2 that differ by an integer and $\lambda_1 > \lambda_2$, then

12. $y_1(x) = x^{\lambda_1} \displaystyle\sum_{n=0}^{\infty} a_n x^n$

and

13. $y_2(x) = K y_1(x) \ln|x| + x^{\lambda_2} \displaystyle\sum_{n=0}^{\infty} b_n x^n,$

where K may be zero.

   In all three cases the coefficients a_n in the solutions $y_1(x)$ are found recursively in terms of the arbitrary constant $a_0 \neq 0$, as already indicated. In Case 1 the solution $y_2(x)$ is found in the same manner, with the coefficients b_n being found recursively in terms of the arbitrary constant $b_0 \neq 0$. However, the solutions $y_2(x)$ in Cases 2 and 3 are more difficult to obtain. One technique for finding $y_2(x)$ involves using a variant of the method of variation of parameters (see 22.10.1).

   If $y_1(x)$ is a known solution of

14. $\dfrac{d^2 y}{dx^2} + p(x)\dfrac{dy}{dx} + q(x)y = 0,$

a second linearly independent solution can be shown to have the form

15. $y_2(x) = y_1(x)v(x),$

where

16. $v(x) = \displaystyle\int \dfrac{\exp[-\int p(x)dx]}{[y_1(x)]^2} dx.$

   This is called the **integral method** for the determination of a second linearly independent solution in terms of a known solution. Often the integral determining $v(x)$ cannot be evaluated analytically, but if the numerator and denominator of the integrand are expressed in terms of power series, the method of 1.11.1.4 may be used to determine their quotient, which may then be integrated term by term to obtain $v(x)$, and hence $y_2(x)$ in the form $y_2(x) = y_1(x)v(x)$. This method will generate the logarithmic term automatically when it is required.

EXAMPLE 1 (distinct roots λ_1, λ_2 not differing by an integer).   The differential equation

$$x\frac{d^2y}{dx^2} + \left(\frac{1}{2} - x\right)\frac{dy}{dx} + 2y = 0$$

has a regular singular point at the origin.

The indicial equation is

$$\lambda\left(\lambda - \tfrac{1}{2}\right) = 0,$$

which corresponds to Case 1 because

$$\lambda_1 = 0, \lambda_2 = \tfrac{1}{2}.$$

The coefficient of $x^{n+\lambda-1}$ obtained when the Frobenius method is used is

$$a_n(n + \lambda)\left(n + \lambda - \tfrac{1}{2}\right) + a_{n-1}(n + \lambda - 3) = 0,$$

so the recursion formula relating a_n to a_{n-1} is

$$a_n = \frac{(n + \lambda - 3)}{(n + \lambda)\left(n + \lambda - \tfrac{1}{2}\right)}a_{n-1}.$$

Setting $\lambda = \lambda_1 = 0$ we find that

$$a_1 = -4a_0, \quad a_2 = \tfrac{4}{3}a_0 \quad \text{and} \quad a_n = 0 \text{ for } n > 2.$$

The corresponding solution

$$y_1(x) = a_0\left(1 - 4x + \tfrac{4}{3}x^2\right)$$

is simply a second-degree polynomial.

Setting $\lambda = \lambda_2 = \tfrac{1}{2}$, a routine calculation shows the coefficient b_n is given by

$$b_n = \frac{b_1}{(4n^2 - 1)(2n - 3)n!},$$

so the second linearly independent solution is

$$y_2(x) = b_0 \sum_{n=0}^{\infty} \frac{x^{n+1/2}}{(4n^2 - 1)(2n - 3)n!}.$$

EXAMPLE 2 (equal roots $\lambda_1 = \lambda_2 = \lambda$).   The differential equation

$$x\frac{d^2y}{dx^2} + (1 - x)\frac{dy}{dx} + y = 0$$

has a regular singular point at the origin.

The indicial equation is

$$\lambda^2 = 0,$$

which corresponds to Case 2 since $\lambda_1 = \lambda_2 = 0$ is a repeated root.

The recursion formula obtained as in Example 1 is

$$a_n = \frac{(n + \lambda - 2)}{(n + \lambda)^2}a_{n-1}.$$

Setting $\lambda = 0$ then shows that one solution is

$$y_1(x) = a_0(1 - x).$$

Using the integral method to find $y_2(x)$ leads to the second linearly independent solution

$$y_2(x) = b_0(1 - x)\left[\ln|x| + 3x + \tfrac{11}{4}x^2 + \tfrac{49}{18}x^3 + \cdots\right],$$

in which there is no simple formula for the general term in the series.

EXAMPLE 3 (roots λ_1, λ_2 that differ by an integer).   The differential equation

$$x\frac{d^2 y}{dx^2} - x\frac{dy}{dx} + 2y = 0$$

has a regular singular point at the origin.

The indicial equation is

$$\lambda(\lambda - 1) = 0,$$

which corresponds to Case 3 because $\lambda_1 = 1$ and $\lambda_2 = 0$.

The recursion formula obtained as in Example 1 is

$$a_n = \frac{(n + \lambda - 3)}{(n + \lambda)(n + \lambda - 1)} \qquad [n \geq 1].$$

Setting $\lambda = 1$ shows one solution to be

$$y_1(x) = a_0\left(x - \tfrac{1}{2}x^2\right).$$

Using the integral method to find $y_2(x)$ leads to the second linearly independent solution

$$y_2(x) = 2\left(x - \tfrac{1}{2}x^2\right)\ln|x| + \left(-1 + \tfrac{1}{2}x + \tfrac{9}{4}x^2 + \cdots\right),$$

where again there is no simple formula for the general term in the series.

22.17 The Hypergeometric Equation

22.17.1

The **hypergeometric equation** due to Gauss contains, as special cases, many of the differential equations whose solutions are the special functions that arise in applications.

The general **hypergeometric equation** has the form

1. $$x(1 - x)\frac{d^2 y}{dx^2} - [(a + b + 1)x - c]\frac{dy}{dx} - aby = 0,$$

in which a, b, and c are real constants.

Using the Frobenius method the equation can be shown to have the general solution

2. $$y(x) = Ay_1(x) + By_2(x),$$

where, for $|x| < 1$ and $c \neq 0, 1, 2, \ldots,$

3. $y_1(x) = F(a, b, c; x)$

$$= 1 + \frac{ab}{c}x + \frac{a(a+1)\,b(b+1)}{c(c+1)}\frac{x^2}{2!}$$

$$+ \frac{a(a+1)(a+2)\,b(b+1)(b+2)}{c(c+1)(c+2)}\frac{x^3}{3!} + \cdots$$

4. $y_2(x) = x^{1-c}F(a - c + 1, b - c + 1, 2 - c; x)$;

and for $|x| > 1, a - b \neq 0, 1, 2, \ldots,$

5. $y_1(x) = x^{-a}F(a, a - c + 1, a - b + 1; x^{-1})$
6. $y_2(x) = x^{-b}F(b, b - c + 1, b - a + 1; x^{-1})$;

while for $(x - 1) < 1, a + b - c \neq 0, 1, 2, \ldots,$

7. $y_1(x) = F(a, b, a + b - c + 1; 1 - x)$
8. $y_2(x) = (1 - x)^{c-a-b}F(c - b, c - a, c - a - b + 1; 1 - x)$.

The **confluent hypergeometric equation** has the form

9. $x\dfrac{d^2y}{dx^2} + (c - x)\dfrac{dy}{dx} - by = 0,$

with b and c real constants, and the general solution obtained by the Frobenius method

10. $y(x) = Ay_1(x) + By_2(x),$

where for $c \neq 0, 1, 2, \ldots,$

11. $y_1(x) = F(b, c; x) = 1 + \dfrac{b}{c}x + \dfrac{b(b+1)}{c(c+1)}\dfrac{x^2}{2!} + \dfrac{b(b+1)(b+2)}{c(c+1)(c+2)}\dfrac{x^3}{3!} + \cdots$

12. $y_2(x) = x^{1-c}F(b - c + 1, 2 - c; x)$.

22.18 Numerical Methods

22.18.1

When numerical solutions to initial value problems are required that cannot be obtained by analytical means it is necessary to use numerical methods. From the many methods that exist we describe in order of increasing accuracy only Euler's method, the modified Euler method, the fourth-order Runge–Kutta method and the Runge–Kutta–Fehlberg method. The section concludes with a brief discussion of how methods for the solution of initial value problems can be used to solve two-point boundary value problems for second-order equations.

Euler's method.   The **Euler method** provides a numerical approximation to the solution of the initial value problem

1. $\dfrac{dy}{dx} = f(x, y),$

which is subject to the initial condition

2. $y(x_0) = y_0$.

Let each increment (step) in x be h, so that at the nth step

3. $x_n = x_0 + nh$.

Then the Euler algorithm for the determination of the approximation y_{n+1} to $y(x_{n+1})$ is

4. $y_{n+1} = y_n + f(x_n, y_n)h$.

The method uses a tangent line approximation to the solution curve through (x_n, y_n) in order to determine the approximation y_{n+1} to $y_{(x_{n+1})}$. The local error involved is $O(h^2)$. The method does not depend on equal step lengths at each stage of the calculation, and if the solution changes rapidly the step length may be reduced in order to control the local error.

Modified Euler method.   The **modified Euler method** is a simple refinement of the Euler method that takes some account of the curvature of the solution curve through (x_n, y_n) when estimating y_{n+1}. The modification involves taking as the gradient of the tangent line approximation at (x_n, y_n) the average of the gradients at (x_n, y_n) and (x_{n+1}, y_{n+1}) as determined by the Euler method.

The algorithm for the modified Euler method takes the following form: If

5. $\dfrac{dy}{dx} = f(x, y)$,

subject to the initial condition

6. $y(x_0) = y_0$,

and all steps are of length h, so that after n steps

7. $x_n = x_0 + nh$,

then

8. $y_{n+1} = y_n + \dfrac{h}{2}[f(x_n, y_n) + f(x_n + h, y_n + f(x_n, y_n)h]$.

The local error involved when estimating y_{n+1} from y_n is $O(h^3)$. Here, as in the Euler method, if required the step length may be changed as the calculation proceeds.

Runge–Kutta fourth-order method.   The **Runge–Kutta (R–K) fourth-order method** is an accurate and flexible method based on a Taylor series approximation to the function $f(x, y)$ in the initial value problem

9. $\dfrac{dy}{dx} = f(x, y)$

subject to the initial condition

10. $y(x_0) = y_0$.

The increment h in x may be changed at each step, but it is usually kept constant so that after n steps

11. $x_n = x_0 + nh$.

The Runge–Kutta algorithm for the determination of the approximation y_{n+1} to $y(x_{n+1})$ is

12. $y_{n+1} = y_n + \frac{1}{6}(k_1 + 2k_2 + 2k_3 + k_4)$,

where

13. $k_1 = hf(x_n, y_n)$

$k_2 = hf\left(x_n + \frac{1}{2}h, y_n + \frac{1}{2}k_1\right)$

$k_3 = hf\left(x_n + \frac{1}{2}h, y_n + \frac{1}{2}k_2\right)$

$k_4 = hf(x_{n+1}, y_n + k_3)$.

The local error involved in the determination of y_{n+1} from y_n is $O(h^5)$.

The Runge–Kutta method for systems. The Runge–Kutta method extends immediately to the solution of problems of the type

14. $\dfrac{dy}{dx} = f(x, y, z)$

15. $\dfrac{dz}{dx} = g(x, y, z)$,

subject to the initial conditions

16. $y(x_0) = y_0$ and $z(x_0) = z_0$.

At the nth integration step, using a step of length h, the Runge–Kutta algorithm for the system takes the form

17. $y_{n+1} = y_n + \frac{1}{6}(k_1 + 2k_2 + 2ki_3 + k_4)$

18. $z_{n+1} = z_n + \frac{1}{6}(K_1 + 2K_2 + 2K_3 + K_4)$,

where

19. $k_1 = hf(x_n, y_n, z_n)$

$k_2 = hf\left(x_n + \frac{1}{2}h, y_n + \frac{1}{2}k_1, z_n + \frac{1}{2}K_1\right)$

$k_3 = hf\left(x_n + \frac{1}{2}h, y_n + \frac{1}{2}k_2, z_n + \frac{1}{2}K_2\right)$

$k_4 = hf\left(x_n + h, y_n + k_3, z_n + K_3\right)$

and

20. $K_1 = hg(x_n, y_n, z_n)$

$K_2 = hg\left(x_n + \frac{1}{2}h, y_n + \frac{1}{2}k_1, z_n + \frac{1}{2}K_1\right)$

$K_3 = hg\left(x_n + \frac{1}{2}h, y_n + \frac{1}{2}k_2, z_n + \frac{1}{2}K_2\right)$

$K_4 = hg(x_n + h, y_n + k_3, z_n + K_3)$.

As with the Runge–Kutta method, the local error involved in the determination of y_{n+1} from y_n and z_{n+1} from z_n is $O(h^5)$.

This method may also be used to solve the second-order equation

21. $\dfrac{d^2 y}{dx^2} = H\left(x, y, \dfrac{dy}{dx}\right)$

subject to the initial conditions

22. $y(x_0) = a$ and $y'(x_0) = b$,

by setting $z = dy/dx$ and replacing the second-order equation by the system

23. $\dfrac{dy}{dx} = z$

24. $\dfrac{dz}{dx} = H(x, y, z)$,

subject to the initial conditions

25. $y(x_0) = a, z(x_0) = b$.

Runge–Kutta–Fehlberg method.   The **Runge–Kutta–Fehlberg** (R–K–F) method is an adaptive technique that uses a Runge–Kutta method with a local error or order 5 in order to estimate the local error in the Runge–Kutta method of order 4. The result is then used to adjust the step length h so the magnitude of the global error is bounded by some given tolerance ε. In general, the step length is changed after each step, but high accuracy is attained if ε is taken to be suitably small, though this may be at the expense of (many) extra steps in the calculation. Because of the computation involved, the method is only suitable for implementation on a computer, though the calculation is efficient because only six evaluations of the $f(x, y)$ are required at each step compared with the four required for the Runge–Kutta method of order 4.

The R–K–F algorithm for the determination of the approximation $\tilde{y}_{n+1}$ to $y(x_{n+1})$ is as follows: It is necessary to obtain a numerical solution to the initial value problem

26. $\dfrac{dy}{dx} = f(x, y)$,

subject to the initial condition

27. $y(x_0) = y_0$,

in which the magnitude of the global error is to be bounded by a given tolerance ε.

The approximation $\tilde{y}_{n+1}$ to $y(x_{n+1})$ is given by

28. $\tilde{y}_{n+1} = y_n \frac{16}{135}k_1 + \frac{6656}{12825}k_3 + \frac{28561}{56430}k_4 - \frac{9}{50}k_5 + \frac{2}{55}k_6$,

where y_{n+1} used in the determination of the step length is

29. $y_{n+1} = y_n + \frac{25}{216}k_1 + \frac{1408}{2565}k_3 + \frac{2197}{4104}k_4 - \frac{1}{5}k_s$

and

30. $k_1 = hf(x_n, y_n)$
       $k_2 = hf\left(x_n + \frac{1}{4}h, y_n + \frac{1}{4}k_1\right)$
       $k_3 = hf\left(x_n + \frac{3}{8}h, y_n + \frac{3}{32}k_1 + \frac{9}{32}k_2\right)$
       $k_4 = hf\left(x_n + \frac{12}{13}h, y_n + \frac{1932}{2197}k_1 - \frac{7200}{2197}k_2 + \frac{7296}{2197}k_3\right)$
       $k_5 = hf\left(x_n + h, y_n + \frac{439}{216}k_1 - 8k_2 + \frac{3680}{513}k_3 - \frac{845}{4104}k_4\right)$
       $k_6 = hf\left(x_n + \frac{1}{2}h, y_n - \frac{8}{27}k_1 + 2k_2 - \frac{3544}{2565}k_3 + \frac{1859}{4104}k_4 - \frac{11}{40}k_5\right)$.

The factor μ by which the new step length μh is to be determined so that the global error bound ϵ is maintained is usually taken to be given

31. $\mu = 0.84\left(\dfrac{\epsilon h}{|\tilde{y}_{n+1} - y_{n+1}|}\right)^{1/4}.$

The relative accuracy of these methods is illustrated by their application to the following two examples.

EXAMPLE 1.   Solve the initial value problem

$$x\frac{dy}{dx} = -6y + 3xy^{4/3},$$

subject to the initial condition $y(1) = 2$.

After division by x, the differential equation is seen to be a Bernoulli equation, and a routine calculation shows its solution to be

$$y = \left[x + x^2\left(\tfrac{1}{2}\,2^{2/3} - 1\right)\right]^{-3}.$$

The numerical solutions obtained by the methods described above are as follows:

$\langle y_n \rangle$

x_n	Euler	Modified Euler	R–K	R–K–F	Exact
1.0	2	2	2	2	2
1.1	1.555953	1.624075	1.626173	1.626165	1.626165
1.2	1.24815	1.354976	1.358449	1.358437	1.358437
1.3	1.027235	1.156977	1.161392	1.161378	1.161378
1.4	0.86407	1.008072	1.013169	1.013155	1.013155
1.5	0.740653	0.894173	0.899801	0.899786	0.899786
1.6	0.645428	0.805964	0.812040	0.812025	0.812025
1.7	0.570727	0.737119	0.743608	0.743593	0.743593
1.8	0.511317	0.683251	0.690149	0.690133	0.690133
1.9	0.46354	0.641264	0.648593	0.648578	0.648578
2.0	0.424781	0.608957	0.616762	0.616745	0.616745

EXAMPLE 2.   Solve Bessel's equation

$$x^2\frac{d^2 y}{dx^2} + x\frac{dy}{dx} + (x^2 - n^2)y = 0$$

in the interval $1 \le x \le 5$ for the case $n = 0$, subject to the initial conditions $y(1) = 1$ and $y'(1) = 0$.

This has the analytical solution

$$y(x) = \frac{Y_1(1)J_0(x)}{J_0(1)Y_1(1) - J_1(1)Y_0(1)} - \frac{J_1(1)Y_0(x)}{J_0(1)Y_1(1) - J_1(1)Y_0(1)}.$$

The numerical results obtained by the Runge–Kutta method with a uniform step length $h = 0.4$ are shown in the second column of the following table, while the third and fourth columns show the results obtained by the R–K–F method with $\epsilon = 10^{-6}$

and the exact result, respectively. It is seen that in the interval $1 \leq x \leq 5$, the R–K–F method and the exact result agree to six decimal places.

| | | $\langle y_n \rangle$ | |
x_n	R–K	R–K–F	Exact
1.0	1	1.000000	1.000000
1.4	0.929215	0.929166	0.929166
1.8	0.74732	0.747221	0.747221
2.2	0.495544	0.495410	0.495410
2.6	0.214064	0.213918	0.213918
3.0	−0.058502	−0.058627	−0.058627
3.4	−0.288252	−0.288320	−0.288320
3.8	−0.449422	−0.449401	−0.449401
4.2	−0.527017	−0.526887	−0.526887
4.6	−0.518105	−0.517861	−0.517861
5.0	−0.431532	−0.431190	−0.431190

Two-point boundary value problems—the shooting method.   **A two-point boundary value problem** for the second-order equation,

32. $\dfrac{d^2 y}{dx^2} = f(x, y, y')$ $[a \leq x \leq b]$,

which may be either linear or nonlinear, involves finding the solution $y(x)$ that satisfies the **boundary conditions**

33. $y(a) = \alpha$ and $y(b) = \beta$

at the end points of the interval $a \leq x \leq b$. Numerical methods for solving *initial value problems* for second-order equations cannot be used to solve this problem directly, because instead of specifying y and y' at an initial point $x = a$, y alone is specified at two distinct points $x = a$ and $x = b$.

    To understand the **shooting method** used to solve 22.18.1.31 subject to the boundary conditions of 22.18.1.32, let us suppose that one of the methods for solving initial value problems, say, the Runge–Kutta method, is applied twice to the equation with two slightly different sets of initial conditions. Specifically, suppose the *same* initial condition $y(a) = \alpha$ is used in each case, but that two *different* initial gradients $y'(a)$ are used, so that

     (I) $y(a) = \alpha$ and $y'(a) = \gamma$

and

     (II) $y(a) = \alpha$ and $y'(a) = \delta$,

where $\gamma \neq \delta$ are chosen arbitrarily. Let the two different solutions $y_1(x)$ and $y_2(x)$ correspond, respectively, to the initial conditions (I) and (II). Typical solutions are illustrated in Figure 22.1, in which $y_1(b)$ and $y_2(b)$ differ from the required result $y(b) \equiv \beta$.

    The approach used to obtain the desired solution is called the *shooting method* because if each solution curve is considered as the trajectory of a particle shot from

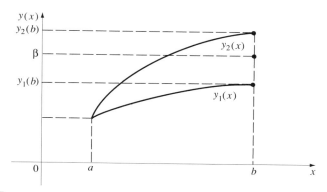

FIGURE 22.1 ∎

the point $x = a$ at the elevation $y(a) = \alpha$, provided there is a unique solution, the required terminal value $y(b) = \beta$ will be attained when the trajectory starts with the correct gradient $y'(a)$.

When 22.18.1.31 is a linear equation, and so can be written

34. $y'' = p(x)y' + q(x)y + r(x)$ $[a \le x \le b]$,

the appropriate choice for $y'(a)$ is easily determined. Setting

35. $y(x) = k_1 y_1(x) + k_2 y_2(x)$,

with

36. $k_1 + k_2 = 1$,

and substituting into 22.18.1.35 leads to the result

37. $\dfrac{d^2}{dx^2}(k_1 y_1 + k_2 y_2) = p(x)\dfrac{d}{dx}(k_1 y_1 + k_2 y_2) + q(x)(k_1 y_1 + k_2 y_2) + r(x)$,

which shows that $y(x) = k_1 y_1(x) + k_2 y_2(x)$ is itself a solution. Furthermore, $y(x) = k_1 y_1(x) + k_2 y_2(x)$ satisfies the left-hand boundary condition $y(a) = \alpha$.

Setting $x = b$, $y(b) = \beta$ in 22.18.1.35 gives

$\beta = k_1 y_1(b) + k_2 y_2(b)$,

which can be solved in conjunction with $k_1 + k_2 = 1$ to give

38. $k_1 = \dfrac{\beta - y_2(b)}{y_1(b) - y_2(b)}$ and $k_2 = 1 - k_1$.

The solution is then seen to be given by

39. $y(x) = \left[\dfrac{\beta - y_2(b)}{y_1(b) - y_2(b)}\right] y_1(x) + \left[\dfrac{y_1(b) - \beta}{y_1(b) - y_2(b)}\right] y_2(x)$ $[a \le x \le b]$.

A variant of this method involves starting from the two quite different initial value problems; namely, the original equation

40. $y'' = p(x)y' + q(x)y + r(x)$,

with the initial conditions

(III) $y(a) = \alpha$ and $y'(a) = 0$

and the corresponding *homogeneous* equation

41. $y'' = p(x)y' + q(x)y$,

with the initial conditions

(IV) $y(a) = 0$ and $y'(a) = 1$.

Using the fact that adding to the solution of the homogeneous equation any solution of the inhomogeneous equation will give rise to the general solution of the inhomogeneous equation, a similar form of argument to the one used above shows the required solution to be given by

42. $y(x) = y_3(x) + \left[\dfrac{\beta - y_3(b)}{y_4(b)} \right] y_4(x)$,

where $y_3(x)$ and $y_4(x)$ are, respectively, the solutions of 22.18.1.40 with boundary conditions (III) and 22.18.1.41 with boundary conditions (IV).

The method must be modified when 22.18.1.31 is **nonlinear,** because solutions of homogeneous equations are no longer additive. It then becomes necessary to use an iterative method to adjust repeated estimates of $y'(a)$ until the terminal value $y(b) = \beta$ is attained to the required accuracy.

We mention only the iterative approach based on the *secant method* of interpolation, because this is the simplest to implement. Let the two-point boundary value problem be

43. $y'' = f(x, y, y')$ $[a \le x \le b]$,

subject to the boundary conditions

44. $y(a) = \alpha$ and $y(b) = \beta$.

Let k_0 and k_1 be two estimates of the initial gradient $y'(a)$, and denote the solution of

$$y'' = f(x, y, y'), \qquad \text{with } y(a) = \alpha \quad \text{and} \quad y'(a) = k_0,$$

by $y_0(x)$, and the solution of

$$y'' = f(x, y, y'), \qquad \text{with } y(a) = \alpha \quad \text{and} \quad y'(a) = k_1,$$

by $y_1(x)$.

The iteration then proceeds using successive values of the gradient k_i, for $i = 2, 3, \ldots$, and the corresponding terminal values $y_i(b)$, in the scheme

$$k_i = k_{i-1} - \frac{(y_{i-1}(b) - \beta)(k_{i-1} - k_{i-2})}{y_{i-1}(b) - y_{i-2}(b)},$$

starting from k_0 and k_1, until for some $i = N$, $|y_N(b) - y_{N-1}(b)| < \varepsilon$, where ε is a preassigned tolerance. The required approximate solution is then given by $y_N(x)$, for $a \le x \le b$.

In particular, it is usually necessary to experiment with the initial estimates k_0 and k_1 to ensure the convergence of the iterative scheme.

23

Vector Analysis

23.1 Scalars and Vectors

23.1.1 Basic definitions

23.1.1.1   A **scalar** quantity is completely defined by a single real number (positive or negative) that measures its *magnitude*. Examples of scalars are length, mass, temperature, and electric potential. In print, scalars are represented by Roman or Greek letters like r, m, T, and ϕ.

A **vector** quantity is defined by giving its *magnitude* (a nonnegative scalar), and *its line of action* (a line in space) together with its **sense** (direction) along the line. Examples of vectors are velocity, acceleration, angular velocity, and electric field. In print, vector quantities are represented by Roman and Greek boldface letters like ν, $\mathbf{a}$, $\mathbf{\Omega}$, and $\mathbf{E}$. By convention, the magnitudes of vectors ν, $\mathbf{a}$, and $\mathbf{\Omega}$ are usually represented by the corresponding ordinary letters v, a, and Ω, etc. The magnitude of a vector $\mathbf{r}$ is also denoted by $|\mathbf{r}|$, so that

1. $r = |\mathbf{r}|$.

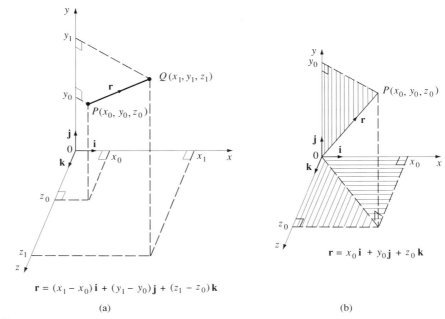

FIGURE 23.1 ▮

A vector of unit magnitude in the direction of **r**, called a **unit vector**, is denoted by $\mathbf{e}_r$, so that

2. $\mathbf{r} = r\mathbf{e}_r$.

The **null vector (zero vector) 0** is a vector with zero magnitude and no direction.

A geometrical interpretation of a vector is obtained by using a straight-line segment parallel to the line of action of the vector, whose length is equal (or proportional) to the magnitude of the vector, with the sense of the vector being indicated by an arrow along the line segment. The end of the line segment from which the arrow is directed is called the **initial point** of the vector, while the other end (toward which the arrow is directed) is called the **terminal point** of the vector.

A **right-handed system** of rectangular cartesian coordinate axes $0\{x, y, z\}$ is one in which the positive direction along the z-axis is determined by the direction in which a right-handed screw advances when rotated from the x- to the y-axis. In such a system the signed lengths of the projections of a vector **r** with initial point $P(x_0, y_0, z_0)$ and terminal point $Q(x_1, y_1, z_1)$ onto the x-, y-, and z-axes are called the x, y, and z **components** of the vector. Thus the x, y, and z components of **r** directed from P to Q are $x_1 - x_0$, $y_1 - y_0$, and $z_1 - z_0$, respectively (Figure 23.1(a)). A vector directed from the origin 0 to the point $P(x_0, y_0, z_0)$ has x_0, y_0, and z_0 as its respective x, y, and z components [Figure 23.1(b)]. Special unit vectors directed along the x-, y-, and z-axes are denoted by **i**, **j**, and **k**, respectively.

The cosines of the angles α, β, and γ between **r** and the respective x-, y-, and z-axes shown in Figure 23.2 are called the **direction cosines** of the vector **r**. If the

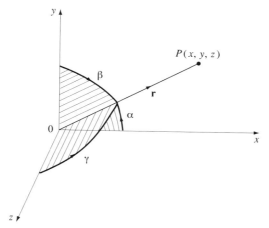

FIGURE 23.2 ∎

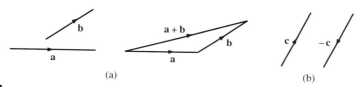

(a)

(b)

FIGURE 23.3 ∎

components of $\mathbf{r}$ are x, y, and z, then the respective direction cosines of $\mathbf{r}$, denoted by l, m, and n, are

3. $l = \dfrac{x}{r}$, $\quad m = \dfrac{y}{r}$, $\quad n = \dfrac{z}{r}$ $\quad$ with $r = (x^2 + y^2 + z^2)^{1/2}$.

The direction cosines are related by

4. $l^2 + m^2 + n^2 = 1$.

Numbers u, v, and w proportional to l, m, and n, respectively, are called **direction ratios**.

23.1.2 Vector addition and subtraction

23.1.2.1 **Vector addition** of vectors $\mathbf{a}$ and $\mathbf{b}$, denoted by $\mathbf{a} + \mathbf{b}$, is performed by first translating vector $\mathbf{b}$, without rotation, so that its initial point coincides with the terminal point of $\mathbf{a}$. The vector sum $\mathbf{a} + \mathbf{b}$ is then defined as the vector whose initial point is the initial point of $\mathbf{a}$, and whose terminal point is the new terminal point of $\mathbf{b}$ (the **triangle rule** for vector addition) (Figure 23.3(a)).

The **negative** of vector $\mathbf{c}$, denoted by $-\mathbf{c}$, is obtained from $\mathbf{c}$ by reversing its sense, as in Fig. 23.3(b), and so

1. $c = |\mathbf{c}| = |-\mathbf{c}|$.

The **difference** $\mathbf{a} - \mathbf{b}$ of vectors $\mathbf{a}$ and $\mathbf{b}$ is defined as the vector sum $\mathbf{a} + (-\mathbf{b})$. This corresponds geometrically to translating vector $-\mathbf{b}$, without rotation, until its

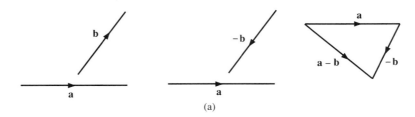

(a)

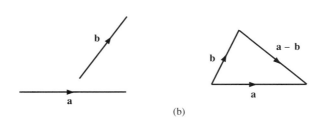

(b)

FIGURE 23.4 ∎

initial point coincides with the terminal point of **a**, when the vector **a** − **b** is the vector drawn from the initial point of **a** to the new terminal point of −**b** (Figure 23.4(a)). Equivalently, **a** − **b** is obtained by bringing into coincidence the initial points of **a** and **b** and defining **a** − **b** as the vector drawn from the terminal point of **b** to the terminal point of **a** [Figure 23.4(b)].

Vector addition obeys the following algebraic rules:

2. $\mathbf{a} + (-\mathbf{a}) = \mathbf{a} - \mathbf{a} = \mathbf{0}$

3. $\mathbf{a} + \mathbf{b} + \mathbf{c} = \mathbf{a} + \mathbf{c} + \mathbf{b} = \mathbf{b} + \mathbf{c} + \mathbf{a}$ (commutative law)

4. $(\mathbf{a} + \mathbf{b}) + \mathbf{c} = \mathbf{a} + (\mathbf{b} + \mathbf{c})$ (associative law)

The geometrical interpretations of laws 3 and 4 are illustrated in Figures 23.5(a) and 23.5(b).

23.1.3 Scaling vectors

23.1.3.1   A vector **a** may be **scaled** by the scalar λ to obtain the new vector $\mathbf{b} = \lambda\mathbf{a}$. The magnitude $b = |\mathbf{b}| = |\lambda\mathbf{a}| = |\lambda|a$. The sense of **b** is the same as that of **a** if $\lambda > 0$, but it is *reversed* if $\lambda < 0$. The scaling operation performed on vectors obeys the laws:

1. $\lambda\mathbf{a} = \mathbf{a}\lambda$ (commutative law)

2. $(\lambda + \mu)\mathbf{a} = \lambda\mathbf{a} + \mu\mathbf{a}$ (distributive law)

3. $\lambda(\mu\mathbf{a}) = \mu(\lambda\mathbf{a}) = (\lambda\mu)\mathbf{a}$ (associative law)

4. $\lambda(\mathbf{a} + \mathbf{b}) = \lambda\mathbf{a} + \lambda\mathbf{b}$ (distributive law)

where λ, μ are scalars and **a, b** are vectors.

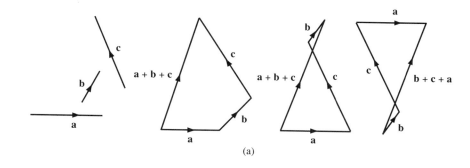

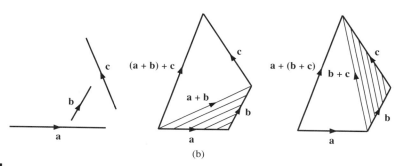

FIGURE 23.5 ∎

23.1.4 Vectors in component form

23.1.4.1 If **a**, **b**, and **c** are any three noncoplanar vectors, an arbitrary vector **r** may always be written in the form

1. $\mathbf{r} = \lambda_1 \mathbf{a} + \lambda_2 \mathbf{b} + \lambda_3 \mathbf{c}$,

where the scalars λ_1, λ_2, and λ_3 are the components of **r** in the **triad** of reference vectors **a**, **b**, **c**. In the important special case of rectangular Cartesian coordinates $0\{x, y, z\}$, with unit vectors **i**, **j**, and **k** along the x-, y-, and z-axes, respectively, the vector **r** drawn from point $P(x_0, y_0, z_0)$ to point $Q(x_1, y_1, z_1)$ can be written (Figure 23.1(a))

2. $\mathbf{r} = (x_1 - x_0)\mathbf{i} + (y_1 - y_0)\mathbf{j} + (z_1 - z_0)\mathbf{k}$.

Similarly, the vector drawn from the origin to the point $P(x_0, y_0, z_0)$ becomes (Figure 23.1(b))

3. $\mathbf{r} = x_0\mathbf{i} + y_0\mathbf{j} + z_0\mathbf{k}$.

For 23.1.4.1.2 the magnitude of **r** is

4. $r = |\mathbf{r}| = [(x_1 - x_0)^2 + (y_1 - y_0)^2 + (z_1 - z_0)^2]^{1/2}$,

whereas for 23.1.4.1.3 the magnitude of **r** is

$$r = |\mathbf{r}| = \left(x_0^2 + y_0^2 + z_0^2\right)^{1/2}.$$

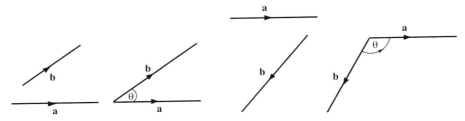

FIGURE 23.6 ▮

In terms of the direction cosines l, m, n (see 23.1.1.1.3) the vector $\mathbf{r} = x\mathbf{i} + y\mathbf{j} + z\mathbf{k}$ becomes

$$\mathbf{r} = \mathbf{r}(l\mathbf{i} + m\mathbf{j} + n\mathbf{k}),$$

where $l\mathbf{i} + m\mathbf{j} + n\mathbf{k}$ is the unit vector in the direction of $\mathbf{r}$.

If $\mathbf{a} = a_1\mathbf{i} + a_2\mathbf{j} + a_3\mathbf{k}$, $\mathbf{b} = b_1\mathbf{i} + b_2\mathbf{j} + b_3\mathbf{k}$, and λ and μ are scalars, then

5. $\lambda\mathbf{a} = \lambda a_1\mathbf{i} + \lambda a_2\mathbf{j} + \lambda a_3\mathbf{k}$,
6. $\mathbf{a} + \mathbf{b} = (a_1 + b_1)\mathbf{i} + (a_2 + b_2)\mathbf{j} + (a_3 + b_3)\mathbf{k}$,
7. $\lambda\mathbf{a} + \mu\mathbf{b} = (\lambda a_1 + \mu b_1)\mathbf{i} + (\lambda a_2 + \mu b_2)\mathbf{j} + (\lambda a_3 + \mu b_3)\mathbf{k}$,

which are equivalent to the results in 23.1.3.

23.2 Scalar Products

23.2.1

The **scalar product (dot product** or **inner product)** of vectors $\mathbf{a} = a_1\mathbf{i} + a_2\mathbf{j} + a_3\mathbf{k}$ and $\mathbf{b} = b_1\mathbf{i} + b_2\mathbf{j} + b_3\mathbf{k}$ inclined at an angle θ to one another and written $\mathbf{a} \cdot \mathbf{b}$ is defined as the scalar (Figure 23.6)

1. $\mathbf{a} \cdot \mathbf{b} = |\mathbf{a}||\mathbf{b}| \cos\theta$
    $= ab\cos\theta$
    $= a_1b_1 + a_2b_2 + a_3b_3$.

If required, the angle between $\mathbf{a}$ and $\mathbf{b}$ may be obtained from

2. $\cos\theta = \dfrac{\mathbf{a} \cdot \mathbf{b}}{|\mathbf{a}||\mathbf{b}|} = \dfrac{a_1b_1 + a_2b_2 + a_3b_3}{\left(a_1^2 + a_2^2 + a_3^2\right)^{1/2}\left(b_1^2 + b_2^2 + b_3^2\right)^{1/2}}.$

Properties of the scalar product.   If $\mathbf{a}$ and $\mathbf{b}$ are vectors and λ and μ are scalars, then:

3. $\mathbf{a} \cdot \mathbf{b} = \mathbf{b} \cdot \mathbf{a}$ (commutative property)
4. $(\lambda\mathbf{a}) \cdot (\mu\mathbf{b}) = \lambda\mu\mathbf{a} \cdot \mathbf{b}$ (associative property)
5. $\mathbf{a} \cdot (\mathbf{b} + \mathbf{c}) = \mathbf{a} \cdot \mathbf{b} + \mathbf{a} \cdot \mathbf{c}$ (distributive property)

Special cases.

6. $\mathbf{a} \cdot \mathbf{b} = 0$ if $\mathbf{a}, \mathbf{b}$ are orthogonal $(\theta = \pi/2)$

7. $\mathbf{a} \cdot \mathbf{b} = |\mathbf{a}||\mathbf{b}| = ab$ if $\mathbf{a}$ and $\mathbf{b}$ are parallel $(\theta = 0)$

8. $\mathbf{i} \cdot \mathbf{i} = \mathbf{j} \cdot \mathbf{j} = \mathbf{k} \cdot \mathbf{k} = 1$ and $\mathbf{i} \cdot \mathbf{j} = \mathbf{j} \cdot \mathbf{k} = \mathbf{k} \cdot \mathbf{i} = 0.$

23.3 Vector Products

23.3.1

The **vector product (cross product)** of vectors $\mathbf{a} = a_1\mathbf{i} + a_2\mathbf{j} + a_3\mathbf{k}$ and $\mathbf{b} = b_1\mathbf{i} + b_2\mathbf{j} + b_3\mathbf{k}$ inclined at an angle θ to one another and written $\mathbf{a} \times \mathbf{b}$ is defined as the vector

1. $\mathbf{a} \times \mathbf{b} = |\mathbf{a}||\mathbf{b}|\sin\theta\mathbf{n} = ab\,\sin\theta\mathbf{n},$

where $\mathbf{n}$ is a unit vector normal to the plane containing $\mathbf{a}$ and $\mathbf{b}$ directed in the sense in which a right-handed screw would advance if rotated from $\mathbf{a}$ to $\mathbf{b}$ (Figure 23.7).

An alternative and more convenient definition of $\mathbf{a} \times \mathbf{b}$ is

2. $\mathbf{a} \times \mathbf{b} = \begin{vmatrix} \mathbf{i} & \mathbf{j} & \mathbf{k} \\ a_1 & a_2 & a_3 \\ b_1 & b_2 & b_3 \end{vmatrix}.$

If required, the angle θ between $\mathbf{a}$ and $\mathbf{b}$ follows from

3. $\sin\theta = \dfrac{|\mathbf{a} \times \mathbf{b}|}{ab},$

though the result 23.2.1.2 is usually easier to use.

Properties of the vector product.   If $\mathbf{a}$ and $\mathbf{b}$ are vectors and λ and μ are scalars, then

4. $\mathbf{a} \times \mathbf{b} = -\mathbf{a} \times \mathbf{b}$ (noncommutative)

5. $(\lambda\mathbf{a}) \times (\mu\mathbf{b}) = \lambda\mu\mathbf{a} \times \mathbf{b}$ (associative property)

6. $\mathbf{a} \times (\mathbf{b} + \mathbf{c}) = \mathbf{a} \times \mathbf{b} + \mathbf{a} \times \mathbf{c}$ (distributive property)

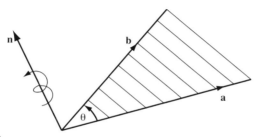

FIGURE 23.7 ∎

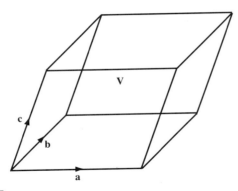

FIGURE 23.8 ▮

Special cases.

7. $\mathbf{a} \times \mathbf{b} = \mathbf{0}$ if $\mathbf{a}$ and $\mathbf{b}$ are parallel $(\theta = 0)$

8. $\mathbf{a} \times \mathbf{b} = ab\mathbf{n}$ if $\mathbf{a}$ and $\mathbf{b}$ are orthogonal $(\theta = \pi/2)$

9. $\mathbf{i} \times \mathbf{j} = \mathbf{k}, \quad \mathbf{j} \times \mathbf{k} = \mathbf{i}, \quad \mathbf{k} \times \mathbf{i} = \mathbf{j}$

10. $\mathbf{i} \times \mathbf{i} = \mathbf{j} \times \mathbf{j} = \mathbf{k} \times \mathbf{k} = \mathbf{0}$

23.4 Triple Products

23.4.1

The **scalar triple product** of the three vectors $\mathbf{a} = a_1\mathbf{i} + a_2\mathbf{j} + a_3\mathbf{k}$, $\mathbf{b} = b_1\mathbf{i} + b_2\mathbf{j} + b_3\mathbf{k}$, and $\mathbf{c} = c_1\mathbf{i} + c_2\mathbf{j} + c_3\mathbf{k}$, written $\mathbf{a} \cdot (\mathbf{b} \times \mathbf{c})$, is the scalar

1. $\mathbf{a} \cdot (\mathbf{b} \times \mathbf{c}) = \mathbf{b} \cdot (\mathbf{c} \times \mathbf{a}) = \mathbf{c} \cdot (\mathbf{a} \times \mathbf{b}).$

In terms of components

2. $\mathbf{a} \cdot (\mathbf{b} \times \mathbf{c}) = \begin{bmatrix} a_1 & a_2 & a_3 \\ b_1 & b_2 & b_3 \\ c_1 & c_2 & c_3 \end{bmatrix}.$

The alternative notation $[\mathbf{abc}]$ is also used for the scalar triple product in place of $\mathbf{a} \cdot (\mathbf{b} \times \mathbf{c})$.

In geometrical terms the absolute value of $\mathbf{a} \cdot (\mathbf{b} \times \mathbf{c})$ may be interpreted as the volume V of a parallelepiped in which $\mathbf{a}$, $\mathbf{b}$, and $\mathbf{c}$ form three adjacent edges meeting at a corner (Figure 23.8). This interpretation provides a useful test for the linear independence of any three vectors. The vectors $\mathbf{a}$, $\mathbf{b}$, and $\mathbf{c}$ are *linearly dependent* if $\mathbf{a} \cdot (\mathbf{b} \times \mathbf{c}) = 0$, because $V = 0$ implies that the vectors are coplanar, and so $\mathbf{a} = \lambda\mathbf{b} + \mu\mathbf{c}$ for some scalars λ and μ; whereas they are *linearly independent* if $\mathbf{a} \cdot (\mathbf{b} \times \mathbf{c}) \neq 0$.

The **vector triple product** of the three vectors $\mathbf{a}$, $\mathbf{b}$, and $\mathbf{c}$, denoted by $\mathbf{a} \times (\mathbf{b} \times \mathbf{c})$, is given by

3. $\mathbf{a} \times (\mathbf{b} \times \mathbf{c}) = (\mathbf{a} \cdot \mathbf{c})\mathbf{b} - (\mathbf{a} \cdot \mathbf{b})\mathbf{c}.$

The parentheses are essential in a vector triple product to avoid ambiguity, because $\mathbf{a} \times (\mathbf{b} \times \mathbf{c}) \neq (\mathbf{a} \times \mathbf{b}) \times \mathbf{c}$.

23.5 Products of Four Vectors

23.5.1

Two other products arise that involve the four vectors $\mathbf{a}$, $\mathbf{b}$, $\mathbf{c}$, and $\mathbf{d}$. The first is the *scalar product*

1. $(\mathbf{a} \times \mathbf{b}) \cdot (\mathbf{c} \times \mathbf{d}) = (\mathbf{a} \cdot \mathbf{c})(\mathbf{b} \cdot \mathbf{d}) - (\mathbf{a} \cdot \mathbf{d})(\mathbf{b} \cdot \mathbf{c})$,

and the second is the *vector product*

2. $(\mathbf{a} \times \mathbf{b}) \times (\mathbf{c} \times \mathbf{d}) = \mathbf{a} \cdot (\mathbf{b} \times \mathbf{d})\mathbf{c} - \mathbf{a} \cdot (\mathbf{b} \times \mathbf{c})\mathbf{d}$.

23.6 Derivatives of Vector Functions of a Scalar *t*

23.6.1

Let $x(t)$, $y(t)$, and $z(t)$ be continuous functions of t that are differentiable as many times as necessary, and let $\mathbf{i}$, $\mathbf{j}$, and $\mathbf{k}$ be the triad of fixed unit vectors introduced in 23.1.4. Then the vector $\mathbf{r}(t)$ given by

1. $\mathbf{r}(t) = x(t)\mathbf{i} + y(t)\mathbf{j} + z(t)\mathbf{k}$

is a vector function of the scalar variable t that has the same continuity and differentiability properties as its components. The first- and second-order derivatives of $\mathbf{r}(t)$ with respect to t are

2. $\dfrac{d\mathbf{r}}{dt} = \dot{\mathbf{r}} = \dfrac{dx}{dt}\mathbf{i} + \dfrac{dy}{dt}\mathbf{j} + \dfrac{dz}{dt}\mathbf{k}$

and

3. $\dfrac{d^2\mathbf{r}}{dt^2} = \ddot{\mathbf{r}} = \dfrac{d^2x}{dt^2}\mathbf{i} + \dfrac{d^2y}{dt^2}\mathbf{j} + \dfrac{d^2z}{dt^2}\mathbf{k}$.

Higher order derivatives are defined in similar fashion so that, in general,

4. $\dfrac{d^n\mathbf{r}}{dt^n} = \dfrac{d^nx}{dt^n}\mathbf{i} + \dfrac{d^ny}{dt^n}\mathbf{j} + \dfrac{d^nz}{dt^n}\mathbf{k}$.

If $\mathbf{r}$ is the **position vector** of a point in space of time t, then $\dot{\mathbf{r}}$ is its *velocity* and $\ddot{\mathbf{r}}$ is its *acceleration* (Figure 23.9).

Differentiation of combinations of vector functions of a scalar t. Let $\mathbf{u}$ and $\mathbf{v}$ be continuous functions of the scalar variable t that are differentiable as many times as necessary, and let $\phi(t)$ be a scalar function of t with the same continuity and differentiability properties as the components of the vector functions. Then the following differentiability results hold:

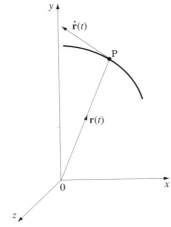

FIGURE 23.9 ∎

1. $\dfrac{d}{dt}(\mathbf{u} + \mathbf{v}) = \dfrac{d\mathbf{u}}{dt} + \dfrac{d\mathbf{v}}{dt}$

2. $\dfrac{d}{dt}(\phi\mathbf{u}) = \dfrac{d\phi}{dt}\mathbf{u} + \phi\dfrac{d\mathbf{u}}{dt}$

3. $\dfrac{d}{dt}(\mathbf{u} \cdot \mathbf{v}) = \dfrac{d\mathbf{u}}{dt} \cdot \mathbf{v} + \mathbf{u} \cdot \dfrac{d\mathbf{v}}{dt}$

4. $\dfrac{d}{dt}(\phi\mathbf{u} \cdot \mathbf{v}) = \dfrac{d\phi}{dt}\mathbf{u} \cdot \mathbf{v} + \phi\dfrac{d\mathbf{u}}{dt} \cdot \mathbf{v} + \phi\mathbf{u} \cdot \dfrac{d\mathbf{v}}{dt}$

5. $\dfrac{d}{dt}(\mathbf{u} \times \mathbf{v}) = \dfrac{d\mathbf{u}}{dt} \times \mathbf{v} + \mathbf{u} \times \dfrac{d\mathbf{v}}{dt}$

6. $\dfrac{d}{dt}(\phi\mathbf{u} \times \mathbf{v}) = \dfrac{d\phi}{dt}\mathbf{u} \times \mathbf{v} + \phi\dfrac{d\mathbf{u}}{dt} \times \mathbf{v} + \phi\mathbf{u} \times \dfrac{d\mathbf{v}}{dt}$

23.7 Derivatives of Vector Functions of Several Scalar Variables

23.7.1

Let $u_i(x, y, z)$ and $v_i(x, y, z)$ for $i = 1, 2, 3$ be continuous functions of the scalar variables x, y, and z, and let them have as many partial derivatives as necessary. Define

1. $\mathbf{u}(x, y, z) = u_1\mathbf{i} + u_2\mathbf{j} + u_3\mathbf{k}$
2. $\mathbf{v}(x, y, z) = v_1\mathbf{i} + v_2\mathbf{j} + v_3\mathbf{k},$

where **i**, **j**, and **k** are the triad of fixed unit vectors introduced in 23.1.4. Then the following differentiability results hold:

3. $\dfrac{\partial \mathbf{u}}{\partial x} = \dfrac{\partial u_1}{\partial x}\mathbf{i} + \dfrac{\partial u_2}{\partial x}\mathbf{j} + \dfrac{\partial u_3}{\partial x}\mathbf{k}$

4. $\dfrac{\partial \mathbf{u}}{\partial y} = \dfrac{\partial u_1}{\partial y}\mathbf{i} + \dfrac{\partial u_2}{\partial y}\mathbf{j} + \dfrac{\partial u_3}{\partial y}\mathbf{k}$

5. $\dfrac{\partial \mathbf{u}}{\partial z} = \dfrac{\partial u_1}{\partial z}\mathbf{i} + \dfrac{\partial u_2}{\partial z}\mathbf{j} + \dfrac{\partial u_3}{\partial z}\mathbf{k},$

with corresponding results of $\partial \mathbf{v}/\partial x$, $\partial \mathbf{v}/\partial y$, and $\partial \mathbf{v}/\partial z$.

Second-order and higher derivatives of **u** and **v** are defined in the obvious manner:

6. $\dfrac{\partial^2 \mathbf{u}}{\partial x^2} = \dfrac{\partial}{\partial x}\left(\dfrac{\partial \mathbf{u}}{\partial x}\right), \quad \dfrac{\partial^2 \mathbf{u}}{\partial x \partial y} = \dfrac{\partial}{\partial x}\left(\dfrac{\partial \mathbf{u}}{\partial y}\right), \quad \dfrac{\partial^2 \mathbf{u}}{\partial x \partial z} = \dfrac{\partial}{\partial x}\left(\dfrac{\partial \mathbf{u}}{\partial z}\right), \dots$

7. $\dfrac{\partial^3 \mathbf{u}}{\partial x^3} = \dfrac{\partial}{\partial x}\left(\dfrac{\partial^2 \mathbf{u}}{\partial x^2}\right), \quad \dfrac{\partial^3 \mathbf{u}}{\partial x^2 \partial y} = \dfrac{\partial}{\partial x}\left(\dfrac{\partial^2 \mathbf{u}}{\partial x \partial y}\right), \quad \dfrac{\partial^3 \mathbf{u}}{\partial x \partial z^2} = \dfrac{\partial}{\partial x}\left(\dfrac{\partial^2 \mathbf{u}}{\partial z^2}\right), \dots$

8. $\dfrac{\partial}{\partial x}(\mathbf{u} \cdot \mathbf{v}) = \dfrac{\partial \mathbf{u}}{\partial x} \cdot \mathbf{v} + \mathbf{u} \cdot \dfrac{\partial \mathbf{v}}{\partial x}$

9. $\dfrac{\partial}{\partial x}(\mathbf{u} \times \mathbf{v}) = \dfrac{\partial \mathbf{u}}{\partial x} \times \mathbf{v} + \mathbf{u} \times \dfrac{\partial \mathbf{v}}{\partial x},$

with corresponding results for derivatives with respect to *y* and *z* and for higher order derivatives.

10. $d\mathbf{u} = \dfrac{\partial \mathbf{u}}{\partial x}dx + \dfrac{\partial \mathbf{u}}{\partial y}dy + \dfrac{\partial \mathbf{u}}{\partial z}dz$ (total differential)

and if $x = x(t)$, $y = y(t)$, $z = z(t)$,

11. $d\mathbf{u} = \left(\dfrac{\partial \mathbf{u}}{\partial x}\dfrac{dx}{dt} + \dfrac{\partial \mathbf{u}}{\partial y}\dfrac{dy}{dt} + \dfrac{\partial \mathbf{u}}{\partial z}\dfrac{dz}{dt}\right)dt$ (chain rule)

23.8 Integrals of Vector Functions of a Scalar Variable *t*

23.8.1

Let the vector function $f(t)$ of the scalar variable *t* be

1. $\mathbf{f}(t) = f_1(t)\mathbf{i} + f_2(t)\mathbf{j} + f_3(t)\mathbf{k},$

where f_1, f_2, and f_3 are scalar functions of *t* for which a function $\mathbf{F}(t)$ exists such that

2. $\mathbf{f}(t) = \dfrac{d\mathbf{F}}{dt}.$

Then

3. $\displaystyle\int \mathbf{f}(t)\, dt = \int \frac{d\mathbf{F}}{dt}\, dt = \mathbf{F}(t) + \mathbf{c},$

where $\mathbf{c}$ is an arbitrary vector constant. The function $\mathbf{F}(t)$ is called an **antiderivative** of $\mathbf{f}(t)$, and result 23.8.1.3 is called an **indefinite integral** of $\mathbf{f}(t)$. Expressed differently, 23.8.1.3 becomes

4. $\displaystyle \mathbf{F}(t) = \mathbf{i} \int f_1(t)\, dt + \mathbf{j} \int f_2(t)\, dt = \mathbf{k} \int f_3(t)\, dt + \mathbf{c}.$

The **definite integral** of $f(t)$ between the scalar limits $t = t_1$ and $t = t_2$ is

5. $\displaystyle \int_{t_1}^{t_2} \mathbf{f}(t)\, dt = \mathbf{F}(t_2) - \mathbf{F}(t_1).$

Properties of the definite integral.   If λ is a scalar constant, t_3 is such that $t_1 < t_3 < t_2$, and $\mathbf{u}(t)$ and $\mathbf{v}(t)$ are vector functions of the scalar variable t, then

1. $\displaystyle \int_{t_1}^{t_2} \lambda \mathbf{u}(t)\, dt = \lambda \int_{t_1}^{t_2} \mathbf{u}(t)\, dt$ (homogeneity)

2. $\displaystyle \int_{t_1}^{t_2} [\mathbf{u}(t) + \mathbf{v}(t)]\, dt = \int_{t_1}^{t_2} \mathbf{u}(t)\, dt + \int_{t_1}^{t_2} \mathbf{v}(t)\, dt$ (linearity)

3. $\displaystyle \int_{t_1}^{t_2} \mathbf{u}(t)\, dt = - \int_{t_2}^{t_1} \mathbf{u}(t)\, dt$ (interchange of limits)

4. $\displaystyle \int_{t_1}^{t_2} \mathbf{u}(t)\, dt = \int_{t_1}^{t_3} \mathbf{u}(t)\, dt + \int_{t_3}^{t_2} \mathbf{u}(t)\, dt$

(integration over contiguous intervals)

23.9 Line Integrals

23.9.1

Let $\mathbf{F}$ be a continuous and differentiable vector function of position $P(x, y, z)$ in space, and let C be a **path (arc)** joining points $P_1(x_1, y_1, z_1)$ and $P_2(x_2, y_2, z_2)$. Then the **line integral** of $\mathbf{F}$ taken along the path C from P_1 to P_2 is defined as (Figure 23.10)

1. $\displaystyle \int_C \mathbf{F} \cdot d\mathbf{r} = \int_{P_1}^{P_2} \mathbf{F} \cdot d\mathbf{r} = \int_C (F_1\, dx + F_2\, dy + F_3\, dz),$

where

2. $\mathbf{F} = F_1 \mathbf{i} + F_2 \mathbf{j} + F_3 \mathbf{k},$

and

3. $d\mathbf{r} = dx\, \mathbf{i} + dy\, \mathbf{j} + dz\, \mathbf{k}$

is a differential vector displacement along the path C. It follows that

4. $\displaystyle \int_{P_1}^{P_2} \mathbf{F} \cdot d\mathbf{r} = - \int_{P_2}^{P_1} \mathbf{F} \cdot dr,$

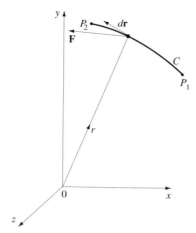

FIGURE 23.10 ∎

while for three points P_1, P_2, and P_3 on C,

5. $\displaystyle\int_{P_1}^{P_2} \mathbf{F} \cdot d\mathbf{r} = \int_{P_1}^{P_3} \mathbf{F} \cdot d\mathbf{r} + \int_{P_3}^{P_2} \mathbf{F} \cdot d\mathbf{r}.$

A special case of a line integral occurs when $\mathbf{F}$ is given by

6. $\mathbf{F} = \operatorname{grad}\phi = \nabla\phi,$

where in rectangular Cartesian coordinates

7. $\operatorname{grad}\phi = \mathbf{i}\dfrac{\partial\phi}{\partial x} + \mathbf{j}\dfrac{\partial\phi}{\partial y} + \mathbf{k}\dfrac{\partial\phi}{\partial z},$

for

8. $\displaystyle\int_C \mathbf{F} \cdot d\mathbf{r} = \int_{P_1}^{P_2} \mathbf{F} \cdot d\mathbf{r} = \phi(P_2) - \phi(P_1),$

and the line integral is *independent* of the path C, depending only on the initial point P_1 and terminal point P_2 of C. A **vector field** of the form

9. $\mathbf{F} = \operatorname{grad}\phi$

is called a **conservative field**, and ϕ is then called a **scalar potential**. For the definition of $\operatorname{grad}\phi$ in terms of other coordinate systems see 24.2.1 and 24.3.1.

In a conservative field, if C is a *closed curve*, it then follows that

10. $\displaystyle\int_C \mathbf{F} \cdot d\mathbf{r} = \oint_C \mathbf{F} \cdot d\mathbf{r} = 0,$

where the symbol $\oint$ indicates that the **curve** (contour) C is closed.

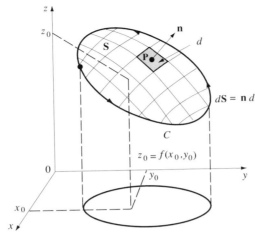

$$dS = \mathbf{n}\,d$$

FIGURE 23.11 ∎

23.10 Vector Integral Theorems

23.10.1

Let a surface S defined by $z = f(x, y)$ that is bounded by a closed space curve C have an element of surface area $d\sigma$, and let $\mathbf{n}$ be a unit vector normal to S at a representative point P (Figure 23.11).

Then the **vector element of surface area** $d\mathbf{S}$ of surface S is defined as

1. $d\mathbf{S} = d\sigma\,\mathbf{n}$.

The **surface integral** of a vector function $\mathbf{F}(x, y, z)$ over the surface S is defined as

2. $\displaystyle \int_S \mathbf{F} \cdot d\mathbf{S} = \int_S \mathbf{F} \cdot \mathbf{n}\,d\sigma.$

The **Gauss divergence theorem** states that if S is a closed surface containing a volume V with volume element dV, and if the vector element of surface area $d\mathbf{S} = \mathbf{n}\,d\sigma$, where $\mathbf{n}$ is the unit normal directed out of V and $d\sigma$ is an element of surface area of S, then

3. $\displaystyle \int_V \operatorname{div} \mathbf{F}\,dV = \int_S \mathbf{F} \cdot d\mathbf{S} = \int_S \mathbf{F} \cdot \mathbf{n}\,d\sigma.$

The Gauss divergence theorem relates the volume integral of div $\mathbf{F}$ to the surface integral of the normal component of $\mathbf{F}$ over the closed surface S.

In terms of the rectangular Cartesian coordinates $0\{x, y, z\}$, the **divergence** of the vector $\mathbf{F} = F_1\mathbf{i} + F_2\mathbf{j} + F_3\mathbf{k}$, written div $\mathbf{F}$, is defined as

4. $\displaystyle \operatorname{div} \mathbf{F} = \frac{\partial F_1}{\partial x} + \frac{\partial F_2}{\partial y} + \frac{\partial F_3}{\partial z}.$

For the definitions of div $\mathbf{F}$ in terms of other coordinate systems see 24.2.1 and 24.3.1.

Stokes's theorem states that if C is a closed curve spanned by an open surface S, and $\mathbf{F}$ is a vector function defined on S, then

5. $$\oint_C \mathbf{F} \cdot d\mathbf{r} = \int_S (\nabla \times \mathbf{F}) \cdot d\mathbf{S} = \int_S (\nabla \times \mathbf{F}) \cdot \mathbf{n} \, d\sigma.$$

In this theorem the direction of unit normal $\mathbf{n}$ in the vector element of surface area $d\mathbf{S} = d\sigma \mathbf{n}$ is chosen such that it points in the direction in which a right-handed screw would advance when rotated in the sense in which the closed curve C is traversed. A surface for which the normal is defined in this manner is called an **oriented surface**.

The surface S shown in Figure 23.11 is oriented in this manner when C is traversed in the direction shown by the arrows. In terms of rectangular Cartesian coordinates $0\{x, y, z\}$, the **curl** of the vector $\mathbf{F} = F_1\mathbf{i} + F_2\mathbf{j} + F_3\mathbf{k}$, written either $\nabla \times \mathbf{F}$, or curl $\mathbf{F}$, is defined as

6. $$\nabla \times \mathbf{F} = \left(\mathbf{i}\frac{\partial}{\partial x} + \mathbf{j}\frac{\partial}{\partial y} + \mathbf{k}\frac{\partial}{\partial z} \right) \times \mathbf{F}$$

$$= \left(\frac{\partial F_3}{\partial y} - \frac{\partial F_2}{\partial z} \right)\mathbf{i} + \left(\frac{\partial F_1}{\partial z} - \frac{\partial F_3}{\partial x} \right)\mathbf{j} + \left(\frac{\partial F_2}{\partial x} - \frac{\partial F_1}{\partial y} \right)\mathbf{k}.$$

For the definition of $\nabla \times \mathbf{F}$ in terms of other coordinate systems see 24.2.1 and 24.3.1.

Green's first and second theorems (identities). Let U and V be scalar functions of position defined in a volume V contained within a simple closed surface S, with an outward-drawn vector element of surface area $d\mathbf{S}$. Suppose further that the **Laplacians** $\nabla^2 U$ and $\nabla^2 V$ are defined throughout V, except on a finite number of surfaces inside V, across which the second-order partial derivatives of U and V are bounded but discontinuous. **Green's first theorem** states that

7. $$\int (U \nabla V) \cdot d\mathbf{S} = \int_V [U \nabla^2 V + (\nabla U) \cdot (\nabla V)] \, dV,$$

where in rectangular Cartesian coordinates

8. $$\nabla^2 U = \frac{\partial^2 U}{\partial x^2} + \frac{\partial^2 U}{\partial y^2} + \frac{\partial^2 U}{\partial z^2}.$$

The Laplacian operator ∇^2 is also often denoted by Δ, or by Δ_n if it is necessary to specify the number n of space dimensions involved, so that in Cartesian coordinates $\Delta_2 U = \partial^2 U/\partial x^2 + \partial^2 U/\partial y^2$.

For the definition of the Laplacian in terms of other coordinate systems see 24.2.1 and 24.3.1.

Green's second theorem states that

9. $$\int_V (U \nabla^2 V - V \nabla^2 U) \, dV = \int_S (U \nabla V - V \nabla U) \cdot d\mathbf{S}.$$

In two dimensions $0\{x, y\}$, **Green's theorem in the plane** takes the form

10. $$\oint_C (P \, dx + Q \, dy) = \int_A \left(\frac{\partial Q}{\partial x} - \frac{\partial P}{\partial y} \right) dx \, dy,$$

where the scalar functions $P(x, y)$ and $Q(x, y)$ are defined and differentiable in some plane area A bounded by a simple closed curve C except, possibly, across an arc γ in A joining two distinct points of C, and the integration is performed counterclockwise around C.

23.11 A Vector Rate of Change Theorem

Let u be a continuous and differentiable scalar function of position and time defined throughout a moving volume $V(t)$ bounded by a surface $S(t)$ moving with velocity **v**. Then the rate of change of the volume integral of **u** is given by

1. $$\frac{d}{dt} \int_{V(t)} u \, dV = \int_{V(t)} \left\{ \frac{\partial u}{\partial t} + \text{div}(u\mathbf{v}) \right\} dV.$$

23.12 Useful Vector Identities and Results

23.12.1

In each identity that follows the result is expressed first in terms of grad, div, and curl, and then in operator notation. **F** and **G** are suitably differentiable vector functions and V and W are suitably differentiable scalar functions.

1. $\text{div}(\text{curl}\,\mathbf{F}) \equiv \nabla \cdot (\nabla \times F) \equiv \mathbf{0}$

2. $\text{curl}(\text{grad}\,V) \equiv \nabla \times (\nabla V) \equiv \mathbf{0}$

3. $\text{grad}(VW) \equiv V\,\text{grad}\,W + W\,\text{grad}\,V \equiv V\nabla W + W\nabla V$

4. $\text{curl}(\text{curl}\,\mathbf{F}) \equiv \text{grad}(\text{div}\,\mathbf{F}) - \nabla^2\mathbf{F} \equiv \nabla(\nabla \cdot \mathbf{F}) - \nabla^2\mathbf{F}$

5. $\text{div}(\text{grad}\,V) \equiv \nabla \cdot (\nabla V) \equiv \nabla^2 V$

6. $\text{div}(V\mathbf{F}) \equiv V\,\text{div}\,\mathbf{F} + \mathbf{F} \cdot \text{grad}\,V \equiv \nabla \cdot (V\mathbf{F}) \equiv V\nabla \cdot \mathbf{F} + \mathbf{F} \cdot \nabla V$

7. $\text{curl}(V\mathbf{F}) \equiv V\,\text{curl}\,\mathbf{F} - \mathbf{F} \times \text{grad}\,V \equiv V\nabla \times \mathbf{F} - \mathbf{F} \times \nabla V$

8. $\text{grad}(\mathbf{F} \cdot \mathbf{G}) \equiv \mathbf{F} \times \text{curl}\,\mathbf{G} + \mathbf{G} \times \text{curl}\,\mathbf{F} + (\mathbf{F} \cdot \nabla)\mathbf{G} + (\mathbf{G} \cdot \nabla)\mathbf{F}$

   $\equiv \mathbf{F} \times (\nabla \times \mathbf{G}) + \mathbf{G} \times (\nabla \times \mathbf{F}) + (\mathbf{F} \cdot \nabla)\mathbf{G} + (\mathbf{G} \cdot \nabla)\mathbf{F}$

9. $\text{div}(\mathbf{F} \times \mathbf{G}) \equiv \mathbf{G} \cdot \text{curl}\,\mathbf{F} - \mathbf{F} \cdot \text{curl}\,\mathbf{G} \equiv \mathbf{G} \cdot (\nabla \times \mathbf{F}) - \mathbf{F} \cdot (\nabla \times \mathbf{G})$

10. $\text{curl}(\mathbf{F} \times \mathbf{G}) \equiv \mathbf{F}\,\text{div}\,\mathbf{G} - \mathbf{G}\,\text{div}\,\mathbf{F} + (\mathbf{G} \cdot \nabla)\mathbf{F} - (\mathbf{F} \cdot \nabla)\mathbf{G}$

   $\equiv \mathbf{F}(\nabla \cdot \mathbf{G}) - \mathbf{G}(\nabla \cdot \mathbf{F}) + (\mathbf{G} \cdot \nabla)\mathbf{F} - (\mathbf{F} \cdot \nabla)\mathbf{G}$

11. $\mathbf{F} \cdot \text{grad}\,V = \mathbf{F} \cdot (\nabla V) = (\mathbf{F} \cdot \nabla)V$ is proportional to the *directional derivative* of V in the direction **F** and it becomes the directional derivative of V in the direction **F** when **F** is a unit vector.

12. $\mathbf{F} \cdot \text{grad}\,\mathbf{G} = (\mathbf{F} \cdot \nabla)\mathbf{G}$ is proportional to the directional derivative of **G** in the direction of **F** and it becomes the directional derivative of **G** in the direction of **F** when **F** is a unit vector.

24

Systems of Orthogonal Coordinates

24.1 Curvilinear Coordinates

24.1.1 Basic definitions

24.1.1.1   Let (x, y, z) and (ξ_1, ξ_2, ξ_3) be related by the equations

1. $x = X(\xi_1, \xi_2, \xi_3), \quad y = Y(\xi_1, \xi_2, \xi_3), \quad z = Z(\xi_1, \xi_2, \xi_3),$

where the functions X, Y, and Z are continuously differentiable functions of their arguments, and the *Jacobian* determinant

2. $J = \begin{vmatrix} \dfrac{\partial X}{\partial \xi_1} & \dfrac{\partial X}{\partial \xi_2} & \dfrac{\partial X}{\partial \xi_3} \\[2ex] \dfrac{\partial Y}{\partial \xi_1} & \dfrac{\partial Y}{\partial \xi_2} & \dfrac{\partial Y}{\partial \xi_3} \\[2ex] \dfrac{\partial Z}{\partial \xi_1} & \dfrac{\partial Z}{\partial \xi_2} & \dfrac{\partial Z}{\partial \xi_3} \end{vmatrix}$

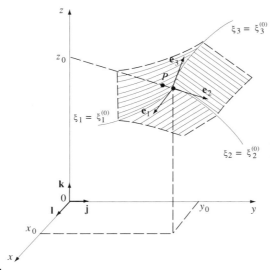

FIGURE 24.1 ∎

does not vanish throughout some region R of space. Then in R 24.1.1.1.1 can be solved uniquely for ξ_1, ξ_2, and ξ_3 in terms of x, y, and z to give

3. $\xi_1 = \Xi_1(x, y, z), \quad \xi_2 = \Xi_2(x, y, z), \quad \xi_3 = \Xi_3(x, y, z).$

The **position vector r** of a point in space with the rectangular Cartesian coordinates (x, y, z) can be written

4. $\mathbf{r} = \mathbf{r}(x, y, z).$

Then the point P with the rectangular Cartesian coordinates (x_0, y_0, z_0) corresponds to the point with the corresponding coordinates $(\xi_1^{(0)}, \xi_2^{(0)}, \xi_3^{(0)})$ in the new coordinate system, and so to the intersection of the three one-parameter curves (Figure 24.1) defined by

5. $\mathbf{r} = \mathbf{r}\left(\xi_1, \xi_2^{(0)}, \xi_3^{(0)}\right), \quad \mathbf{r} = \mathbf{r}\left(\xi_1^{(0)}, \xi_2, \xi_3^{(0)}\right), \quad \mathbf{r} = \mathbf{r}\left(\xi_1^{(0)}, \xi_2^{(0)}, \xi_3\right).$

In general, the coordinates ξ_1, ξ_2, ξ_3 are called **curvilinear coordinates,** and they are said to be **orthogonal** when the unit tangent vectors $\mathbf{e}_1$, $\mathbf{e}_2$, and $\mathbf{e}_3$ to the curves $\xi_1 = \xi_1^{(0)}, \xi_2 = \xi_2^{(0)}, \xi_3 = \xi_3^{(0)}$ through $[\xi_1^{(0)}, \xi_2^{(0)}, \xi_3^{(0)}]$ identifying point $P(x_0, y_0, z_0)$ are all mutually orthogonal.

The vectors $\mathbf{e}_1$, $\mathbf{e}_2$, and $\mathbf{e}_3$ are defined by

6. $\dfrac{\partial \mathbf{r}}{\partial \xi_1} = h_1 \mathbf{e}_1, \quad \dfrac{\partial \mathbf{r}}{\partial \xi_2} = h_2 \mathbf{e}_2, \quad \dfrac{\partial \mathbf{r}}{\partial \xi_3} = h_3 \mathbf{e}_3,$

where the **scale factors** h_1, h_2, and h_3 are given by

7. $h_1 = \left| \dfrac{\partial \mathbf{r}}{\partial \xi_1} \right|, \quad h_2 = \left| \dfrac{\partial \mathbf{r}}{\partial \xi_2} \right|, \quad h_3 = \left| \dfrac{\partial \mathbf{r}}{\partial \xi_3} \right|.$

The following general results are valid for orthogonal curvilinear coordinates:

8. $d\mathbf{r} = h_1 d\xi_1 \mathbf{e}_1 + h_2 d\xi_2 \mathbf{e}_2 + h_3 d\xi_3 \mathbf{e}_3$,

and if ds is an element of arc length, then

$$(ds)^2 = d\mathbf{r} \cdot d\mathbf{r} = h_1^2 d\xi_1^2 + h_2^2 d\xi_2^2 + h_3^2 d\xi_3^2.$$

24.2 Vector Operators in Orthogonal Coordinates

24.2.1

Let V be a suitably differentiable scalar function of position, and $\mathbf{F} = F_1 \mathbf{e}_1 + F_2 \mathbf{e}_2 + F_3 \mathbf{e}_3$ be a vector function defined in terms of the orthogonal curvilinear coordinates ξ_1, ξ_2, ξ_3 introduced in 24.1.1. Then the vector operations of **gradient, divergence,** and **curl** and the scalar **Laplacian** operator $\nabla^2 V$ take the form:

1. $\operatorname{grad} V = \nabla V = \dfrac{1}{h_1} \dfrac{\partial V}{\partial \xi_1} \mathbf{e}_1 + \dfrac{1}{h_2} \dfrac{\partial V}{\partial \xi_2} \mathbf{e}_2 + \dfrac{1}{h_3} \dfrac{\partial V}{\partial \xi_3} \mathbf{e}_3$

2. $\operatorname{div} \mathbf{F} = \nabla \cdot \mathbf{F} = \dfrac{1}{h_1 h_2 h_3} \left[\dfrac{\partial}{\partial \xi_1} (h_2 h_3 F_1) + \dfrac{\partial}{\partial \xi_2} (h_3 h_1 F_2) + \dfrac{\partial}{\partial \xi_3} (h_1 h_2 F_3) \right]$

3. $\operatorname{curl} \mathbf{F} = \nabla \times \mathbf{F} = \dfrac{1}{h_1 h_2 h_3} \begin{vmatrix} h_1 \mathbf{e}_1 & h_2 \mathbf{e}_2 & h_3 \mathbf{e}_3 \\ \dfrac{\partial}{\partial \xi_1} & \dfrac{\partial}{\partial \xi_2} & \dfrac{\partial}{\partial \xi_3} \\ h_1 F_1 & h_2 F_2 & h_3 F_3 \end{vmatrix}$

4. $\nabla^2 V = \dfrac{1}{h_1 h_2 h_3} \left[\dfrac{\partial}{\partial \xi_1} \left(\dfrac{h_2 h_3}{h_1} \dfrac{\partial V}{\partial \xi_1} \right) + \dfrac{\partial}{\partial \xi_2} \left(\dfrac{h_3 h_1}{h_2} \dfrac{\partial V}{\partial \xi_2} \right) + \dfrac{\partial}{\partial \xi_3} \left(\dfrac{h_1 h_2}{h_3} \dfrac{\partial V}{\partial \xi_3} \right) \right]$

The above operations have the following properties:

5. $\nabla(V + W) = \nabla V + \nabla W$

6. $\nabla \cdot (\mathbf{F} + \mathbf{G}) = \nabla \cdot \mathbf{F} + \nabla \cdot \mathbf{G}$

7. $\nabla \cdot (V\mathbf{F}) = V(\nabla \cdot \mathbf{F}) + (\nabla V) \cdot \mathbf{F}$

8. $\nabla \cdot (\nabla V) = \nabla^2 V$

9. $\nabla \times (\mathbf{F} + \mathbf{G}) = \nabla \times \mathbf{F} + \nabla \times \mathbf{G}$

10. $\nabla \times (V\mathbf{F}) = V(\nabla \times \mathbf{F}) + (\nabla V) \times \mathbf{F}$,

where V, W are differentiable scalar functions and $\mathbf{F}, \mathbf{G}$ are differentiable vector functions.

24.3 Systems of Orthogonal Coordinates

24.3.1

The following are the most frequently used systems of orthogonal curvilinear coordinates. In each case the relationship between the curvilinear coordinates and

Cartesian coordinates is given together with the scale factors h_1, h_2, h_3 and the forms taken by $\nabla V, \nabla \cdot \mathbf{F}, \nabla \times \mathbf{F}$, and $\nabla^2 V$.

1. In terms of a right-handed set of **rectangular Cartesian coordinates** $0\{x, y, z\}$ with the fixed unit vectors $\mathbf{i}, \mathbf{j}$, and $\mathbf{k}$ along the x-, y-, and z-axes, respectively, and scalar V and vector $\mathbf{F} = F_1 \mathbf{i} + F_2 \mathbf{j} + F_3 \mathbf{k}$:

(a) $\operatorname{grad} V = \nabla V = \dfrac{\partial V}{\partial x}\mathbf{i} + \dfrac{\partial V}{\partial y}\mathbf{j} + \dfrac{\partial V}{\partial z}\mathbf{k}$

(b) $\operatorname{div} \mathbf{F} = \nabla \cdot \mathbf{F} = \dfrac{\partial F_1}{\partial x} + \dfrac{\partial F_2}{\partial y} + \dfrac{\partial F_3}{\partial z}$

(c) $\operatorname{curl} \mathbf{F} = \nabla \times \mathbf{F} = \begin{vmatrix} \mathbf{i} & \mathbf{j} & \mathbf{k} \\ \dfrac{\partial}{\partial x} & \dfrac{\partial}{\partial y} & \dfrac{\partial}{\partial z} \\ F_1 & F_2 & F_3 \end{vmatrix}$

(d) $\nabla^2 V = \dfrac{\partial^2 V}{\partial x^2} + \dfrac{\partial^2 V}{\partial y^2} + \dfrac{\partial^2 V}{\partial z^2}$

2. **Cylindrical polar coordinates** (r, θ, z) are three-dimensional coordinates defined as in Figure 24.2.

(a) $x = r \cos\theta, \quad y = r \sin\theta, \quad z = z \qquad [0 \le \theta \le 2\pi]$

(b) $h_1^2 = 1, \quad h_2^2 = r^2, \quad h_3^2 = 1$

(c) $\mathbf{F} = F_1 \mathbf{e}_r + F_2 \mathbf{e}_\theta + F_3 \mathbf{e}_z$

(d) $\operatorname{grad} V = \nabla V = \dfrac{\partial V}{\partial r}\mathbf{e}_r + \dfrac{1}{r}\dfrac{\partial V}{\partial \theta}\mathbf{e}_\theta + \dfrac{\partial V}{\partial z}\mathbf{e}_z$

(e) $\operatorname{div} \mathbf{F} = \nabla \cdot \mathbf{F} = \dfrac{1}{r}\left[\dfrac{\partial}{\partial r}(r F_1) + \dfrac{\partial F_2}{\partial \theta} + r\dfrac{\partial F_3}{\partial z} \right]$

(f) $\operatorname{curl} \mathbf{F} = \nabla \times \mathbf{F} = \dfrac{1}{r}\begin{vmatrix} \mathbf{e}_r & r\mathbf{e}_\theta & \mathbf{e}_z \\ \dfrac{\partial}{\partial r} & \dfrac{\partial}{\partial \theta} & \dfrac{\partial}{\partial z} \\ F_1 & r F_2 & F_3 \end{vmatrix}$

(g) $\nabla^2 V = \dfrac{1}{r}\left[\dfrac{\partial}{\partial r}\left(r\dfrac{\partial V}{\partial r} \right) + \dfrac{1}{r}\dfrac{\partial}{\partial \theta}\left(\dfrac{\partial V}{\partial \theta} \right) + \dfrac{\partial}{\partial z}\left(r\dfrac{\partial V}{\partial z} \right) \right]$

The corresponding expressions for **plane polar coordinates** follow from (a) to (g) above by omitting the terms involving z, and by confining attention to the plane $z = 0$ in Figure 24.2.

3. **Spherical coordinates** (r, θ, ϕ) are three-dimensional coordinates defined as in Figure 24.3.

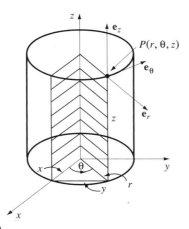

FIGURE 24.2 ∎

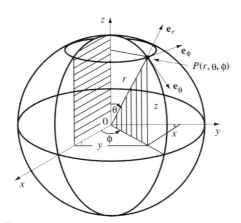

FIGURE 24.3 ∎

(a) $x = r \sin\theta \cos\phi$, $\quad y = r \sin\theta \sin\phi$, $\quad z = r \cos\theta$,
$$[0 \le \theta \le \pi, 0 \le \phi < 2\pi]$$

(b) $h_1^2 = 1$, $\quad h_2^2 = r^2$, $\quad h_3^2 = r^2 \sin^2\theta$

(c) $\mathbf{F} = F_1 \mathbf{e}_r + F_2 \mathbf{e}_\theta + F_3 \mathbf{e}_\phi$

(d) $\operatorname{grad} V = \dfrac{\partial V}{\partial r} \mathbf{e}_r + \dfrac{1}{r} \dfrac{\partial V}{\partial \theta} \mathbf{e}_\theta + \dfrac{1}{r \sin\theta} \dfrac{\partial V}{\partial \phi} \mathbf{e}_\phi$

(e) $\operatorname{div} \mathbf{F} = \nabla \cdot \mathbf{F} = \dfrac{1}{r^2} \dfrac{\partial}{\partial r}(r^2 F_1) + \dfrac{1}{r \sin\theta} \dfrac{\partial}{\partial \theta}(\sin\theta\, F_2) + \dfrac{1}{r \sin\theta} \dfrac{\partial F_3}{\partial \phi}$

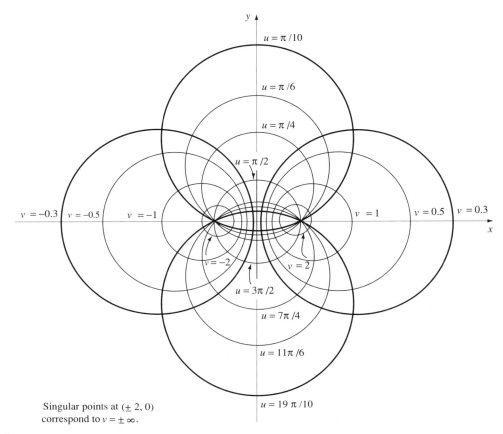

Singular points at $(\pm\, 2, 0)$
correspond to $v = \pm\,\infty$.

FIGURE 24.4 ∎

(f) $\operatorname{curl} \mathbf{F} = \nabla \times \mathbf{F} = \dfrac{1}{r^2 \sin\theta} \begin{vmatrix} \mathbf{e}_r & r\,\mathbf{e}_\theta & r\sin\theta\,\mathbf{e}_\phi \\ \dfrac{\partial}{\partial r} & \dfrac{\partial}{\partial \theta} & \dfrac{\partial}{\partial \phi} \\ F_1 & r F_2 & r\sin\theta\, F_3 \end{vmatrix}$

(g) $\nabla^2 V = \dfrac{1}{r^2}\dfrac{\partial}{\partial r}\left(r^2 \dfrac{\partial V}{\partial r}\right) + \dfrac{1}{r^2 \sin\theta}\dfrac{\partial}{\partial \theta}\left(\sin\theta \dfrac{\partial V}{\partial \theta}\right) + \dfrac{1}{r^2 \sin^2\theta}\dfrac{\partial^2 V}{\partial \phi^2}$

4. **Bipolar coordinates** (u, v, z) are three-dimensional coordinates defined as in Figure 24.4, which shows the coordinate system in the plane $z = 0$. In three dimensions the translation of these coordinate curves parallel to the z-axis generates cylindrical coordinate surfaces normal to the plane $z = 0$.

(a) $x = \dfrac{a \sinh v}{\cosh v - \cos u}, \quad y = \dfrac{a \sin u}{\cosh v - \cos u}, \quad z = z$

$[0 \le u < 2\pi,\ -\infty < v < \infty,\ -\infty < z < \infty]$

(b) $\quad h_1^2 = h_2^2 = R^2 = \dfrac{a^2}{(\cosh v - \cos u)^2}, \quad h_3^2 = 1$

(c) $\quad \mathbf{F} = F_1\mathbf{e}_u + F_2\mathbf{e}_v + F_3\mathbf{e}_z$

(d) $\quad \text{grad } V = \dfrac{1}{R}\left(\dfrac{\partial V}{\partial u}\mathbf{e}_u + \dfrac{\partial V}{\partial v}\mathbf{e}_v\right) + \dfrac{\partial V}{\partial z}\mathbf{e}_z$

(e) $\quad \text{div } \mathbf{F} = \nabla \cdot \mathbf{F} = \dfrac{1}{R^2}\left[\dfrac{\partial}{\partial u}(RF_1) + \dfrac{\partial}{\partial v}(RF_2) + \dfrac{\partial}{\partial z}(R^2F_3)\right]$

(f) $\quad \text{curl } \mathbf{F} = \nabla \times \mathbf{F} = \dfrac{1}{R^2}\begin{vmatrix} R\mathbf{e}_u & R\mathbf{e}_v & \mathbf{e}_z \\ \dfrac{\partial}{\partial u} & \dfrac{\partial}{\partial v} & \dfrac{\partial}{\partial z} \\ RF_1 & RF_2 & F_3 \end{vmatrix}$

(g) $\quad \nabla^2 V = \dfrac{1}{R^2}\left(\dfrac{\partial^2 V}{\partial u^2} + \dfrac{\partial^2 V}{\partial v^2}\right) + \dfrac{\partial^2 V}{\partial z^2}$

5. **Toroidal coordinates** (u, v, ϕ) are three-dimensional coordinates defined in terms of two-dimensional bipolar coordinates. They are obtained by relabeling the y-axis in Figure 24.4 as the z-axis, and then rotating the curves $u = \text{const.}$ and $v = \text{const.}$ about the new z-axis so that each curve $v = \text{const.}$ generates a **torus.** The angle ϕ is measured about the z-axis from the (x, z)-plane, with $0 \le \phi < 2\pi$.

(a) $\quad x = \dfrac{a \sinh v \cos \phi}{\cosh v - \cos u}, \quad y = \dfrac{a \sinh v \sin \phi}{\cosh v - \cos u}, \quad z = \dfrac{a \sinh u}{\cosh v - \cos u},$

$$[0 \le u < 2\pi, -\infty < v < \infty, 0 \le \phi < 2\pi]$$

(b) $\quad h_1^2 = h_2^2 = R^2 = \dfrac{a^2}{(\cosh v - \cos u)^2}, \quad h_3^2 = \dfrac{a^2 \sinh^2 v}{(\cosh v - \cos u)^2}$

(c) $\quad \mathbf{F} = F_1\mathbf{e}_u + F_2\mathbf{e}_v + F_3\mathbf{e}_\phi$

where $\mathbf{e}_u, \mathbf{e}_v, \mathbf{e}_\phi$ are the unit vectors in the toroidal coordinates.

(d) $\quad \text{grad } V = \nabla V = \dfrac{1}{R}\left(\dfrac{\partial V}{\partial u}\mathbf{e}_u + \dfrac{\partial V}{\partial v}\mathbf{e}_v\right) + \dfrac{1}{R \sinh v}\dfrac{\partial V}{\partial \phi}\mathbf{e}_\phi$

(e) $\quad \text{div } \mathbf{F} = \nabla \cdot \mathbf{F}$

$$= \dfrac{1}{R^3 \sinh v}\left[\dfrac{\partial}{\partial u}(R^2 \sinh v F_1) + \dfrac{\partial}{\partial v}(R^2 \sinh v F_2) + \dfrac{\partial}{\partial \phi}(R^2 F_3)\right]$$

(f) $\quad \text{curl } \mathbf{F} = \nabla \times \mathbf{F} = \dfrac{1}{\sinh v}\begin{vmatrix} \mathbf{e}_u & \mathbf{e}_v & \sinh v\,\mathbf{e}_\phi \\ \dfrac{\partial}{\partial u} & \dfrac{\partial}{\partial v} & \dfrac{\partial}{\partial \phi} \\ F_1 & F_2 & \sinh v\,F_3 \end{vmatrix}$

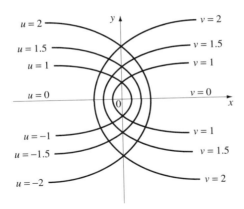

FIGURE 24.5 ▮

$$\text{(g)} \quad \nabla^2 V = \frac{1}{R^3 \sinh v} \left[\frac{\partial}{\partial u} \left(R \sinh v \frac{\partial V}{\partial u} \right) \right.$$

$$\left. + \frac{\partial}{\partial v} \left(R \sinh v \frac{\partial V}{\partial v} \right) + \frac{\partial}{\partial \phi} \left(\frac{R}{\sinh v} \frac{\partial V}{\partial \phi} \right) \right]$$

6. **Parabolic cylindrical coordinates** (u, v, z) are three-dimensional coordinates defined as in Figure 24.5, which shows the coordinate system in the plane $z = 0$. In three dimensions the translation of these coordinate curves parallel to the z-axis generates parabolic cylindrical coordinate surfaces normal to the plane $z = 0$.

   (a) $x = \frac{1}{2}(u^2 - v^2), \quad y = uv, \quad z = z$

   (b) $h_1^2 = h_2^2 = h^2 = u^2 + v^2, \quad h_3^2 = 1$

   (c) $\mathbf{F} = F_1 \mathbf{e}_u + F_2 \mathbf{e}_v + F_3 \mathbf{e}_z$

   (d) $\operatorname{grad} V = \nabla V = \frac{1}{h} \left(\frac{\partial V}{\partial u} \mathbf{e}_u + \frac{\partial V}{\partial v} \mathbf{e}_v \right) + \frac{1}{uv} \frac{\partial V}{\partial z} \mathbf{e}_z$

   (e) $\operatorname{div} \mathbf{F} = \nabla \cdot \mathbf{F} = \frac{1}{h^2} \left[\frac{\partial}{\partial u} (h F_1) + \frac{\partial}{\partial v} (h F_2) + \frac{\partial}{\partial z} (h^2 F_3) \right]$

   (f) $\operatorname{curl} \mathbf{F} = \nabla \times \mathbf{F} = \frac{1}{h^2} \begin{vmatrix} h\mathbf{e}_u & h\mathbf{e}_v & \mathbf{e}_z \\ \dfrac{\partial}{\partial u} & \dfrac{\partial}{\partial v} & \dfrac{\partial}{\partial z} \\ h F_1 & h F_2 & F_3 \end{vmatrix}$

   (g) $\nabla^2 V = \frac{1}{h^2} \left(\frac{\partial^2 V}{\partial u^2} + \frac{\partial^2 V}{\partial v^2} \right) + \frac{\partial^2 V}{\partial z^2}$

7. **Paraboloidal coordinates** (u, v, ϕ) are three-dimensional coordinates defined in terms of two-dimensional parabolic cylindrical coordinates. They are obtained by relabeling the x- and y-axes in Figure 24.5 as the z- and x-axes, respectively, and then rotating the curves about the new z-axis, so that each parabola generates a paraboloid. The angle ϕ is measured about the z-axis from the (x, z)-plane, with $0 \leq \phi < 2\pi$.

(a) $x = uv \cos \phi, \quad y = uv \sin \phi, \quad z = \frac{1}{2}(u^2 - v^2)$

$$[u \geq 0, v \geq 0, 0 \leq \phi < 2\pi]$$

(b) $h_1^2 = h_2^2 = h^2 = u^2 + v^2, \quad h_3^2 = u^2 v^2$

(c) $\mathbf{F} = F_1 \mathbf{e}_u + F_2 \mathbf{e}_v + F_3 \mathbf{e}_\phi$

where $\mathbf{e}_u, \mathbf{e}_v, \mathbf{e}_\phi$ are the unit vectors in the paraboloidal coordinates.

(d) $\operatorname{grad} V = \nabla V = \dfrac{1}{h}\left(\dfrac{\partial V}{\partial u}\mathbf{e}_u + \dfrac{\partial V}{\partial v}\mathbf{e}_v\right) + \dfrac{1}{uv}\dfrac{\partial V}{\partial \phi}\mathbf{e}_\phi$

(e) $\operatorname{div} \mathbf{F} = \nabla \cdot \mathbf{F} = \dfrac{1}{h^2 uv}\left[\dfrac{\partial}{\partial u}(huv F_1) + \dfrac{\partial}{\partial v}(huv F_2) + \dfrac{\partial}{\partial \phi}(h^2 F_3)\right]$

(f) $\operatorname{curl} \mathbf{F} = \nabla \times \mathbf{F} = \begin{vmatrix} \mathbf{e}_u & \mathbf{e}_v & \dfrac{uv}{h}\mathbf{e}_\phi \\[2mm] \dfrac{\partial}{\partial u} & \dfrac{\partial}{\partial v} & \dfrac{\partial}{\partial \phi} \\[2mm] F_1 & F_2 & \dfrac{uv}{h}F_3 \end{vmatrix}$

(g) $\nabla^2 V = \dfrac{1}{h^2 u}\dfrac{\partial}{\partial u}\left(u\dfrac{\partial V}{\partial u}\right) + \dfrac{1}{h^2 v}\dfrac{\partial}{\partial v}\left(v\dfrac{\partial V}{\partial v}\right) + \dfrac{1}{u^2 v^2}\dfrac{\partial^2 V}{\partial \phi^2}$

8. **Elliptic cylindrical coordinates** (u, v, z) are three-dimensional coordinates defined as in Figure 24.6, which shows the coordinate system in the plane $z = 0$. In three-dimensions the translation of the coordinate curves parallel to the z-axis generates elliptic cylinders corresponding to the curves $u = $ const., and parabolic cylinders corresponding to the curves $v = $ const.

(a) $x = a \cosh u \cos v, \quad y = a \sinh u \sin v, \quad z = z$

$$[u \geq 0, 0 \leq v < 2\pi, -\infty < z < \infty]$$

(b) $h_1^2 = h_2^2 = h^2 = a^2(\sinh^2 u + \sin^2 v), \quad h_3^2 = 1$

(c) $\mathbf{F} = F_1 \mathbf{e}_u + F_2 \mathbf{e}_v + F_3 \mathbf{e}_z$

(d) $\operatorname{grad} V = \dfrac{1}{h}\dfrac{\partial V}{\partial u}\mathbf{e}_u + \dfrac{1}{h}\dfrac{\partial V}{\partial v}\mathbf{e}_v + \dfrac{\partial V}{\partial z}\mathbf{e}_z$

(e) $\operatorname{div} \mathbf{F} = \nabla \cdot \mathbf{F} = \dfrac{1}{h}\left[\dfrac{\partial}{\partial u}(h F_1) + \dfrac{\partial}{\partial v}(h F_2)\right] + \dfrac{\partial F_3}{\partial z}$

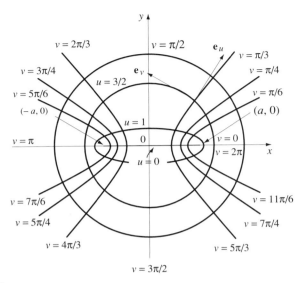

FIGURE 24.6 ∎

(f) $\mathrm{curl}\,\mathbf{F} = \nabla \times \mathbf{F} = \dfrac{1}{h^2} \begin{vmatrix} h\mathbf{e}_u & h\mathbf{e}_v & \mathbf{e}_z \\[4pt] \dfrac{\partial}{\partial u} & \dfrac{\partial}{\partial v} & \dfrac{\partial}{\partial z} \\[4pt] hF_1 & hF_2 & F_3 \end{vmatrix}$

(g) $\nabla^2 V = \dfrac{1}{h^2}\left(\dfrac{\partial^2 V}{\partial u^2} + \dfrac{\partial^2 V}{\partial v^2} \right) + \dfrac{\partial^2 V}{\partial z^2}$

9. **Prolate spheroidal coordinates** (ξ, η, ϕ) are three-dimensional coordinates defined in terms of two-dimensional elliptic cylindrical coordinates. They are obtained by relabeling the x-axis in Figure 24.6 as the z-axis, rotating the coordinate curves about the new z-axis, and taking as the family of coordinate surfaces planes containing this axis. The curves $u = $ const. then generate prolate spheroidal surfaces.

(a) $x = a \sinh \xi \, \sin \eta \, \cos \phi, \quad y = a \sinh \xi \, \sin \eta \, \sin \phi, \quad z = a \cosh \xi \, \cos \eta$
$$[\xi \geq 0, 0 \leq \eta \leq \pi, 0 \leq \phi < 2\pi]$$

(b) $h_1^2 = h_2^2 = h^2 = a^2(\sinh^2 \xi + \sin^2 \eta), \quad h_3^2 = a^2 \sinh^2 \xi \, \sin^2 \eta$

(c) $\mathbf{F} = F_1 \mathbf{e}_\xi + F_2 \mathbf{e}_\eta + F_3 \mathbf{e}_\phi$

(d) $\mathrm{grad}\,V = \dfrac{1}{h}\dfrac{\partial V}{\partial \xi}\mathbf{e}_\xi + \dfrac{1}{h}\dfrac{\partial V}{\partial \eta}\mathbf{e}_\eta + \dfrac{1}{h_3}\dfrac{\partial V}{\partial \phi}\mathbf{e}_\phi$

(e) $\mathrm{div}\,\mathbf{F} = \nabla \cdot \mathbf{F} = \dfrac{1}{h^2 h_3}\left[\dfrac{\partial}{\partial \xi}(hh_3 F_1) + \dfrac{\partial}{\partial \eta}(hh_3 F_2) \right] + \dfrac{1}{h_3}\dfrac{\partial F_3}{\partial \phi}$

(f) $\quad \text{curl } \mathbf{F} = \nabla \times \mathbf{F} = \dfrac{1}{h^2 h_3} \begin{vmatrix} h\mathbf{e}_\xi & h\mathbf{e}_\eta & h_3\mathbf{e}_\phi \\ \dfrac{\partial}{\partial \xi} & \dfrac{\partial}{\partial \eta} & \dfrac{\partial}{\partial \phi} \\ hF_1 & hF_2 & h_3F_3 \end{vmatrix}$

(g) $\quad \nabla^2 V = \dfrac{1}{h^2 \sinh \xi} \dfrac{\partial}{\partial \xi}\left(\sinh \xi \dfrac{\partial V}{\partial \xi} \right) + \dfrac{1}{h^2 \sin \eta} \dfrac{\partial}{\partial \eta}\left(\sin \eta \dfrac{\partial V}{\partial \eta} \right)$

$$+ \dfrac{1}{a^2 \sinh^2 \xi \sin^2 \eta} \dfrac{\partial^2 V}{\partial \phi^2}$$

10. **Oblate spheroidal coordinates** (ξ, η, ϕ) are three-dimensional coordinates defined in terms of two-dimensional elliptic cylindrical coordinates. They are obtained by relabeling the y-axis in Figure 24.6 as the z-axis, rotating the coordinate curves about the new z-axis, and taking as the third family of coordinate surfaces planes containing this axis. The curves $u = \text{const.}$ then generate oblate spheroidal surfaces.

(a) $\quad x = a \cosh \xi \cos \eta \cos \phi, \qquad y = a \cosh \xi \cos \eta \sin \phi, \qquad z = a \sinh \xi \sin \eta$

$$[\xi \geq 0, -\pi/2 \leq \eta \leq \pi/2, 0 \leq \phi < 2\pi]$$

(b) $\quad h_1^2 = h_2^2 = h^2 = a^2(\sinh^2 \xi + \sin^2 \eta), \quad h_3^2 = a^2 \cosh^2 \xi \cos^2 \eta$

(c) $\quad \mathbf{F} = F_1 \mathbf{e}_\xi + F_2 \mathbf{e}_\eta + F_3 \mathbf{e}_\phi$

(d) $\quad \text{grad } V = \nabla V = \dfrac{1}{h} \dfrac{\partial V}{\partial \xi} \mathbf{e}_\xi + \dfrac{1}{h} \dfrac{\partial V}{\partial \eta} \mathbf{e}_\eta + \dfrac{1}{h_3} \dfrac{\partial V}{\partial \phi} \mathbf{e}_\phi$

(e) $\quad \text{div } \mathbf{F} = \nabla \cdot \mathbf{F} = \dfrac{1}{h^2 h_3}\left[\dfrac{\partial}{\partial \xi}(hh_3 F_1) + \dfrac{\partial}{\partial \eta}(hh_3 F_2) \right] + \dfrac{1}{h_3} \dfrac{\partial F_3}{\partial \phi}$

(f) $\quad \text{curl } \mathbf{F} = \nabla \times \mathbf{F} = \dfrac{1}{h^2 h_3} \begin{vmatrix} h\mathbf{e}_\xi & h\mathbf{e}_\eta & h_3\mathbf{e}_\phi \\ \dfrac{\partial}{\partial \xi} & \dfrac{\partial}{\partial \eta} & \dfrac{\partial}{\partial \phi} \\ hF_1 & hF_2 & h_3F_3 \end{vmatrix}$

(g) $\quad \nabla^2 V = \dfrac{1}{h^2 \cosh \xi} \dfrac{\partial}{\partial \xi}\left(\cosh \xi \dfrac{\partial V}{\partial \xi} \right) + \dfrac{1}{h^2 \cos \eta} \dfrac{\partial}{\partial \eta}\left(\cos \eta \dfrac{\partial V}{\partial \eta} \right)$

$$+ \dfrac{1}{a^2 \cosh^2 \xi \cos^2 \eta} \dfrac{\partial^2 V}{\partial \phi^2}$$

25

Partial Differential Equations and Special Functions

25.1 Fundamental Ideas

25.1.1 Classification of equations

25.1.1.1 A **partial differential equation** (PDE) of **order** n, for an unknown function Φ of the m independent variables $x_1, x_2, \ldots, x_m$ ($m \geq 2$), is an equation that relates one or more of the nth-order partial derivatives of Φ to some or all of Φ, $x_1, x_2, \ldots, x_m$ and the partial derivatives of Φ or order less than n.

The most general second-order PDE can be written

1. $F\left(x_1, x_2, \ldots, x_m, \Phi, \dfrac{\partial \Phi}{\partial x_1}, \ldots, \dfrac{\partial \Phi}{\partial x_m}, \dfrac{\partial^2 \Phi}{\partial x_1^2}, \ldots, \dfrac{\partial^2 \Phi}{\partial x_i \partial x_j}, \ldots, \dfrac{\partial^2 \Phi}{\partial x_m^2}\right) = 0,$

where F is an arbitrary function of its arguments. A **solution** of 25.1.1.1.1 in a region R of the space to which the independent variables belong is a twice differentiable function that satisfies the equation throughout R.

A **boundary value problem** for a PDE arises when its solution is required to satisfy conditions on a boundary in space. If, however, one of the independent variables is the time t and the solution is required to satisfy certain conditions when $t = 0$, this

375

leads to an **initial value problem** for the PDE. Many physical situations involve a combination of both of these situations, and they then lead to an **initial boundary value problem.**

A **linear second-order PDE** for the function $\Phi(x_1, x_2, \ldots, x_m)$ is an equation of the form

2. $\displaystyle \sum_{i,j=1}^{m} A_{ij} \frac{\partial^2 \Phi}{\partial x_i \partial x_j} + \sum_{i=1}^{m} B_i \frac{\partial \Phi}{\partial x_i} + C\Phi = f,$

where the A_{ij}, B_i, C, and f are functions of the m independent variables $x_1, x_2, \ldots, x_m$. The equation is said to be **homogeneous** if $f \equiv 0$, otherwise it is **inhomogeneous (nonhomogeneous).**

The most general linear second-order PDE for the function $\Phi(x, y)$ of the two independent variables x and y is

3. $\displaystyle A(x, y)\frac{\partial^2 \Phi}{\partial x^2} + 2B(x, y)\frac{\partial^2 \Phi}{\partial x \partial y} + C(x, y)\frac{\partial^2 \Phi}{\partial y^2}$

$\displaystyle \qquad + d(x, y)\frac{\partial \Phi}{\partial x} + e(x, y)\frac{\partial \Phi}{\partial y} + f(x, y)\Phi = g(x, y),$

where x, y may be two spatial variables, or one space variable and the time (usually denoted by t).

An important, more general second-order PDE that is related to 25.1.1.1.1 is

4. $\displaystyle A\frac{\partial^2 \Phi}{\partial x^2} + 2B\frac{\partial^2 \Phi}{\partial x \partial y} + C\frac{\partial^2 \Phi}{\partial y^2} = H\left(x, y, \Phi, \frac{\partial \Phi}{\partial x}, \frac{\partial \Phi}{\partial y}\right),$

where A, B, and C, like H, are functions of $x, y, \Phi, \partial\Phi/\partial x$, and $\partial\Phi/\partial y$. A PDE of this type is said to be **quasilinear** (linear in its second (highest) order derivatives).

Linear homogeneous PDEs such as 25.1.1.1.2 and 25.1.1.1.3 have the property that if Φ_1 and Φ_2 are solutions, then so also is $c_1\Phi_1 + c_2\Phi_2$, where c_1 and c_2 are arbitrary constants. This behavior of solutions of PDEs is called the **linear superposition property,** and it is used for the construction of solutions to initial or boundary value problems.

The second-order PDEs 25.1.1.1.3 and 25.1.1.1.4 are classified throughout a region R of the (x, y)-plane according to certain of their mathematical properties that depend on the sign of $\Delta = B^2 - AC$. The equations are said to be of **hyperbolic type (hyperbolic)** whenever $\Delta > 0$, to be of **parabolic type (parabolic)** whenever $\Delta = 0$, and to be of **elliptic type (elliptic)** whenever $\Delta < 0$.

The most important linear homogeneous second-order PDEs in one, two, or three space dimensions and time are:

5. $\displaystyle \frac{1}{c^2}\frac{\partial^2 \Phi}{\partial t^2} = \nabla^2 \Phi$ (wave equation: hyperbolic)

6. $\displaystyle \kappa \frac{\partial \Phi}{\partial t} = \nabla^2 \Phi$ [diffusion (heat) equation: parabolic]

7. $\nabla^2 \Phi = 0$ (Laplace's equation: elliptic),

where c and κ are constants, and the form taken by the **Laplacian** $\nabla^2\Phi$ is determined by the coordinate system that is used. Laplace's equation is independent of the time and may be regarded as the **steady-state** form of the two previous equations, in the sense that they reduce to it if, after a suitably long time, their time derivatives may be neglected.

Only in exceptional cases is it possible to find general solutions of PDEs, so instead it becomes necessary to develop techniques that enable them to be solved subject to auxiliary conditions (initial and boundary conditions) that identify specific problems. The most frequently used initial and boundary conditions for second-order PDEs are those of Cauchy, Dirichlet, Neumann, and Robin. For simplicity these conditions are now described for second-order PDEs involving two independent variables, although they can be extended to the case of more independent variables in an obvious manner.

When the time t enters as an independent variable, a problem involving a PDE is said to be a **pure initial value problem** if it is completely specified by describing how the solution starts at time $t = 0$, and no spatial boundaries are involved. If only a first-order time derivative $\partial\Phi/\partial t$ of the solution Φ occurs in the PDE, as in the heat equation, the initial condition involving the specification of Φ at $t = 0$ is called a **Cauchy condition.** If, however, a second-order time derivative $\partial^2\Phi/\partial t^2$ occurs in the PDE, as in the wave equation, the associated **Cauchy conditions** involve the specification of both Φ and $\partial\Phi/\partial t$ at $t = 0$. In each of these cases, the determination of the solution Φ that satisfies both the PDE and the Cauchy condition(s) is called a **Cauchy problem.**

In other problems only the spatial variables x and y are involved in a PDE that contains both the terms $\partial^2\Phi/\partial x^2$ and $\partial^2\Phi/\partial y^2$ and governs the behavior of Φ in a region D of the (x, y)-plane. The region D will be assumed to lie within a closed boundary curve Γ that is smooth at all but a finite number of points, at which sharp corners occur (region D is *bounded*). A **Dirichlet condition** for such a PDE involves the specification of Φ on Γ. If $\partial\Phi/\partial n = \mathbf{n} \cdot \text{grad } \Phi$, with $\mathbf{n}$ the inward drawn normal to Γ, a **Neumann condition** involves the specification of $\partial\Phi/\partial n$ on Γ. A **Robin condition** arises when Φ is required to satisfy the condition $\alpha(x, y)\Phi + \beta(x, y)\partial\Phi/\partial n = h(x, y)$ on Γ, where h may be identically zero.

The determinations of the solutions Φ satisfying both the PDE and Dirichlet, Neumann, or Robin conditions on Γ are called **boundary value problems of the Dirichlet, Neumann,** or **Robin types,** respectively.

When a PDE subject to auxiliary conditions gives rise to a solution that is unique (except possibly for an arbitrary additive constant), and depends continuously on the data in the auxiliary conditions, it is said to be **well posed, or properly posed.** An **ill-posed** problem is one in which the solution does not possess the above properties.

Well-posed problems for the **Poisson equation** (the inhomogeneous Laplace equation)

8. $\nabla^2\Phi(x, y) = H(x, y)$

involving the above conditions are as follows:

The Dirichlet problem.

9. $\nabla^2 \Phi(x, y) = H(x, y)$ in D with $\Phi = f(x, y)$ on Γ

yields a unique solution that depends continuously on the inhomogeneous term $H(x, y)$ and the boundary data $f(x, y)$.

The Neumann problem.

10. $\nabla^2 \Phi(x, y) = H(x, y)$ in D with $\partial \Phi / \partial n = g(x, y)$ on Γ

yields a unique solution, apart from an arbitrary additive constant, that depends continuously on the inhomogeneous term $H(x, y)$ and the boundary data $g(x, y)$, provided that H and g satisfy the **compatibility condition**

11. $\displaystyle \int_D H(x, y) \, dA = \int_\Gamma g \, d\sigma,$

where dA is the element of area in D and $d\sigma$ is the element of arc length along Γ. No solution exists if the compatibility condition is not satisfied.

The Robin problem.

12. $\nabla^2 \Phi(x, y) = H(x, y)$ in D with $\alpha \Phi + \beta \dfrac{\partial \Phi}{\partial n} = h$ on Γ

yields a unique solution that depends continuously on the inhomogeneous term $H(x, y)$ and the boundary data $\alpha(x, y)$, $\beta(x, y)$, and $h(x, y)$.

   If the PDE holds in a region D that is *unbounded*, the above conditions that ensure well-posed problems for the Poisson equation must be modified as follows:

Dirichlet conditions.   Add the requirement that Φ is bounded in D.

Neumann conditions.   Delete the compatibility condition and add the requirement that $\Phi(x, y) \to 0$ as $x^2 + y^2 \to \infty$.

Robin conditions.   Add the requirement that Φ is bounded in D.

   The variability of the coefficients $A(x, y)$, $B(x, y)$, and $C(x, y)$ in 25.1.1.1.3 can cause the equation to change its type in different regions of the (x, y)-plane. An equation exhibiting this property is said to be of **mixed type.** One of the most important equations of mixed type is the **Tricomi equation**

$$y \frac{\partial^2 \Phi}{\partial x^2} + \frac{\partial^2 \Phi}{\partial y^2} = 0,$$

which first arose in the study of transonic flow. This equation is elliptic for $y > 0$, hyperbolic for $y < 0$, and degenerately parabolic along the line $y = 0$. Such equations are difficult to study because the appropriate auxiliary conditions vary according to the type of the equation. When a solution is required in a region within which the parabolic degeneracy occurs, the matching of the solution across the degeneracy gives rise to considerable mathematical difficulties.

25.2 Method of Separation of Variables

25.2.1 Application to a hyperbolic problem

25.2.1.1   The method of **separation of variables** is a technique for the determination of the solution of a boundary value or an initial value problem for a linear homogeneous equation that involves attempting to separate the spatial behavior of a solution from its time variation (**temporal behavior**). It will suffice to illustrate the method by considering the special linear homogeneous second-order hyperbolic equation

1. $\quad \operatorname{div}(k\nabla\phi) - h\phi = \rho\dfrac{\partial^2\phi}{\partial t^2}$,

in which $k > 0, h \geq 0$, and $\phi(x, t)$ is a function of position vector $\mathbf{x}$, and the time t. A typical **homogeneous boundary condition** to be satisfied by ϕ on some fixed surface S in space bounding a volume V is

2. $\quad \left(k_1\dfrac{\partial\phi}{\partial n} + k_2\phi\right)_S = 0$,

where k_1, k_2 are constants and $\partial/\partial n$ denotes the derivative of ϕ normal to S. The appropriate *initial conditions* to be satisfied by ϕ when $t = 0$ are

3. $\quad \phi(S, 0) = \phi_1(S)$ and $\dfrac{\partial\phi}{\partial t}(S, 0) = \phi_2(S)$.

The homogeneity of both 25.2.1.1.1 and the boundary condition 25.2.1.1.2 means that if $\tilde{\phi}_1$ and $\tilde{\phi}_2$ are solutions of 25.2.1.1.1 satisfying 25.2.1.1.2, then the function $c_1\tilde{\phi}_1 + c_2\tilde{\phi}_2$ with c_1, c_2 being arbitrary constants will also satisfy the same equation and boundary condition.

The method of **separation of variables** proceeds by seeking a solution of the form

4. $\quad \phi(x, t) = U(\mathbf{x})T(t)$,

in which $U(\mathbf{x})$ is a function only of the spatial position vector $\mathbf{x}$, and $T(t)$ is a function only of the time t.

Substitution of 25.2.1.1.4 into 25.2.1.1.1, followed by division by $\rho U(\mathbf{x})T(t)$, gives

5. $\quad \dfrac{L[U]}{\rho U} = \dfrac{T''}{T}$,

where $T'' = d^2T/dt^2$, and we have set

6. $\quad L[U] = \operatorname{div}(k\nabla U) - hU$.

The spatial vector $\mathbf{x}$ has been *separated* from the time variable t in 25.2.1.1.5, so the left-hand side is a function only of $\mathbf{x}$, whereas the right-hand side is a function only of t. It is only possible for these functions of $\mathbf{x}$ and t to be equal for all $\mathbf{x}$ and t if

7. $\quad \dfrac{L[U]}{\rho U} = \dfrac{T''}{T} = -\lambda$,

with λ an absolute constant called the **separation constant.** This result reduces to the equation

8. $L[U] + \lambda \rho U = 0,$

with the boundary condition

9. $\left(k_1 \dfrac{\partial U}{\partial n} + k_2 U \right)_S = 0$

governing the spatial variation of the solution, and the equation

10. $T'' + \lambda T = 0,$

with

11. $T(0) = \phi_1$ and $T'(0) = \phi_2,$

governing the time variation of the solution.

Problem 25.2.1.1.8 subject to boundary condition 25.2.1.1.9 is called a **Sturm-Liouville problem,** and it only has **nontrivial solutions** (not identically zero) for special values $\lambda_1, \lambda_2, \ldots$ of λ called the **eigenvalues** of the problem. The solutions $U_1, U_2, \ldots$, corresponding to the eigenvalues $\lambda_1, \lambda_2, \ldots$, are called the **eigenfunctions** of the problem.

The solution of 25.2.1.1.10 for each $\lambda_1, \lambda_2, \ldots$, subject to the initial conditions of 25.2.1.1.11, may be written

12. $T_n(t) = C_n \cos \sqrt{\lambda_n} t + D_n \sin \sqrt{\lambda_n} t$ $[n = 1, 2, \ldots]$

so that

$\Phi_n(x, t) = U_n(x) T_n(t).$

The solution of the original problem is then sought in the form of the linear combination

13. $\phi(x, t) = \displaystyle\sum_{n=1}^{\infty} U_n(x)[C_n \cos \sqrt{\lambda_n} t + D_n \sin \sqrt{\lambda_n} t].$

To determine the constants C_n and D_n it is necessary to use a fundamental property of eigenfunctions. Setting

14. $\|U_n\|^2 = \displaystyle\int_D \rho(\mathbf{x}) U_n^2(\mathbf{x}) \, dV,$

and using the Gauss divergence theorem, it follows that the eigenfunctions $U_n(\mathbf{x})$ are orthogonal over the volume V with respect to the **weight function** $\rho(\mathbf{x})$, so that

15. $\displaystyle\int_D \rho(\mathbf{x}) U_m(\mathbf{x}) U_n(\mathbf{x}) \, dV = \begin{cases} 0, & m \neq n, \\ \|U_n\|^2, & m = n. \end{cases}$

The constants C_n follow from 25.2.1.1.13 by setting $t = 0$, replacing $\phi(\mathbf{x}, t)$, by $\phi_1(\mathbf{x})$, multiplying by $U_m(x)$, integrating over D, and using 25.2.1.1.15 to obtain

16. $C_m = \dfrac{1}{\|U_m\|^2} \displaystyle\int_D \rho(\mathbf{x}) \phi_1(\mathbf{x}) U_m(\mathbf{x}) \, dV.$

The constants D_m follow in similar fashion by differentiating $\phi(\mathbf{x}, t)$ with respect to t and then setting $t = 0$, replacing $\partial\phi/\partial t$ by $\phi_2(\mathbf{x})$ and proceeding as in the determination of C_m to obtain

17. $\quad D_m = \dfrac{1}{\|U_m\|^2} \displaystyle\int_D \rho(\mathbf{x})\phi_2(\mathbf{x})U_m(\mathbf{x})\,dV.$

The required solution to 25.2.1.1.1 subject to the boundary condition 25.2.1.1.2 and the initial conditions 25.2.1.1.3 then follows by substituting for C_n, D_n in 25.2.1.1.13.

If in 25.2.1.1.1 the term $\rho\partial^2\phi/\partial t^2$ is replaced by $\rho\partial\phi/\partial t$ the equation becomes *parabolic*. The method of separation of variables still applies, though in place of 25.2.1.1.10, the equation governing the time variation of the solution becomes

18. $\quad T' + \lambda T = 0,$

and so

19. $\quad T(t) = e^{-\lambda t}.$

Apart from this modification, the argument leading to the solution proceeds as before.

Finally, 25.2.1.1.1 becomes elliptic if $\rho \equiv 0$, for which only an appropriate boundary condition is needed. In this case, separation of variables is only performed on the spatial variables, though the method of approach is essentially the same as the one already outlined. The boundary conditions lead first to the eigenvalues and eigenfunctions, and then to the solution.

25.3 The Sturm–Liouville Problem and Special Functions

25.3.1

In the Sturm–Liouville problem 25.2.1.1.8, subject to the boundary condition 25.2.1.1.9, the operator $L[\phi]$ is a purely spatial operator that may involve any number of space dimensions. The coordinate system that is used determines the form of $L[\phi]$ (see 24.2.1) and the types of special functions that enter into the eigenfunctions.

To illustrate matters we consider as a representative problem 25.2.1.1.1 in cylindrical polar coordinates (r, θ, z) with $k = $ const., $\rho = $ const., and $h \equiv 0$, when the equation takes the form

1. $\quad \dfrac{\partial^2\phi}{\partial r^2} + \dfrac{1}{r}\dfrac{\partial\phi}{\partial r} + \dfrac{1}{r^2}\dfrac{\partial^2\phi}{\partial\theta^2} + \dfrac{\partial^2\phi}{\partial z^2} = \dfrac{1}{c^2}\dfrac{\partial^2\phi}{\partial t^2},$

where $c^2 = k/\rho$.

The first step in separating variables involves separating out the time variation by writing

2. $\quad \phi(r, \theta, z, t) = U(r, \theta, z)T(t).$

Substituting for ϕ in 25.3.1.1 and dividing by UT gives

3. $\quad \dfrac{1}{U}\left[\dfrac{\partial^2 U}{\partial r^2} + \dfrac{1}{r}\dfrac{\partial U}{\partial r} + \dfrac{1}{r^2}\dfrac{\partial^2 U}{\partial\theta^2} + \dfrac{\partial^2 U}{\partial z^2}\right] = \dfrac{1}{c^2 T}\dfrac{d^2 T}{dt^2}.$

Introducing a separation constant $-\lambda^2$ by setting $1/(c^2 T)\, d^2 T/dt^2 = -\lambda^2$ reduces 25.3.1.3 to the two differential equations

4. $\quad \dfrac{d^2 T}{dt^2} + c^2\lambda^2 T = 0$ $\hspace{4cm}$ (time variation equation)

and

5. $\quad \dfrac{\partial^2 U}{\partial r^2} + \dfrac{1}{r}\dfrac{\partial U}{\partial r} + \dfrac{1}{r^2}\dfrac{\partial^2 U}{\partial\theta^2} + \dfrac{\partial^2 U}{\partial z^2} + \lambda^2 U = 0.$ $\hspace{1cm}$ (Sturm–Liouville equation)

The separation constant has been chosen to be negative because solutions of wave-like equations must be oscillatory with respect to time (see 25.3.1.4).

To separate the variables in this last equation, which is the Sturm–Liouville equation, it is necessary to set

6. $\quad U(r, \theta, z) = R(r)\Theta(\theta)Z(z)$

and to introduce two further separation constants. However, a considerable simplification results if fewer independent variables are involved. This happens, for example, in the cylindrical polar coordinate system when the solution in planes parallel to $z = 0$ is the same, so there is no variation with z. This reduces the mathematical study of the three-dimensional problem to a two-dimensional one involving the plane polar coordinates (r, θ). Removing the term $\partial^2 U/\partial z^2$ in the Sturm–Liouville equation reduces it to

7. $\quad \dfrac{\partial^2 U}{\partial r^2} + \dfrac{1}{r}\dfrac{\partial U}{\partial r} + \dfrac{1}{r^2}\dfrac{\partial^2 U}{\partial\theta^2} + \lambda^2 U = 0.$

To separate the variables in 25.3.1.7 we now set

8. $\quad U(r, \theta) = R(r)\Theta(\theta),$

substitute for U, and multiply by $r^2/(R\Theta)$ to obtain

9. $\quad \dfrac{1}{R}\left(r^2\dfrac{\partial^2 R}{\partial r^2} + r\dfrac{\partial R}{\partial r} + \lambda^2 r^2 R\right) = -\dfrac{1}{\Theta}\dfrac{d^2\Theta}{d\theta^2},$

where the expression on the left is a function only of r, whereas the one on the right is a function only of θ. Introducing a new separation constant μ^2 by writing $\mu^2 = -(1/\Theta)d^2\Theta/d\theta^2$ shows that

10. $\quad \dfrac{d^2\Theta}{d\theta^2} + \mu^2\Theta = 0$

and

11. $\quad r^2\dfrac{d^2 R}{dr^2} + r\dfrac{dR}{dr} + (\lambda^2 r^2 - \mu^2)R = 0.$

To proceed further we must make use of the boundary condition for the problem which, as yet, has not been specified. It will be sufficient to consider the solution of 25.3.1.1 in a circular cylinder of radius a with its axis coinciding with the z-axis, inside of which the solution ϕ is finite, while on its surface $\phi = 0$. Because the solution ϕ will be of the form $\phi(r, \theta, z, t) = R(r)\Theta(\theta)T(t)$, it follows directly from this that the boundary condition $\phi(a, \theta, z, t) = 0$ corresponds to the condition

12. $R(a) = 0$.

The solution of 25.3.1.10 is

13. $\Theta(\theta) = \tilde{A} \cos \mu\theta + \tilde{B} \sin \mu\theta$,

where $\tilde{A}$, $\tilde{B}$ are arbitrary constants. Inside the cylinder $r = a$ the solution must be invariant with respect to a rotation through 2π, so that $\Theta(\theta + 2\pi) = \Theta(\theta)$, which shows that μ must be an integer $n = 0, 1, 2, \ldots$. By choosing the reference line from which θ is measured 25.3.1.3 may be rewritten in the more convenient form

14. $\Theta_n(\theta) = A_n \cos n\theta$.

The use of integral values of n in 25.3.1.11 shows the variation of the solution ϕ with r to be governed by **Bessel's equation** of integral order n

15. $r^2 \dfrac{d^2 R}{dr^2} + r \dfrac{dR}{dr} + (\lambda^2 r^2 - n^2) R = 0$,

which has the general solution (see 17.1.1.1.8)

16. $R(r) = B J_n(\lambda r) + C Y_n(\lambda r)$.

The condition that the solution ϕ must be finite inside the cylinder requires us to set $C = 0$, because $Y_n(\lambda r)$ is infinite at the origin (see 17.2.22.2), while the condition $R(a) = 0$ in 25.3.1.12 requires that

17. $J_n(\lambda a) = 0$,

which can only hold if λa is a zero of $J_n(x)$. Denoting the zeros of $J_n(x)$ by $j_{n,1}, j_{n,2}, \ldots$ (see 17.5.11) it follows that

18. $\lambda_{n,m} = j_{n,m}/a \qquad [n = 0, 1, \ldots; m = 1, 2, \ldots]$.

The possible spatial modes of the solution are thus described by the *eigenfunctions*

19. $U_{n,m}(r, \theta) = J_n\left(\dfrac{j_{n,m}r}{a}\right) \cos n\theta \qquad [n = 0, 1, \ldots; m = 1, 2, \ldots]$.

The solution ϕ is then found by setting

20. $\phi(r, \theta, z, t) = \displaystyle\sum_{\substack{n=0 \\ m=1}}^{\infty} J_n\left(\dfrac{j_{n,m}r}{a}\right) \cos n\theta [A_{nm} \cos(\lambda_{n,m}ct) + B_{nm} \sin(\lambda_{n,m}ct)]$,

and choosing the constants A_{nm}, B_{nm} so that specified initial conditions are satisfied. In this result the multiplicative constant A_n in 25.3.1.14 has been incorporated into the arbitrary constants A_{nm}, B_{nm} introduced when 25.3.1.4 was solved.

The eigenfunctions 25.3.1.19 may be interpreted as the possible *spatial modes* of an oscillating uniform circular membrane that is clamped around its circumference $r = a$. Each term in 25.3.1.20 represents the time variation of a possible mode, with the response to specified initial conditions comprising a suitable sum of such terms.

Had 25.2.1.1 been expressed in terms of spherical polar coordinates (r, θ, ϕ), with $k = \text{const.}$, $h \equiv 0$, and $\rho = 0$, the equation would have reduced to Laplace's equation $\nabla^2 \phi = 0$. In the case where the solution is required to be finite and independent of the azimuthal angle ϕ, separation of variables would have led to eigenfunctions of the form

$$U_n(r, \theta) = Ar^n P_n(\cos \theta),$$

where $P_n(\cos \theta)$ is the *Legendre polynomial* of degree n.

Other choices of coordinate systems may lead to different and less familiar special functions, the properties of many of which are only to be found in more advanced reference works.

25.4 A First-Order System and the Wave Equation

25.4.1

Although linear second-order partial differential equations govern the behavior of many physical systems, they are not the only type of equation that is of importance. In most applications that give rise to second-order equations, the equations occur as a result of the elimination of a variable, sometimes a vector, between a coupled system of more fundamental **first-order equations.** The study of such first-order systems is of importance because when variables cannot be eliminated in order to arrive at a single higher order equation the underlying first-order system itself must be solved.

A typical example of the derivation of a single second-order equation from a coupled system of first-order equations is provided by considering **Maxwell's equations** in a vacuum. These comprise a first-order linear system of the form

1. $\dfrac{\partial \mathbf{E}}{\partial t} = \text{curl } \mathbf{H}, \qquad \dfrac{\partial \mathbf{H}}{\partial t} = -\text{curl } \mathbf{E}, \qquad \text{with div } \mathbf{E} = 0.$

Here $\mathbf{E} = (E_1, E_2, E_3)$ is the electric vector, $\mathbf{H} = (H_1, H_2, H_3)$ is the magnetic vector, and the third equation is the condition that no distributed charge is present.

Differentiation of the first equation with respect to t followed by substitution for $\partial \mathbf{H}/\partial t$ from the second equation leads to the following second-order vector differential equation for $\mathbf{E}$:

2. $\dfrac{\partial^2 \mathbf{E}}{\partial t^2} = -\text{curl (curl } \mathbf{E}).$

Using vector identity 23.1.1.4 together with the condition div $\mathbf{E} = 0$ shows $\mathbf{E}$ satisfies the vector wave equation

3. $\dfrac{\partial^2 \mathbf{E}}{\partial t^2} = \nabla^2 \mathbf{E}.$

The linearity of this equation then implies that each of the components of $\mathbf{E}$ separately must satisfy this same equation, so

4. $\dfrac{\partial^2 U}{\partial t^2} = \nabla^2 U,$

where U may be any component of $\mathbf{E}$.

An identical argument shows that the components of **H** satisfy this same scalar wave equation. Thus the study of solutions of the first-order system involving Maxwell's equations reduces to the study of the single second-order scalar wave equation.

An example of a first-order quasilinear system of equations, the study of which cannot be reduced to the study of a single higher order equation, is provided by the equations governing **compressible gas flow**

5. $\dfrac{\partial \rho}{\partial t} + \operatorname{div}(\rho \mathbf{u}) = 0$

6. $\dfrac{\partial \mathbf{u}}{\partial t} + \mathbf{u} \cdot \operatorname{grad} \mathbf{u} + \dfrac{1}{\rho} \operatorname{grad} p = 0$

7. $p = f(\rho),$

where ρ is the gas density, $\mathbf{u}$ is the gas velocity, p is the gas pressure, and $f(\rho)$ is a known function of ρ (it is the **constitutive equation** relating the pressure and density). Only in the case of linear acoustics, in which the pressure variations are small enough to justify linearization of the equations, can they be reduced to the study of the scalar wave equation.

25.5 Conservation Equations (Laws)

25.5.1

A type of first-order equation that is of fundamental importance in applications is the **conservation equation,** sometimes called a **conservation law.** In the one-dimensional case let $u = u(x, t)$ represent the density per unit volume of a quantity of physical interest. Then in a cylindrical volume of cross-sectional area A normal to the x-axis and extending from $x = a$ to $x = b$, the amount present at time t is

1. $Q = A \displaystyle\int_a^b u(x, t)\, dx.$

Let $f(x, t)$ at position x and time t be the amount of u that is flowing through a unit area normal to the x-axis per unit time. The quantity $f(x, t)$ is called the **flux** of u at position x and time t. Considering the flux at the ends of the cylindrical volume we see that

2. $Q = A[f(a, t) - f(b, t)],$

because in a one-dimensional problem there is no flux normal to the axis of the cylinder (through the curved walls).

If there is an internal source for u distributed throughout the cylinder it is necessary to take account of its effect on u before arriving at a final *balance equation* (*conservation law*) for u. Suppose u is created (or removed) at a rate $h(x, t, u)$ at position x and time t. Then the rate of production (or removal) of u throughout the volume $= A \int_a^b h(x, t, u)\, dx$. Balancing all three of these results to find the rate of change of Q with respect to t gives

3. $\dfrac{d}{dt} \displaystyle\int_a^b u(x, t)\, dx = f(a, t) - f(b, t) + \int_a^b h(x, t, u)\, dx.$

This is the **integral form of the conservation equation** for u.

Provided $u(x, t)$ and $f(x, t)$ are differentiable, this may be rewritten as

4. $\displaystyle\int_a^b \left[\dfrac{\partial u}{\partial t} + \dfrac{\partial f}{\partial x} - h(x, t, u) \right] dx = 0,$

for arbitrary a and b. The result can only be true for all a and b if

5. $\dfrac{\partial u}{\partial t} + \dfrac{\partial f}{\partial x} = h(x, t, u),$

which is the **differential equation form of the conservation equation** for u.

In more space dimensions the differential form of the conservation equation becomes the equation in **divergence form**

6. $\dfrac{\partial u}{\partial t} + \operatorname{div} \mathbf{f} = h(\mathbf{x}, t, u).$

25.6 The Method of Characteristics

25.6.1

Because the fundamental properties of first-order systems are reflected in the behavior of single first-order scalar equations, the following introductory account will be restricted to this simpler case. Consider the single first-order quasilinear equation

1. $a(x, t, u) \dfrac{\partial u}{\partial t} + b(x, t, u) \dfrac{\partial u}{\partial x} = h(x, t, u),$

subject to the initial condition $u(x, 0) = g(x)$.

Let a curve Γ in the (x, t)-plane be defined parametrically in terms of σ by $t = t(\sigma)$, $x = x(\sigma)$. The tangent vector $\mathbf{T}$ to Γ has components $dx/d\sigma, dt/d\sigma$, so the directional derivative of u with respect to σ along Γ is

2. $\dfrac{du}{d\sigma} = \mathbf{T} \cdot \operatorname{grad} u = \dfrac{dx}{d\sigma} \dfrac{\partial u}{\partial x} + \dfrac{dt}{d\sigma} \dfrac{\partial u}{\partial t}.$

Comparison of this result with the left-hand side of 25.6.1.1 shows that it may be rewritten in the **first characteristic form**

3. $\dfrac{du}{d\sigma} = h(x, t, u),$

along the **characteristic curves** in the (x, t)-plane obtained by solving

4. $\dfrac{dt}{d\sigma} = a(x, t, u) \quad \text{and} \quad \dfrac{dx}{d\sigma} = b(x, t, u).$

On occasion it is advantageous to retain the parameter σ, but it is often removed by multiplying $du/d\sigma$ and $dx/d\sigma$ by $d\sigma/dt$ to obtain the equivalent **second char-**

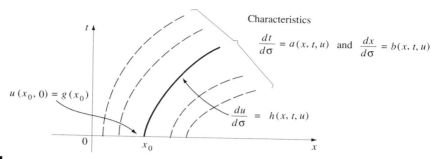

FIGURE 25.1 ∎

acteristic form for the partial differential equation 25.6.1.1.

5. $\dfrac{du}{dt} = \dfrac{h(x, t, u)}{a(x, t, u)}$,

along the characteristic curves obtained by solving

6. $\dfrac{dx}{dt} = \dfrac{b(x, t, u)}{a(x, t, u)}$.

This approach, called solution by the **method of characteristics,** has replaced the original partial differential equation by the solution of an ordinary differential equation that is valid along each member of the family of characteristic curves in the (x, t)-plane. If a characteristic curve C_0 intersects the initial line at $(x_0, 0)$, it follows that at this point the initial condition for u along C_0 must be $u(x_0, 0) = g(x_0)$.

This situation is illustrated in Figure 25.1, which shows typical members of a family of characteristics in the (x, t)-plane together with the specific curve C_0.

When the partial differential equation is linear, the characteristic curves can be found independently of the solution, but in the quasilinear case they must be determined simultaneously with the solution on which they depend. This usually necessitates the use of numerical methods.

EXAMPLE 1. This example involves a constant coefficient first-order equation. Consider the equation

7. $\dfrac{\partial u}{\partial t} + c\dfrac{\partial u}{\partial x} = 0$ [$c > 0$ a const.],

sometimes called the **advection equation,** and subject to the initial condition $u(x, 0) = g(x)$. When written in the second characteristic form this becomes

8. $\dfrac{du}{dt} = 0$

along the characteristic curves given by integrating $dx/dt = c$. Thus the characteristic curves are the family of parallel straight lines $x = ct + \xi$ that intersects the initial line

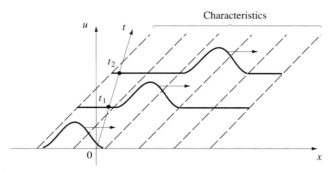

FIGURE 25.2 ∎

$t = 0$ (the x-axis) at the point $(\xi, 0)$. The first equation shows that $u = $ const. along a characteristic, but at the point $(\xi, 0)$ we have $u(\xi, 0) = g(\xi)$, so $u(x, t) = g(\xi)$ along this characteristic. Using the fact that $\xi = x - ct$ it then follows that the required solution is

9. $u(x, t) = g(x - ct)$.

The solution represents a **traveling wave** with initial shape $g(x)$ that moves to the right with constant speed c without change of shape. The nature of the solution relative to the characteristics is illustrated in Figure 25.2.

EXAMPLE 2.   In this example we solve the linear variable coefficient first-order equation

10. $\dfrac{\partial u}{\partial t} + t\dfrac{\partial u}{\partial x} = xt,$

subject to the initial condition $u(x, 0) = \sin x$. Again expressing the equation in the second characteristic form gives

11. $\dfrac{du}{dt} = xt$

along the characteristics given by integrating $dx/dt = t$. Integration shows that the characteristic curves are

12. $\displaystyle\int dx = \int t\,dt \quad \text{or} \quad x = \frac{1}{2}t^2 + \xi,$

where the constant of integration ξ defines the point $(\xi, 0)$ on the initial line through which the characteristic curve passes, while from the initial condition $u(\xi, 0) = \sin \xi$.
   Substituting for x in $du/dt = xt$ and integrating gives

13. $\displaystyle\int du = \int \left(\frac{1}{2}t^3 + \xi t\right)dt \quad \text{or} \quad u = \frac{1}{8}t^4 + \frac{1}{2}\xi t^2 + k(\xi),$

where for the moment the "constant of integration" $k(\xi)$ is an unknown function of ξ. The function $k(\xi)$ is constant along each characteristic, but it differs from characteristic to characteristic, depending on the value of ξ associated with each characteristic.

Because $\xi = x - \frac{1}{2}t^2$, it follows that

14. $u(x, t) = \frac{1}{8}t^4 + \frac{1}{2}(x - \frac{1}{2}t^2)t^2 + k(x - \frac{1}{2}t^2)$.

Setting $t = 0$ and using the initial condition $u(x, 0) = \sin x$ reduces this last result to

15. $\sin x = k(x)$,

so the unknown function k has been determined. Replacing x in $k(x)$ by $x - \frac{1}{2}t^2$, and using the result in the expression for $u(x, t)$ gives

16. $u(x, t) = \frac{1}{8}t^4 + \frac{1}{2}(x - \frac{1}{2}t^2)t^2 + \sin(x - \frac{1}{2}t^2)$.

This expression for $u(x, t)$ satisfies the initial condition and the differential equation, so it is the required solution.

EXAMPLE 3.   This example involves a simple first-order quasilinear equation. We now use the method of characteristics to solve the initial value problem

17. $\dfrac{\partial u}{\partial t} + f(u)\dfrac{\partial u}{\partial x} = 0$,

subject to the initial condition $u(x, 0) = g(x)$, where $f(u)$ and $g(x)$ are known continuous and differentiable functions.

Proceeding as before, the second characteristic form for the equation becomes

18. $\dfrac{du}{dt} = 0$

along the characteristic curves given by integrating $dx/dt = f(u)$. The first equation shows $u = $ constant along a characteristic curve. Using this result in the second equation and integrating shows the characteristic curves to be the family of straight lines

19. $x = tf(u) + \xi$,

where $(\xi, 0)$ is the point on the initial line from which the characteristic emanates. From the initial condition it follows that the value of u transported along this characteristic must be $u = g(\xi)$. Because $\xi = x - tf(u)$, it follows that the solution is given *implicitly* by

20. $u(x, t) = g[x - tf(u)]$.

The implicit nature of this solution indicates that the solution need not necessarily always be unique. This can also be seen by computing $\partial u/\partial x$, which is

21. $\dfrac{\partial u}{\partial x} = \dfrac{g'(x - tf(u))}{1 + tg'(x - tf(u))f'(u)}$.

If the functions f and g are such that the denominator vanishes for some $t = t_c > 0$, the derivative $\partial u/\partial x$ becomes unbounded when $t = t_c$. When this occurs, the differential equation can no longer govern the behavior of the solution, so it ceases to have meaning and the solution may become nonunique.

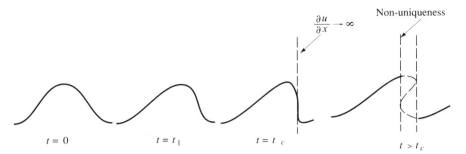

FIGURE 25.3 ∎

The development of the solution when $f(u) = u$ and $g(x) = \sin x$ is shown in Figure 25.3. The wave profile is seen to become steeper due to the influence of non-linearity until $t = t_c$, where the tangent to part of the solution profile becomes infinite. Subsequent to time t_c the solution is seen to become many valued (nonunique). This result illustrates that, unlike the linear case in which $f(u) = c$ (const.), a quasilinear equation of this type cannot describe traveling waves of constant shape.

25.7 Discontinuous Solutions (Shocks)

25.7.1

To examine the nature of **discontinuous solutions** of first-order quasilinear hyperbolic equations, it is sufficient to consider the scalar conservation equation in differential form

1. $\dfrac{\partial u}{\partial t} + \text{div } \mathbf{f} = 0,$

where $u = u(\mathbf{x}, t)$ and $\mathbf{f} = \mathbf{f}(u)$. Let $V(t)$ be an arbitrary volume bounded by a surface $S(t)$ moving with velocity $\boldsymbol{\nu}$. Provided u is differentiable in $V(t)$, it follows from 23.11.1.1 that the rate of change of the volume integral of u is

2. $\dfrac{d}{dt} \displaystyle\int_{V(t)} u\, dV = \int_{V(t)} \left\{ \dfrac{\partial u}{\partial t} + \text{div } (u\boldsymbol{\nu}) \right\} dV.$

Substituting for $\partial u / \partial t$ gives

3. $\dfrac{d}{dt} \displaystyle\int_{V(t)} u\, dV = \int_{V(t)} [\text{div } (u\boldsymbol{\nu}) - \text{div } \mathbf{f}]\, dV,$

and after use of the Gauss divergence theorem 23.9.1.3 this becomes

4. $\dfrac{d}{dt} \displaystyle\int_{V(t)} u\, dV = \int_{S(t)} (u\boldsymbol{\nu} - \mathbf{f}) \cdot d\boldsymbol{\sigma},$

where $d\boldsymbol{\sigma}$ is the outward drawn vector element of surface area of $S(t)$ with respect to $V(t)$.

Now suppose that $V(t)$ is divided into two parts $V_1(t)$ and $V_2(t)$ by a moving surface $\Omega(t)$ across which u is discontinuous, with $u = u_1$ in $V_1(t)$ and $u = u_2$ in $V_2(t)$. Subtracting from this last result the corresponding results when $V(t)$ is replaced first by $V_1(t)$ and then by $V_2(t)$ gives

5. $\quad 0 = \int_{\Omega(t)} (u\boldsymbol{\nu} - \mathbf{f})_1 \cdot d\boldsymbol{\Omega}_1 + \int_{\Omega(t)} (u\boldsymbol{\nu} - \mathbf{f})_2 \cdot d\boldsymbol{\Omega}_2,$

where now $\boldsymbol{\nu}$ is restricted to $\Omega(t)$, and so is the velocity of the discontinuity surface, while $d\boldsymbol{\Omega}_i$ is the outwardly directed vector element of surface area of $\Omega(t)$ with respect to $V_i(t)$ for $i = 1, 2$. As $d\boldsymbol{\Omega}_1 = -d\boldsymbol{\Omega}_2 = \mathbf{n}d\Omega$, say, where $\mathbf{n}$ is the outward drawn unit normal to $\Omega_1(t)$ with respect to $V_1(t)$, it follows that

6. $\quad 0 = \int_{\Omega(t)} [(u\boldsymbol{\nu} - \mathbf{f})_1 \cdot \mathbf{n} - (u\boldsymbol{\nu} - \mathbf{f})_2 \cdot \mathbf{n}] \, d\Omega = 0.$

The arbitrariness of $V(t)$ implies that this result must be true for all $\Omega(t)$, which can only be possible if

7. $\quad (u\boldsymbol{\nu} - \mathbf{f})_1 \cdot \mathbf{n} = (u\boldsymbol{\nu} - \mathbf{f})_2 \cdot \mathbf{n}.$

The speed of the discontinuity surface along the normal $\mathbf{n}$ is the same on either side of $\Omega(t)$, so setting $\boldsymbol{\nu}_1 \cdot n = \boldsymbol{\nu}_2 \cdot \mathbf{n} = s$, leads to the following **algebraic jump condition** that must be satisfied by a discontinuous solution

8. $\quad (u_1 - u_2)s = (\mathbf{f}_1 - \mathbf{f}_2) \cdot \mathbf{n}.$

This may be written more concisely as

9. $\quad [\![u]\!]s = [\![\mathbf{f}]\!] \cdot \mathbf{n},$

where $[\![\alpha]\!] = \alpha_1 - \alpha_2$ denotes the jump in α across $\Omega(t)$.

In general, $\mathbf{f} = \mathbf{f}(u)$ is a *nonlinear* function of u, so for any given s, specifying u on one side of the discontinuity and using the jump condition to find u on the other side may lead to more than one value. This possible nonuniqueness of discontinuous solutions to quasilinear hyperbolic equations is typical, and in physical problems a *criterion* (*selection principle*) must be developed to select the unique physically realizable discontinuous solution from among the set of all mathematically possible but nonphysical solutions.

In gas dynamics the jump condition is called the **Rankine–Hugoniot jump condition,** and a discontinuous solution in which gas flows across the discontinuity is called a **shock,** or a **shock wave.** In the case in which a discontinuity is present, but there is no gas flow across it, the discontinuity is called a **contact discontinuity.** In gas dynamics two shocks are possible mathematically, but one is rejected as nonphysical by appeal to the second law of thermodynamics (the selection principle used there) because of the entropy decrease that occurs across it, so only the compression shock that remains is a physically realizable shock. To discuss discontinuous solutions, in general it is necessary to introduce more abstract selection principles, called **entropy conditions,** which amount to stability criteria that must be satisfied by solutions.

The need to work with conservation equations (equations in divergence form) when considering discontinuous solutions can be seen from the preceding argument, for only then can the Gauss divergence theorem be used to relate the solution on one side of the discontinuity to that on the other.

25.8 Similarity Solutions

25.8.1

When *characteristic scales* exist for length, time, and the dependent variables in a physical problem it is advantageous to free the equations from dependence on a particular choice of scales by expressing them in nondimensional form. Consider, for example, a cylinder of radius ρ_0 filled with a viscous fluid with viscosity ν, which is initially at rest at time $t = 0$. This cylinder is suddenly set into rotation about the axis of the cylinder with angular velocity ω. Although this is a three-dimensional problem, because of radial symmetry about the axis, the velocity **u** at any point in the fluid can only depend on the radius ρ and the time t if gravity is neglected.

Natural length and time scales are ρ_0, which is determined by the geometry of the problem, and the period of a rotation $\tau = 2\pi/\omega$, which is determined by the initial condition. Appropriate nondimensional length and time scales are thus $r = \rho/\rho_0$ and $T = t/\tau$. The dependent variable in the problem is the fluid velocity $\mathbf{u}(r, t)$, which can be made nondimensional by selecting as a convenient characteristic speed $u_0 = \omega\rho_0$, which is the speed of rotation at the wall of the cylinder. Thus, an appropriate nondimensional velocity is $U = u/u_0$.

The only other quantity that remains to be made nondimensional is the viscosity ν. It proves most convenient to work with $1/\nu$ and to define the **Reynolds number** $\mathrm{Re} = u_0\rho_0/\nu$, which is a dimensionless quantity. Once the governing equation has been reexpressed in terms of r, T, U, and Re, and its dimensionless solution has been found, the result may be used to provide the solution appropriate to any choice of ρ_0, ω, and ν for which the governing equations are valid.

Equations that have natural characteristic scales for the independent variables are said to be **scale-similar.** The name arises from the fact that any two different problems with the same numerical values for the dimensionless quantities involved will have the same nondimensional solution.

Certain partial differential equations have no natural characteristic scales for the independent variables. These are equations for which **self-similar** solutions can be found. In such problems the nondimensionalization is obtained not by introducing characteristic scales, but by combining the independent variables into *nondimensional groups*.

The classic example of a self-similar solution is provided by considering the unsteady heat equation

1. $$\frac{\partial^2 T}{\partial x^2} = \frac{1}{\kappa}\frac{\partial T}{\partial t}$$

applied to a semi-infinite slab of heat conducting material with thermal diffusivity κ, with x measured into the slab normal to the bounding face $x = 0$. Suppose that for $t < 0$ the material is all at the initial temperature T_0, and that at $t = 0$ the temperature of the face of the slab is suddenly changed to T_1. Then there is a natural characteristic temperature scale provided by the temperature difference $T_1 - T_0$, so a convenient nondimensional temperature is $\tau = (T - T_0)/(T_1 - T_0)$. However, in this example, no natural length and time scales can be introduced.

If a combination η, say, of variables x and t is to be introduced in place of the separate variables x and t themselves, the one-dimensional heat equation will be simplified if this change of variable reduces it to a second-order ordinary differential equation in terms of the single independent variable η. The consequence of this approach will be that instead of there being a different temperature profile for each time t, there will be a single profile in terms of η from which the temperature profile at any time t can be deduced. It is this scaling of the solution on itself that gives rise to the name *self-similar solution*.

Let us try setting

2. $\quad \eta = Dx/t^n \quad$ and $\quad \tau = f(\eta) \qquad [D = \text{const.}]$

in the heat equation, where a suitable choice for n has still to be made. Routine calculation shows that the heat equation becomes

3. $\quad \dfrac{d^2 f}{d\eta^2} + \dfrac{n}{\kappa D^2} t^{2n-1} \eta \dfrac{df}{d\eta} = 0.$

For this to become an ordinary differential equation in terms of the independent variable η, it is necessary for the equation to be independent of t, which may be accomplished by setting $n = \frac{1}{2}$ to obtain

4. $\quad \dfrac{d^2 f}{d\eta^2} + \dfrac{1}{2\kappa D^2} \eta \dfrac{df}{d\eta} = 0.$

It is convenient to choose D so that $2\kappa D^2 = 1$, which corresponds to

5. $\quad D = 1/\sqrt{2\kappa}.$

The heat equation then reduces to the variable coefficient second-order ordinary differential equation

6. $\quad \dfrac{d^2 f}{d\eta^2} + \eta \dfrac{df}{d\eta} = 0, \quad$ with $\eta = x/\sqrt{2\kappa t}.$

The initial condition shows that $f(0) = 1$, while the temperature variation must be such that for all time t, $T \to T_0$ as $\eta \to \infty$, so another condition on f is $f(\eta) \to 0$ as $\eta \to \infty$. Integration of the equation for f subject to these conditions can be shown to lead to the result

7. $\quad T = T_0 + (T_1 - T_0)\text{erfc}\left(\dfrac{x}{2\sqrt{\kappa t}}\right).$

Sophisticated group theoretic arguments must be used to find the similarity variable in more complicated cases. However, this example illustrates the considerable advantage of this approach when a similarity variable can be found, because it reduces by one the number of independent variables involved in the partial differential equation.

As a final simple example of self-similar solutions, we mention the *cylindrical wave equation*

8. $\dfrac{\partial^2 \Phi}{\partial r^2} + \dfrac{1}{r}\dfrac{\partial \Phi}{\partial r} = \dfrac{1}{c^2}\dfrac{\partial^2 \Phi}{\partial t^2},$

which has the self-similar solution

9. $\Phi(r, t) = rf(\eta), \quad \text{with } \eta = \dfrac{r}{ct},$

where f is a solution of the ordinary differential equation

10. $\eta(1 - \eta^2) f''(\eta) + (3 - 2\eta^2) f' + f/\eta = 0.$

If the wave equation is considered to describe an expanding cylindrically symmetric wave, the radial speed of the wave v_r can be shown to be

11. $v_r = -[f(\eta) + \eta f'(\eta)],$

which in turn can be reduced to

12. $v_r = \begin{cases} A(1 - \eta^2)/\eta, & n \leq 1, \\ 0, & n > 1, \end{cases}$

where A is a constant of integration.

Other solutions of 25.8.1.8 can be found if appeal is made to the fact that solutions are invariant with respect to a time translation, so that in the solution t may be replaced by $t - t^*$, for some constant t^*.

Result 25.8.1.12 then becomes

13. $v_r = \begin{cases} A(t^*)[c^2(t - t^*)^2 - r^2]^{1/2}/r, & t^* \leq t - r/c, \\ 0, & t^* > t - r/c, \end{cases}$

and different choices for $A(t^*)$ will generate different solutions.

25.9   Burgers's Equation, the KdV Equation, and the KdVB Equation

25.9.1

In time-dependent partial differential equations, certain higher order spatial derivatives are capable of interpretation in terms of important physical effects. For example, in **Burgers's equation**

1. $\dfrac{\partial u}{\partial t} + u\dfrac{\partial u}{\partial x} = v\dfrac{\partial^2 u}{\partial x^2} \qquad [v > 0],$

the term on the right may be interpreted as a **dissipative** effect; namely, as the removal of energy from the system described by the equation. In the **Korteweg–de Vries (KdV) equation**

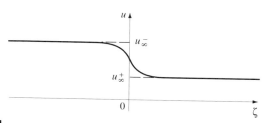

FIGURE 25.4 ■

2. $\dfrac{\partial u}{\partial t} + u\dfrac{\partial u}{\partial x} + \mu\dfrac{\partial^3 u}{\partial x^3} = 0,$

the last term on the left represents a **dispersive** effect; namely, a smoothing effect that causes localized disturbances in waves that are propagated to spread out and disperse.

Burgers's equation serves to model a gas shock wave in which energy dissipation is present ($\nu > 0$). The steepening effect of nonlinearity in the second term on the left can be balanced by the dissipative effect, leading to a traveling wave of constant form, unlike the case examined earlier corresponding to $\nu = 0$ in which a smooth initial condition evolved into a discontinuous solution (shock). The steady traveling wave solution for Burgers's equation that describes the so-called **Burgers's shock wave** is

3. $u(\zeta) = \dfrac{1}{2}\left(u_\infty^- + u_\infty^+\right) - \dfrac{1}{2}\left(u_\infty^- - u_\infty^+\right)\tanh\left[\left(\dfrac{u_\infty^- - u_\infty^+}{4\nu}\right)\zeta\right],$

where $\zeta = x - ct$, with the speed of propagation $c = \frac{1}{2}(u_\infty^- + u_\infty^+)$, $u_\infty^- > u_\infty^+$, and u_∞^- and u_∞^+ denote, respectively, the solutions at $\zeta \to -\infty$ and $\zeta \to +\infty$. This describes a smooth transition from u_∞^- to u_∞^+. The Burgers's shock wave profile is shown in Figure 25.4.

The celebrated KdV equation was first introduced to describe the propagation of long waves in shallow water, but it has subsequently been shown to govern the asymptotic behavior of many other physical phenomena in which nonlinearity and dispersion compete. In the KdV equation, the smoothing effect of the dispersive term can balance the steepening effect of the nonlinearity in the second term to lead to a traveling wave of constant shape in the form of a pulse called a **solitary wave.** The solution for the KdV solitary wave in which $u \to u_\infty$ as $\zeta \to \pm$ is

4. $u(\zeta) = u_\infty + a\,\operatorname{sech}^2\left[\zeta\left(\dfrac{a}{12\mu}\right)^{1/2}\right] \qquad [u_\infty > 0],$

where $\zeta = x - ct$, with the speed of propagation $c = u_\infty + \frac{1}{3}a$.

Notice that, relative to u_∞, the speed of propagation of the solitary wave is proportional to the amplitude a. It has been shown that these solitary wave solutions of the KdV equation have the remarkable property that, although they are solutions of a nonlinear equation, they can *interact* and *preserve their identity* in ways that are similar to those of linear waves. However, unlike linear waves, during the interaction

$t = t_1$ $t = t_2$ $t = t_3$ $t = t_4$

FIGURE 25.5 ▮

the solutions are *not* additive, though after it they have interchanged their positions. This property has led to these waves being called **solitons.** Interaction between two solitons occurs when the amplitude of the one on the left exceeds that of the one on the right, for then overtaking takes place due to the speed of the one on the left being greater than the speed of the one on the right. This interaction is illustrated in Figure 25.5; the waves are *unidirectional* since only a first-order time derivative is present in the KdV equation.

The Korteweg–de Vries–Burgers's (KdVB) equation.

5. $\dfrac{\partial u}{\partial t} + u \dfrac{\partial u}{\partial x} - v \dfrac{\partial^2 u}{\partial x^2} + \mu \dfrac{\partial^3 u}{\partial x^3} = 0 \qquad [v > 0],$

describes wave propagation in which the effects of nonlinearity, dissipation, and dispersion are all present. In steady wave propagation the combined smoothing effects of dissipation and dispersion can balance the steepening effect of nonlinearity and lead to a traveling wave solution moving to the right given by

6. $u(\zeta) = \dfrac{3v^2}{100\mu} [\text{sech}^2(\zeta/2) + 2\tanh(\zeta/2) + 2],$

with

7. $\zeta = \dfrac{-v}{5\mu}\left(x - \dfrac{6v^2}{25\mu}t \right) \quad$ and $\quad$ speed $c = 6v^2/(25\mu),$

or to one moving to the left given by

8. $u(\zeta) = \dfrac{3v^2}{100\mu} [\text{sech}^2(\zeta/2) - 2\tanh(\zeta/2) - 2],$

with

9. $\zeta = \dfrac{v}{5\mu}\left(x + \dfrac{6v^2}{25\mu}t \right) \quad$ and $\quad$ speed $c = -6v^2/(25\mu).$

The wave profile for a KdVB traveling wave is very similar to that of the Burgers's shock wave.

Short Classified Reference List

General tables of integrals

Erdélyi, A., *et. al., Tables of Integral Transforms*, Vols. I and II, McGraw–Hill, New York, 1954,

Gradshteyn, I. S., and Ryzhik, I. M., *in Tables of Integrals, Series and Products* (A. Jeffrey, Ed.), 5th ed., Academic Press, Boston, 1994.

Marichev, O. I., *Handbook of Integral Transforms of Higher Transcendental Functions, Theory and Algorithmic Tables*, Ellis Horwood, Chichester, 1982.

Prudnikov, A. P., Brychkov, Yu. A., and Marichev, O. I., *Integrals and Series*, Vols. 1–4, Gordon and Breach, New York, 1986–1992.

Special functions

Abramowitz, M., and Stegun, I. A., *Handbook of Mathematical Functions*, Dover Publications, New York, 1972.

Erdélyi, A., *et al., Higher Transcendental Functions*, Vols. I–III, McGraw–Hill, New York, 1953–55.

Hobson, E. W., *The Theory of Spherical and Ellipsoidal Harmonics*, Cambridge University Press, London, 1931.

MacRobert, T. M., *Spherical Harmonics*, Methuen, London, 1947.

Magnus, W., Oberhettinger, F., and Soni, R. P., *Formulas and Theorems for the Special Functions of Mathematical Physics*, 3rd ed., Springer-Verlag, Berlin, 1966.

McBride, E. B., *Obtaining Generating Functions*, Springer-Verlag, Berlin, 1971.

Snow, C., *The Hypergeometric and Legendre Functions with Applications to Integral Equations and Potential Theory*, 2nd ed., National Bureau of Standards, Washington, DC, 1952.

Watson, G. N., *A Treatise on the Theory of Bessel Functions*, 2nd ed., Cambridge University Press, London, 1944.

Asymptotics

Copson, E. T., *Asymptotic Expansions*, Cambridge University Press, London, 1965.

DeBruijn, N. G., *Asymptotic Methods in Analysis*, North-Holland, Amsterdam, 1958.

Erdélyi, A., *Asymptotic Expansions*, Dover Publications, New York, 1956.

Olver, F. W. J., *Asymptotics and Special Functions*, Academic Press, New York, 1974.

Elliptic integrals

Abramowitz, M., and Stegun, I. A., *Handbook of Mathematical Functions*, Dover Publications, New York, 1972.

Byrd, P. F., and Friedman, M. D., *Handbook of Elliptic Integrals for Engineers and Physicists*, Springer-Verlag, Berlin, 1954.

Gradshteyn, I. S., and Ryzhik, I. M., *Tables of Integrals, Series, and Products*, (A. Jeffrey, Ed.), 5th ed., Academic Press, Boston, 1994.

Lawden, D. F., *Elliptic Functions and Applications*, Springer-Verlag, Berlin, 1989.

Neville, E. H., *Jacobian Elliptic Functions*, 2nd ed., Oxford University Press, Oxford, 1951.

Prudnikov., A. P., Brychkov, Yu. A., and Marichev, O. I., *Integrals and Series*, Vol. 3, Gordon and Breach, New York, 1990.

Integral transforms

Doetsch, G., *Handbuch der Laplace-Transformation*, Vols. I–IV, Birkhäuser Verlag, Basel, 1950–56.

Doetsch, G., *Theory and Application of the Laplace Transform*, Chelsea, New York, 1965.

Erdélyi, A., *et al.*, *Tables of Integral Transforms*, Vols. I and II, McGraw–Hill, New York, 1954.

Marichev, O. I., *Handbook of Integral Transforms of Higher Transcendental Functions, Theory and Algorithmic Tables*, Ellis Horwood, Chichester, 1982.

Oberhettinger, F., and Badii, L., *Tables of Laplace Transforms*, Springer-Verlag, Berlin, 1973.

Prudnikov, A. P., Brychkov, Yu. A., and Marichev, O. I., *Integrals and Series*, Vol. 4, Gordon and Breach, New York, 1992.

Sneddon, I. N., *Fourier Transforms*, McGraw–Hill, New York, 1951.

Sneddon, I. N., *The Use of Integral Transforms*, McGraw–Hill, New York, 1972.

Widder, D. V., *The Laplace Transforms*, Princeton University Press, Princeton, NJ, 1941.

Orthogonal functions and polynomials

Abramowitz, M., and Stegun, I. A., *Handbook of Mathematical Functions*, Dover Publications, New York, 1972.

Sansone, G., *Orthogonal Functions*, revised English ed., Interscience, New York, 1959.

Szegö, G., *Orthogonal Polynomials*, revised ed., Colloquium Publications XXIII, American Mathematical Society, New York, 1959.

Series

Jolley, I. R. W., *Summation of Series*, Dover Publications, New York, 1962.

Zygmund, A., *Trigonometric Series*, 2nd ed., Vols. I and II, Cambridge University Press, London, 1988.

Numerical tabulations and approximations

Abramowitz, M., and Stegun, I. A., *Handbook of Mathematical Functions*, Dover Publications, New York, 1972.

Hastings, Jr., C., *Approximations for Digital Computers*, Princeton University Press, Princeton, NJ, 1955.

Jahnke, E., and Emde, F., *Tables of Functions with Formulas and Curves*, Dover Publications, New York, 1943.

Jahnke, E., Emde, F., and Lösch, F., *Tables of Higher Functions*, 6th ed., McGraw–Hill, New York, 1960.

Ordinary and partial differential equations

Birkhoff, G., and Gian–Carlo, R., *Ordinary Differential Equations*, 4th ed., Wiley, New York, 1989.

Boyce, W. E., and Di Prima, R. C., *Elementary Differential Equations and Boundary Value Problems*, 5th ed., Wiley, New York, 1992.

Du Chateau, Y., and Zachmann, D., *Applied Partial Differential Equations*, Harper & Row, New York, 1909.

Keener, J. P., *Principles of Applied Mathematics*, Addison–Wesley, New York, 1988.

Logan, J. D., *Applied Mathematics: A Contemporary Approach*, Wiley, New York, 1987.

Strauss, W. A., *Partial Differential Equations*, Wiley, New York, 1992.

Tyn, Myint-U., *Partial Differential Equations of Mathematical Physics*, Elsevier, New York, 1973.

Zachmanoglou, E. C., and Thoe, D. W., *Introduction to Partial Differential Equations and Applications*, William and Wilkins, 1976.

Zauderer, E., *Partial Differential Equations of Applied Mathematics*, 2nd ed., Wiley, New York, 1989.

Zwillinger, D., *Handbook of Differential Equations*, Academic Press, New York, 1989.

Numerical analysis

Ames, W. F., *Nonlinear Partial Differential Equations in Engineering*, Vol. 1, Academic Press, New York, 1965.

Ames, W. F., *Nonlinear Partial Differential Equations in Engineering*, Vol. 2, Academic Press, New York, 1972.

Ames, W. F., *Numerical Methods for Partial Differential Equations*, Nelson, London, 1977.

Atkinson, K. E., *An Introduction to Numerical Analysis*, 2nd ed., Wiley, New York, 1989.

Fröberg, C. E., *Numerical Methods: Theory and Computer Applications*, Addison–Wesley, New York, 1985.

Golub, G. H., and Van Loan, C. E., *Matrix Computations*, Johns Hopkins University Press, Baltimore, 1984.

Henrici, P., *Essentials of Numerical Analysis*, Wiley, New York, 1982.

Johnson, L. W., and Riess, R. D., *Numerical Analysis*, Addison–Wesley, New York, 1982.

Morton, K. W., and Mayers, D. F., *Numerical Solution of Partial Differential Equations*, Cambridge University Press, London, 1994.

Press, W. H., Flannery, B. P., Teukolsky, S. A., and Vellerling, W. T., *Numerical Recipes*, Cambridge University Press, London, 1986.

Schwarz, H. R., *Numerical Analysis: A Comprehensive Introduction*, Wiley, New York, 1989.

Index

NOTES

NOTES

NOTES

NOTES

NOTES

NOTES

NOTES

NOTES

NOTES

NOTES

NOTES

NOTES

NOTES

NOTES